118 Advances in Polymer Science

W0107115

Rigid Polymer Networks

By S. M. Aharoni, S. F. Edwards

With 51 Figures

Springer-Verlag
Berlin Heidelberg GmbH

Dr. S. M. Aharoni
Polymer Science Laboratory
Corporate Research & Technology
AlliedSignal Inc.
Morristown, NJ 07962 / USA

Sir Sam Edwards
Cavendish Professor of Physics
University of Cambridge
Cambridge CB3 OHE / UK

ISBN 978-3-662-14899-0   ISBN 978-3-540-48666-4 (eBook)
DOI 10.1007/978-3-540-48666-4

© Springer-Verlag Berlin Heidelberg 1994
Originally published by Springer-Verlag Berlin Heidelberg New York in 1994
Softcover reprint of the hardcover 1st edition 1994

Library of Congress Catalog Card Number 61-642

Typesetting: Macmillan India Ltd., Bangalore-25

SPIN: 10470558   02/3020 5 4 3 2 1 0 Printed on acid-free paper

Editors

Preface

Synthetic rigid polymer networks comprise part for the thermosetting class of polymers. Some members have now been in use for about a hundred years. These first pioneers, such as the phenol formaldehydes and furfurals, owed their popularity to a combination of several important features: monomer availability, ease and speed of processing, low cost of monomers, process and end-products, ability to be filled by various fillers, and a very desirable combination of good mechanical properties, high gloss surface, and excellent electrical insulation. As time progressed, the potential users elevated the performance requirements of the rigid polymer networks demanding ever-increasing thermal stability and mechanical properties. In response to these increasing expectations, novel rigid polymer networks were created with increasing fractions of thermally stable and mechanically robust moieties, initially as single rings and later as condensed aromatic groups. Laterly, rigid polymer networks whose stiff segments are liquid crystalline in nature were also introduced. In this case, very high modulus and strength may be obtained by first orienting the segments in the desired direction and then setting them in their final form by the addition or in-situ creation of rigid junctions.

As we delved into the subject matter, we were struck by the huge amount of literature dealing with various synthetic aspects of rigid polymer networks, the smaller number of publications dealing with correlations between the molecular structure of the networks and commercially desirable properties, and the far smaller number of publications, devoted to fundamental theoretical description of these networks, their properties, evolution and final structure, and the unique synthetic limitations imposed by the inherent rigidity of the structural units of the growing networks. The remarkably broad range of rigid polymer networks and the great ingenuity invested in their creation stood in stark contrast to the dearth of theoretical interest. It may be that the factor which made them theoretically less than popular is that so much of the literature in the field is patent literature, and the non-Gaussian, non-elastomeric nature of the rigid polymer networks, both alien to most theoreticians.

In this work we aim, therefore, to present the reader with an up-to-date overview of the field of rigid polymer networks with many of its synthetic complexities and challenges, and to whet the appetite of theoreticians and encourage them to solve the many fascinating theoretical problems presently existing in the field.

S.M. Aharoni
S.F. Edwards

Table of Contents

List of Symbols, Abbreviations and Acronyms

a	Persistence length
a	Average tube diameter, network mesh size
a	Coefficient in replica procedure in Sect. 5
A	Kuhn segment length
A	Free energy of a particular member of an ensemble
$\tilde{A}$	Free energy of the deformed state
$\langle \tilde{A} \rangle$	Average free energy of the deformed state
A_{el}	Elastic free energy
Ar	Aromatic unit in chain or pendant group
c	Concentration
C	Equilibrium concentration of gel
C_0	Polymer concentration in "as prepared" gel
C_0*	Critical concentration for "infinite" network formation
C_f	Number density of fractal polymers
C_n	Characteristic ratio of chain with n bonds
C_∞	Characteristic ratio of a chain in the limit of infinity
Cos θ	Angular energy
d	Average diameter of chain or segment
d^2	Effective segment cross-sectional area
D	Bond dissociation energy
D	Fractal dimension
D_s	Fractal surface dimensionality
D_f	Fractal mass dimensionality
DP	Degree of polymerization
$\overline{DP}$	Average degree of polymerization of chain or segment
DABA	4,4'-Diaminobenzanilide
DLA	Diffusion limited aggregation
DMAc	N,N-Dimethylacetamide
DMF	N,N-Dimethylformamide
e_{ij}	Strain
E	Strain modulus
E	Young's modulus of reinforced ensemble
E_f	Young's modulus of reinforcing fiber or segment
E_l	Longitudinal tensile modulus
E_n	Young's modulus of the bulk or matrix
E_r	Rod material modulus
f	Branchpoint or junction functionality
f	Tensile force applied to segment
$\bar{f}$	Average retractive force
F	Force
FP, FPs	Fractal polymer, fractal polymers
$g(r_1 r_2)$	Mathematical description of monomer in Sect. 5

G	Equilibrium shear modulus
H-bond	Hydrogen bond
ΔH_a	Activation energy for glass transition
$\langle h^2 \rangle_0$	Average square unperturbed end-to-end distance
I(q)	Intensity of scattered radiation at q
IR	Infrared
k	Boltzman constant
K_i	Force constants
l	Length of actual or virtual bond; average bond length
l_0	Length of average stiff or flexible segment in network or FP
L	Chain contour length
L	Total length of segments in a FP
L_c	Segment length between entanglements
LALS	Low angle light scattering
M	Molecular weight
M_c	Molecular weight between entanglements
M_n	Number average molecular weight
M_w	Weight average molecular weight
N	Number of repeat units
N	Number density of rods, rodlike segments
N_F	Number of FPs
NF	Network fragment
NMP	N-Methyl-2-pyrrolidinone
NMR	Nuclear magnetic resonance
NTPA	Nitroterephthalic acid
P	Fraction of monomer consumed, probability that monomer was consumed
P	Probability
Py	Pyridine
PPh_3	Triphenylphosphine
q	Scattering vector
r	Displacement length of a chain or segment
r	Chain or segment end-to-end vector
$\langle r^2 \rangle_0$	Mean square magnitude of r
R	Aliphatic or general organic unit in chain, segment or pendant group
R	Gas constant
R	Radius of polymeric species, radius of FP
R_a	Average distance between centers of FPs
R_G	Radius of gyration
R_H	Hydrodynamic radius
R_{GW}	Radius of gyration of worm-like chain
R_1	Length of non-coaxial, actual bond in polyamide segment
R_2	Length of coaxial, virtual bond in polyamide segment

s	Area
S	Entropy
S	Spring constant
SANS	Small angle neutron scattering
SAXS	Small angle X-ray scattering
T	Absolute temperature
T_g	Glass transition temperature
$T_{g\infty}$	Glass transition temperature of uncrosslinked polymer
TPA	Terephthalic acid
TPP	Triphenylphosphite
U_0	Thermal activation energy for bond scission
v	Volume
V	Potential energy of rod
V(l)	Potential energy of bond
V_f	Volume fraction
WAXD	Wide-angle X-ray diffraction
x	Numerical value of DP
x	Axial ratio
$\bar{X}$	Axial ratio of stiff segment of average length
$\bar{X}^*$	Critical axial ratio for the onset of liquid crystallinity
Z_f	Number of reactive groups belonging to all species larger than monomers
Z_m	Number of reactive groups belonging to monomers
α	Valence angle of carbonyl in amide group
α_G, α_L	Coefficients of expansion in the glass, liquid states
β	Valence angle at nitrogen in amide group
γ	Stress concentration factor for brittle failure
ε	Strain
θ	Scattering angle, angle between segment axis of symmetry and network draw direction
λ	Wavelength of scattered beam
λ	Measure of tube deformation, extension
λ	Measure of material deformation
Λ	Deformation matrix
μ	Shear modulus
ν	Number concentration of elastically effective network segments
ρ	Polymer density
σ	Stress
σ_b	Breaking strength
ϕ	Torsional angle around bond between aromatic ring and nitrogen in amide group
ϕ	Volume concentration
ϕ, ϕ^*	Parameters used in integrations in Sect. 5
Φ	1% Extensibility

Ψ	Torsional angle around bond between aromatic ring and carbon in amide group
ω	Torsional angle around central bond in amide group
ω	Stress concentration factor for ductile failure
$\iint \phi g \phi$	Mathematical representation of unreacted monomer in Sect. 5
$\iint \phi^{*2}$	Mathematical representation of difunctional monomer in Sect. 5
$\int \phi^{*3}$	Mathematical representation of trifunctional monomer in Sect. 5

1 Introduction

It took only one generation from Staudinger's proposal [1] that polymers are covalently bonded linear [2] chain molecules [3], through his own description of non-linear network polymers [4] and the application of statistical methods by Guth, Mark [5] and Kuhn [6, 7] to describe the configurational behavior of flexible chain polymers, to a molecular level understanding of many of the important features and properties of polymeric networks, elastomers and gels [8–16]. In these and subsequent papers [17–38] and books [39–46], the theoretical and most of the experimental treatments were confined essentially to systems comprising long flexible chain segments joined to one another in most instances by flexible crosslinking units. Generally, rigid networks were not mentioned at all, or were dealt with in a perfunctory manner. In this monograph we shall concentrate on a particular kind of rigid networks. These are all covalently-bonded permanent networks, completely or mostly aromatic in character.

1.1 Definitions

A network's rigidity is determined by the stiffness and length of its segments and the rigidity and functionality of its junction points. Segmental stiffness is described in terms of the properties exhibited in very dilute solutions by an isolated linear high molecular weight chain analogue. Under such conditions, two types of chains may be distinguished: flexible chains whose work of deformation depends on their large configurational entropy, and stiff or rigid chains whose work of deformation relies on their backbone elasticity. The former chains will be loosely termed here entropic and the latter enthalpic. When appropriate, these terms will carry through to networks and their segments as well. (Entropic and enthalpic chains respectively correspond to the kinetic and static chain flexibilities of the Russian School [47].) The structural features contributing to the enthalpic behavior of polymer chains include rigid cyclic groups, such as individual aromatic and heteroaromatic rings or condensed ring systems preferentially arranged in a coaxial or colinear fashion along the chain axis:

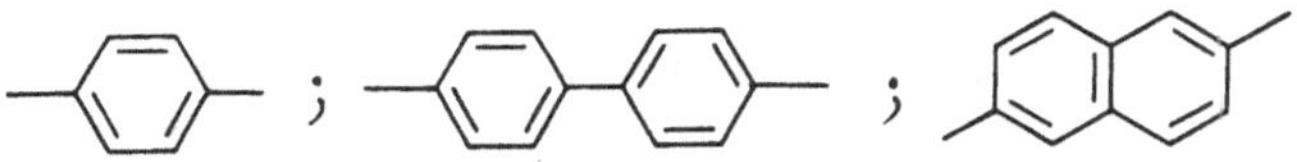

multiple bonds between adjacent atoms along the chain, such as ethynylene, vinylene and imines:

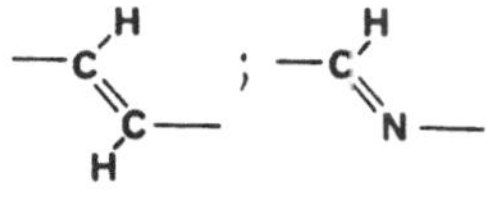

or groups rendered rigid because of electron delocalization, such as the amide residue, or semi-rigid such as the ester group:

$$-C\overset{\displaystyle O}{\underset{\displaystyle \underset{H}{N}-}{\diagup}} \quad ; \qquad -C\overset{\displaystyle O}{\underset{\displaystyle O-}{\diagup}}$$

and systems where conformational interconversions by means of rotational isomerization requires surmounting high energy barriers. Entropic chain rigidity, on the other hand, is strongly affected by temperature, as is clearly manifested by the large decrease in polymer chain flexibility and increase in modulus when the temperature is reduced to below the glass transition temperature, T_g.

In the condensed state and in concentrated solutions, the behavior of polymer chains is affected by the presence of others. In order to undergo conformational transitions many chains perform torsional movements in which chain sections sweep through significant volumes of space. This movement, rather simple for an isolated chain, becomes increasingly hindered with polymer concentration by sections belonging to other chains. Thus, a simple "crankshaft" torsional movement, which is very likely to occur in isolated hydrocarbon chains in which the backbone C-C bonds are all part of the tetrahedral bond structure, becomes practically impossible in the bulk where the chains are close together and may not get out of the way at a sufficient rate to allow the volume sweeping associated with "crankshaft" type rotations. Suppression by crowding or by specific interchain interactions of motions involving long chain sections is more obvious in the case of flexible chains. In the case of rigid or stiff chains, especially the enthalpically rigid ones, even in isolation, chain motional freedom is reduced to a level much lower than for the flexible chains. In the bulk or concentrated solutions the motional freedom of the rigid chains is further reduced by crowding or specific interactions such as interchain hydrogen bonds (H-bonds). In fact, we were recently able to demonstrate by NMR studies that the parallel alignment and interchain H-bonding of linear aromatic polyamides in concentrated solutions do arrest the rotational motions of the amide carbonyl group relative to the aromatic rings, motions which are present in analogous systems where the interchain H-bonds are prevented from being formed [48]. Now, when built into a network, both equilibrium and kinetic flexibilities may be affected by restrictions on chain motions imposed by the junctions holding the segments together. Furthermore, the higher the junction concentration, that is the shorter the segmental contour length, l_0, the stronger the effect of the junctions and the higher the overall network rigidity.

For consistency, and for the purpose of discriminating between the network and its components, we shall describe the network as a whole or its solvent-swollen gel by the adjectives rigid or semi-rigid. The chain segments between junction points will be termed stiff instead of rigid. This removes the possible implication of rod-like behavior (which will be used only in specific instances) and includes worm-like chains and chains where stiffer structural units are connected along the chain by flexible ones. The network junction points will be referred to according to their chemical and structural nature. When their chemical nature is substantially different from the segments they connect, and especially when they are also flexible, they will be called crosslinks. An example is a sulfide or polysulfide crosslink in vulcanized rubber. However, when the junctions are rigid and chemically identical or close in nature to the units comprising the stiff segments, then the junctions will be called branchpoints. An example is a tri- or tetra-functional aromatic ring in an aromatic polymaide network. The number of arms that can emanate from each branchpoint is dictated by the chemical functionality, f, of the branchpoint unit.

The definition of a segment stiffness depends on its nature. If a long stiff segment can be described as a worm-like chain, characterized by a particular smooth and uniform curvature, then the chain persistence length is a good initial descriptor. For a hypothetical worm-like chain of contour length L and statistical mechanical average square of the unperturbed end-to-end distance $\langle h^2 \rangle_0$, it was shown by Porod and Kratky [49, 50], that in the limit of $L \to \infty$,

$$(\langle h^2 \rangle_0 / L) = 2a \tag{1}$$

with a being the persistence length of the chain. For chains treatable by statistical methods,

$$2a = A \tag{2}$$

where A is the length of Kuhn's statistical segment [6, 7]. When the chains are rather flexible, $L/a \gg 1$ and the general relationship [51]

$$6R_G / L = 2a[1 - (L/a)^{-1}] \tag{3}$$

holds, correlating together the chain radius of gyration, R_G, with its contour and persistence lengths. The above relationships lead to the approximation

$$A = 2a \cong 6R_G^2 / L \tag{4}$$

In the case of stiffer chains that conform with the worm-like model, it was shown by Benoit and Doty [52] and Yamakawa [53] that the radius of gyration of such a chain, R_{GW}, relates to L and a in a more complex manner:

$$R_{GW} = [(La/3) - a^2 + (2a^3/L) - (2a^4/L^2)(1 - e^{-L/a})]^{1/2}. \tag{5}$$

The above relationships make it possible to determine the size of a from direct measurements of R_G or R_{GW} by techniques such as light scattering, and from knowledge of L derived from knowledge of the chain structure and its molecular weight. When incorporated in rigid networks, stiff segments are usually shorter than the persistence lengths of the analogous worm-like chains. We may therefore conclude that the flexibility and curvature of the worm-like chain is an upper bound for these properties in the corresponding segments and that, most likely, the flexibility, natural curvature and compliance of the stiff segments decrease as they become shorter. We believe that for the purpose of describing and comparing rigid networks, a descriptor better than the persistence length of an isolated chain may be the segmental axial ratio, $\bar{x}$:

$$\bar{x} = l_0/d \tag{6}$$

in which l_0 is the contour length of an average stiff segment in a network and d is its average diameter. When $\bar{x} > \bar{x}^*$, the critical axial ratio above which the segment is mesogenic in nature, then networks with main-chain stiffness may show liquid-crystal characteristics.

Another kind of enthalpic stiff segment is one where relatively long, stiff or rodlike groups are connected to each other by means of short singly-bonded spacers. The rigid groups can be rigid or stiff mesogenic entities encompassing several aromatic units connected to one another by a single bond or by a rather inflexible bridging group such as ethynylene or amide. The rigid groups may, however, consist of a single aromatic ring or condensed-rings unit. The length of the rigid group is defined in terms of a single virtual bond spanning its whole length. The short spacers may contain a single chain atom serving as a swivel between two rigid groups, such as:

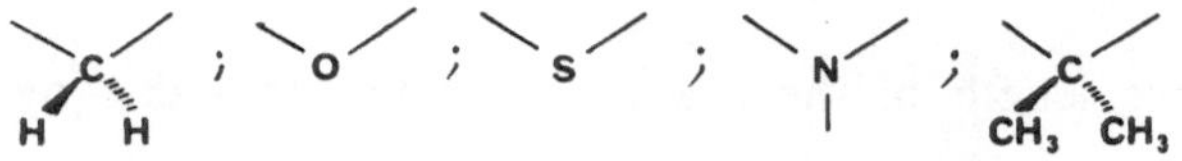

or two chain atoms connecting two rigid groups by a single bond not colinear with the rigid groups' axes, such as

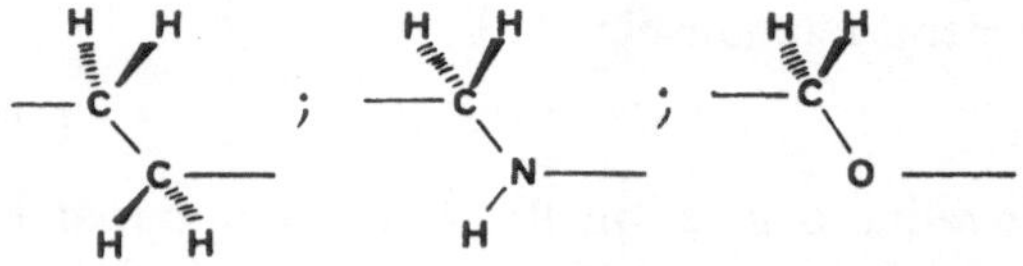

The common feature of all such short spacers is that they allow a measure of torsional motion of each rigid group relative to its neighbors along the chain, but not an independent motion. Similarly, both the single-atom swivel and the two-atom spacer allow a measure of directional reorientation of one rigid group relative to its neighbors, but not an independent directional mobility.

The limitation of the short spacers to only one or two atoms, connected to the rigid groups and to one another by single bonds, rests upon the fact that in the study of liquid crystal polymers it was repeatedly observed [54–60] that spacers with three chain atoms or longer allow for an increasing independence of torsional motions and directional reorientation of the rigid groups along the chain. This independent mobility allows mesogenic units from different chains to align themselves in more or less parallel arrays. The appearance of such arrays and the cooperative behavior of the mesogenic groups are reflected in liquid crystalline behavior. Two-atom spacers and, especially, one-atom swivels, more often than not appear to inhibit parallel array formation [61, 62] and the associated mesomorphic behavior.

The networks discussed in this monograph are, then, of an aromatic nature, with the aromatic groups in the segments and branchpoints or junctions connected to one another directly or through spacers having restricted motional freedom. These spacers contain a very small number of main-chain atoms. The length of the stiff or rodlike segments between branchpoints is not limited and may be as short as a single aromatic ring. In such networks, the overall rigidity or modulus is affected by several structural features. Paramount among the features contributing to increased rigidity are: the shortness of the network segments between branchpoints; the increasing number of reacted functionalities in each branchpoint; the growing stiffness of the branchpoints themselves and of the bridging residues between the aromatic groups in the segments; reduction in the number of chain single bonds in the network, especially those not coaxial or colinear with their segment axis; increases in the height of the energy barriers that segments, or constitutive units in them, have to surmount in order to change torsional angles or escape from the energy minima of conformational isomeric states; increases in the size and rigidity of aromatic groups at the expense of non-aromatic and single bond connections between smaller groups. Among the secondary effects one may mention potential non-covalent bond interactions, such as hydrogen bonds, that may contribute to the rigidity of the system, for instance, in rigid aromatic polyamide networks. The case of ethynylene (acetylene) and diethynylene (diacetylene) bridging groups is interesting. Here, because of symmetry, rotations around the single bonds do not change the rodlike geometry of these groups. An increased number of ethynylene groups increases, however, the kinetic flexibility of a polyethynylene residue.

A reduction in modulus is associated, naturally, with a reversal of all or any of the features listed above. In addition, there are two most important features strongly contributing to the lowering of network modulus. There are dilution and defects. When a dry flexible network imbibes a good solvent, it swells. The ensemble of a network filled with solvent is called a gel. Because of the swelling, the concentration of segment and branchpoints or crosslinks in the gel is lower than in the original network. This reduced concentration is directly reflected in reduced modulus. In the case of rigid, highly aromatic and intensely crosslinked networks, the ability to swell in a good solvent is greatly diminished. This is especially the case when the network is defect free or close to being so. A defect free network may alternatively be called flawless. It is defined here as a network in which every segment is connected at both ends to fully-reacted branchpoints or junctions, and every functionality in the system is consumed. A flawless network does not necessarily means that all segments are identical in length. In a defect free network, the relatively short and substantially inflexible segments cannot stretch much to accomodate the solvent and the network swelling is minimal. Furthermore, because of their chemical nature, many of the rigid aromatic networks must be prepared in solution to start with. When obtained, they are already in the form of a rigid gel consisting mostly of the reaction solvent, residual reagents and reaction by-products. Replacing this liquid by a good solvent leads to some swelling of the gel. This swelling may be limited to only a few percent in the case of low-defect networks having short segments and high branchpoint functionality, f. The swelling may grow up to several hundred percent when the networks have very long segments and only a few branchpoints, when f is low and, especially, when the network is defective. Our experience with rigid networks and gels leads us to believe that network defects are, qualitatively, the most important contributor to the swelling of rigid networks and has a significant bearing on the modulus of such networks and their gels. We shall return to this problem in Sect. 4.2.2 below.

Rigid networks and gels can be prepared in at least three ways, as follows.

(1) By placing all the monomers, from which both the segments and branchpoints are to be constructed, together in the same pot, in the presence or absence of extra solvent, and creating from them the network in a single step. This procedure is routinely used for the creation of networks by polycondensation reactions. In general, it leads to networks with segments not all identical but of average length. We call it one-step synthesis and its products one-step networks or gels.

(2) By preparing stiff telomeric segments in a separate reaction and reacting the reactive end-caps in a subsequent step to form a network. The reaction product of the end-caps may be a new cyclic group, but this is not a necessary condition. This method of network preparation is conceptually similar to the well-known method of chain-ends crosslinking frequently employed in the preparation of flexible networks from long flexible chains by reacting their reactive ends with one another or with an additional multifunctional crosslinking species. We shall refer to this method as end-cap polymerization with the products being end-capped networks.

(3) By preparing in a first step long stiff chains decorated along their length, and not only at their ends, with a multiplicity of reactive groups. The length of these chains is measured in tens of nanometers. In a separate second step these chains are reacted with one another directly or by the introduction of an additional rigid and reactive species. In the resulting network or gelled network, each precursor stiff chain is connected to other such chains by several stiff or rodlike struts holding the ensemble together yet keeping the precursor chains at the junction points at more or less fixed distances from each other. The struts may be chemically identical with the precursor chains such that, in the final network, segments originating from the precursor chains are indistinguishable from those originating from the connecting struts. In other instances the connecting species may be chemically or structurally different from the precursor stiff chains, making the differentiation easy between segments originating from precursor chains and those originating from the connecting species. We call this a two-step synthesis and the products two-step networks or gels.

The creation of a perfect rigid aromatic network will be a remarkable achievement. By a perfect network we mean one with rodlike or stiff segments, all identical in length and all connected at both ends to branchpoints whose functionalities are all satisfied. Such a perfect network will hardly swell when its voids are filled with solvent and it turns to be a gel. Because it is structurally perfect, it will be highly regular and, when studied by an appropriate diffraction method, it will produce a diffraction pattern characteristic of a single crystal. In case of long-segment perfect rigid networks, it is expected that interpenetrating network lattices will be formed to prevent the creation of perfect networks with large repeating voids. Our experience with the preparation of rigid networks in solution instructs us to believe that perfect aromatic rigid networks may not be preparable in solution but may conceivably be created on a solid-solution interface. Here, only monomers or small network components will be deposited from their solution onto the solid growth face of the network. Large network components, such as the pre-gel polymeric species present in all one-step network syntheses, are not allowed to form in solution nor are they allowed to be deposited on the perfect network growth face. There are several descriptions in the literature of single crystal type networks [63–67], most of them interpenetrating in nature [64–67]. The majority of these structures consist of small organic molecules held together by a complicated arrangement of H-bonds [64–66]. The most open system is a framework formed by complexing cuprous ions each with four 4,4′,4″,4‴-tetracyanotetraphenylmethane [63]. None of the reported structures consist of covalently bonded networks whose segments are stiff and aromatic in character.

In reality, when we prepare one-step or end-capped rigid networks they are far from being perfect even when they are flawless. This is because the segments in such networks are hardly ever all identical in length, the latter being a precondition for perfect systems. In both kinds of networks the segment lengths between branchpoints are most often average lengths. The majority of stiff segments may have length identical with the average length of all segments in the network, but many additional segments may be shorter or longer. This results in the networks being disordered and tending towards amorphicity even when all the functionalities in the branchpoints are consumed. An additional factor mitigating against perfection is the problem with the complete utilization of branchpoint functionality in stiff-segment networks. Our experience tells us that this desired outcome is almost impossible to achieve, even in instances where the precursor segments were all identical in length [68]. This leads to defective networks and towards amorphicity. Before the one-step or end-capped rigid aromatic systems become "infinite" networks, that is before they reach the gel point, the growing polymeric species are best described in terms of the fractal model. The growth nature of the two-step rigid systems is less well understood and leaves much work to be done in the future.

1.2 General Classification of Liquid-Crystal Polymers and Networks

Many of the linear stiff polymers described in this work are liquid-crystalline in nature. Several of the permanent networks and gels discussed here possess some properties commonly associated with liquid-crystal polymers (LCPs). A typical such property is the birefringent appearance of the quiescent network or gel when observed in the microscope using cross-polarized light. The birefringence, of course, is a manifestation of segmental anisotropy present in the system. Because they have some liquid-crystal properties, we call such networks and gels liquid-crystal polymer (LCP) networks and gels even though the systems are not truly liquid. The rigid LCP networks are a subset of the much larger family of LCP networks. This family contains both stiff and flexible polymer chains. To appreciate the richness of the LCP networks family and the place the rigid LCP networks take in it, it is useful first to have an overview of LCPs in general, and of the networks prepared from them. Such an overview will be presented in this section. In subsequent sections, however, only the rigid (or rather, stiff) networks will be discussed and the theoretical treatment limited to rigid LCP networks only.

1.2.1 Definition of Liquid-Crystal Polymers

Liquid-crystal polymers are polymers that, in the quiescent state and over characteristic temperature or concentration intervals, exhibit a combination of long-range motion and fluidity together with a degree of order measurable by optical and scattering techniques such as cross-polarized light microscopy, X-ray diffraction and scattering, or small-angle neutron scattering. This order is a reflection of a cooperative behavior of many polymer chains or polymeric segments. With the exception of a very few flexible polymer chains whose cooperative behavior and liquid crystallinity are due to inter- or intra-chain hydrogen bonds [69–76], the vast majority of liquid-crystal polymers owe their unique behavior to the presence, in the chain backbone or pendant to the chain, of relatively stiff and elongated units which, because of their stiffness and shape anisotropy, tend to order themselves in more or less parallel arrays and impart to the system a liquid crystal character. The stiff anisotropic units responsible for polymer liquid crystallinity are called mesogenic units.

The liquid-crystal polymers (LCPs) may be categorized in two broad classes, with several polymers falling in both. One class is that of thermotropic LCPs and the other is that of lyotropic LCPs. When heated in the quiescent state from a crystalline, semi-crystalline or occasionally a glassy state, the thermotropic LCPs pass through one of more phases, mobile yet ordered to varying degrees, before reaching a fully isotropic molten state. The phase or phases between the low temperature crystalline or glassy state and the high temperature isotropic melt, are called mesomorphic phases. Based on their structure, thermotropic LCPs are further divisible into two groups. One includes polymers in which mesogenic units of modest lengths are connected along the chain backbone by flexible "spacers", or are connected to a flexible chain backbone by flexible spacers or tethers, just as pendants to a necklace. The other group contains polymers with rather weak interchain attractive interactions and whose backbones are uniformly stiff to varying degrees, imparting to these polymers rodlike or wormlike character.

Lyotropic LCPs are polymers whose solutions exhibit liquid crystallinity, that is, anisotropic domains in a fluid system, over a characteristic range of concentrations. In more concentrated solutions the system may be multiphasic and contain crystalline particles, amorphous gel particles and anisotropic solution coexisting with one another. Upon dilution, the anisotropic liquid crystalline solution turns biphasic, where anisotropic and isotropic solutions of the same polymer in the same solvent coexist. Upon further dilution, the solution becomes fully isotropic. Polymers that exhibit lyotropic mesomorphicity are either stiff-backbone polymers with strong interchain interaction in the absence of solvent or polymers whose backbones are so extended and rigid that, upon breakup of their crystalline order by the addition of some solvent, the stiff polymer chains retain substantial measure of parallel alignment to remain in mobile anisotropic domains.

A new group of LCPs does not fit the structural attributes presented above. Its liquid crystalline behavior is due to the ability, and often thermodynamic preference, of some relatively flat structural units to stack themselves together, similar to flat discs or coins stacked in tall cylinders. Because of the analogy, such polymers are called discotic LCPs. The discotic LCPs can be divided into two groups. One group consists of arrays of large flat organic molecules such as phthalocyanines, containing a metal fluoride, cyanide or oxide in the center of each flat molecule. Among such groups one finds Al-F [77,78], Ga-F [77,78], Ge-O [79], Co-CN [80] and Si-O [77, 81, 82]. Strong attractive interaction between these groups strings them together in such a fashion that chains such as -Si-O-,Ge-O- or -Co-CN- are formed, threading through the centers of the flat organic molecules and holding them together as long and slender, stiff rods. Aggregates of such rods prefer to pack more or less in parallel arrays and behave as LCPs under appropriately chosen temperature and/or concentration conditions [82]. Additional organometallic variations on this approach, using metal-laynes as the stringing groups, were also reported [83, 84]. The second family of discotic LCPs is that in which flat or saucer-shaped organic groups are connected to one another by either being part of a flexible single-strand or double-strand backbone [85–88], or hanging as pendants attached by a flexible tether to a flexible main-chain [89–91]. In both instances the flexibility of the connecting spacers allows the flat units to stack themselves together in cylindrically-shaped columns. When the system contains a sufficient concentration of sufficiently long stacks, then they are forced, for better packing in space, to align themselves more or less in parallel, imparting to the whole ensemble liquid crystalline characteristics [85–91]. At the writing of these lines we are aware of neither networks comprising discotic LCPs nor any theoretical treatments of such. Therefore, they will not be further referred to below.

A brief description of the structural features of LCPs owing their mesomorphicity to the presence in them of relatively stiff and slender structural units will now be given. It will be followed by a description of LCPs with more or less uniform main-chain mesogenicity. The examples and references below, as well as those throughout this monograph, are not meant to be exhaustive but are given in order to illustrate the large variety and complexity of LCPS presently known.

1.2.2 Flexible Chain Polymers with Mesogenic Groups

These are divided into two broad classes: main-chain LCPs and side-chain LCPs. Because in the latter the mesogenic side-chains are usually rather short and of uniform length, we prefer to name the side-chain LCPs as pendant LCPs. Both main-chain and pendant LCP families have been extensively described in the literature [59, 60, 92 – 96] so that they need not be described below in great detail. However, examples will be given of main-chain and pendant LCPs demonstrating various structural features and chemical families, in order for the reader to appreciate the variety of such LCPs and the synthetic ingenuity invested in their creation.

Main-chain LCPs containing stiff mesogenic groups are schematically shown in Fig. 1. In it the blocks stand for the stiff and slender mesogenic groups and the straight or curly lines stand for the spacers between them. When the axial ratio, x, of the mesogenic groups surpasses about $x = 3$ [54, 55, 97, 98] and when the spacer groups are sufficiently flexible and long [98–102] to allow for motional decoupling of each mesogenic group along the chain from its neighbors so that they may align in parallel arrays with mesogenic groups belonging to the same or other chains, then liquid crystallinity ensues. When the spacers are not flexible or long enough, liquid crystallinity is inhibited [61, 62]. This liquid crystallinity, as a rule, is thermotropic and in solution such polymers generally behave as common flexible coils with longer than usual virtual bonds [61, 62, 70, 103]. Because of this behavior, these polymers may just as well be called flexible polymers with stiff spacers.

Typical main-chain LCPs are described by the general structure

$$-Z-\hspace{-4pt}\left(\hspace{-3pt}\bigcirc\hspace{-3pt}-X-\hspace{-3pt}\bigcirc\hspace{-3pt}-Z\hspace{-3pt}\right)_{\hspace{-3pt}n}\hspace{-4pt}\left(CH_2-Y-CH_2\right)_{\hspace{-3pt}m}$$

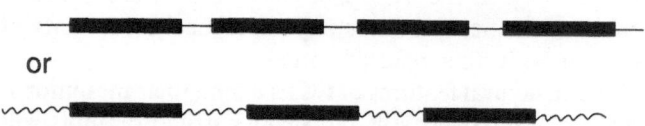

or

Fig. 1. Liquid crystal polymers containing mesogenic groups in the main-chain

in which X may stand for a single bond or groups such as

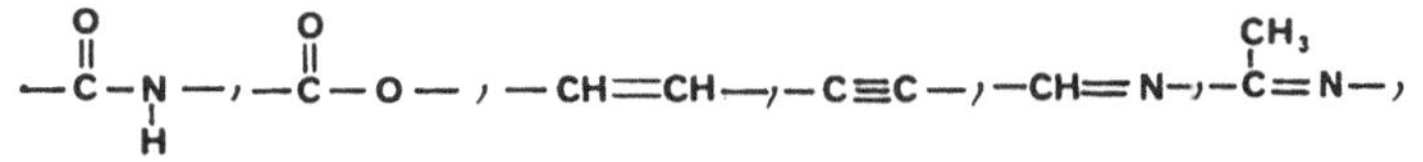

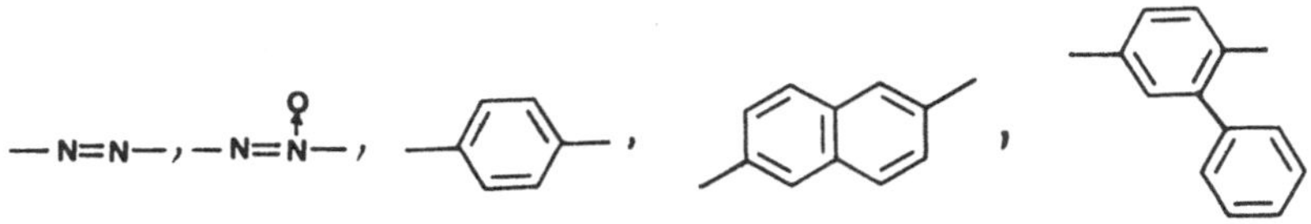

$$-\overset{O}{\overset{\|}{C}}-\overset{H}{\underset{|}{N}}-,\ -\overset{O}{\overset{\|}{C}}-O-,\ -CH{=}CH-,\ -C{\equiv}C-,\ -CH{=}N-,\ -\overset{CH_3}{\underset{|}{C}}{=}N-,$$

$$-N{=}N-,\ -\overset{O}{\overset{\|}{N}}{=}N-,$$

and several others. Y may represent $\left(\ CH_2\ \right)_n$, $\left(\ CH_2CH_2O\ \right)_n$ and similar flexible groups. Relatively short polysiloxane runs, such as polydimethyl-siloxane

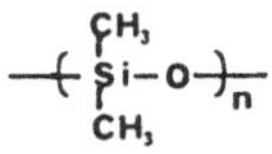

$$-\left(\ \overset{CH_3}{\underset{CH_3}{\overset{|}{\underset{|}{Si}}}}-O\ \right)_n-$$

may be inserted between the mesogenic groups instead of the flexible hydrocarbon or polyether groups shown above. The Z in the above structure may stand for any coupling group connecting the stiff the flexible units. Among such coupling groups one frequently finds ether oxygen, ester and amide groups, and more recently such flexible groups as carbonate and urethane:

$$-O-\overset{O}{\overset{\|}{C}}-O-\ ,\qquad -O-\overset{O}{\overset{\|}{C}}-\overset{H}{\underset{|}{N}}-$$

Combinations of various lengths of m and n together with multiple X,Y and Z groups create a rich menu of main-chain liquid crystal polymers.

In the second class of flexible-chain LCPs, the mesogenic groups are attached to the chain by spacers of various lengths and flexibility such that liquid crystallinity may arise from an appropriate packing of the mesogenic pendants in layer-like structures. In Fig. 2 the case is shown in which the pendants belong to the same chain and packing in the same layer of pendants belonging to different chains does not necessarily take place. On the other hand, cases were reported where the mesogenic pendant groups of several chains are all interdigitated in a single layer, as in Fig. 3. Lyotropic liquid-crystallinity is relatively rare in the case of LCPs with pendant mesogenic groups.

The general structure of pendant LCPs is described by:

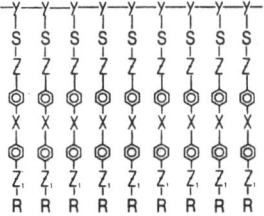

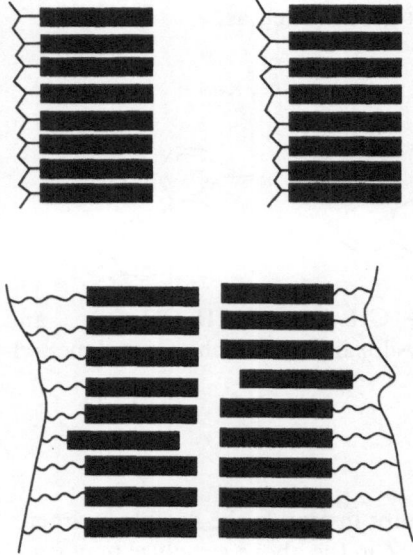

Fig. 2. LCPs containing side-chain mesogens. Organized layers contain pendants attached to the same flexible main-chain

Fig. 3. LCPs containing side-chain mesogens. Layers contain interdigitated pendants connected to several flexible main-chains

where Y is a repeat unit in a flexible main-chain, such as

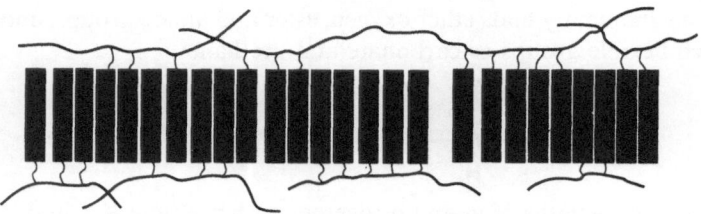

and may contain monomers or oligomers having no mesogenic group attached to them; S is a short or long spacer such as alkylene or alkoxy groups whose flexibility depends on its composition and length. X are stiff groups in the mesogenic units, similar to those listed for the main-chain LCPs. The R groups are generally methyl or longer alkyl moieties. On occasion they may be non-symmetric and optically active, imparting to the LCPs interesting properties such as ferroelectricity, and on other occasions the R group may be missing altogether. The Z and Z_1 groups are connecting groups such as a single bond, -O- and -O-$\overset{\overset{O}{\|}}{C}$-. When R is missing the Z_1 group may contain a single atom such as hydrogen or halogen. The pendant LCPs lend themselves to a broad range of compositions by mixing various liquid crystalline monomers, or mixing meso-genic and non-mesogenic ones, or by using any of the above described mono-mers mixed together with monomers of the same or other copolymerizable unit but bearing a cholesterol-based mesogenic group. In the vast majority of pendant LCPs, the placement of the mesogenic pendant groups is more of less perpendicular to the backbone axis. There exist, however, several examples of pendant LCPs where the mesogenic group is attached in its middle to a flexible spacer such that the mesogenic group adopts a direction more or less parallel to the chain [104, 105]. Copolymers of these and perpendicular pendant mono-mers were also reported.

In flexible main-chain and pendant LCPs, the generally identical length of the mesogenic groups is conducive to their layered arrangement and the appearance of the corresponding smectic liquid crystalline phase. In cases where mesogens not of uniform length are a substantial fraction of all mesogenic species in the system, smectic liquid crystallinity may not be observed. Nematic liquid crystallinity is then often observed. When cholesterol groups serve as the pendant mesogens, cholesteric liquid crystallinity often ensues.

1.2.3 Networks with Flexible Chains and Stiff Mesogenic Groups

Here, we shall first describe networks containing stiff mesogenic groups in the main chain, connected to one another and crosslinked by flexible spacers and junctions. We then describe similar networks in which the junctions are stiff branchpoints. After these, networks with flexible main chains and stiff meso-genic pendants will be discussed. These are also divisible into two: networks in which the junctions consist of flexible units and networks in which flexible chains are connected to one another by means of stiff mesogenic units reacted at both ends. There exist in the literature several LCP networks containing two kinds of mesogens or two kinds of junctions or crosslinks. These are not amenable yet to theoretical description and will not be specifically discussed.

1.2.3.1 Networks with Stiff Main-Chain Mesogens, Flexible Spacers and Flexible junctions

As a rule, networks of this kind are prepared in two steps. In the first, long linear chains containing alternating stiff mesogens units and flexible spacers are prepared. All or many of the spacers contain reactive sites. In a second step, these sites are reacted with difunctional monomers or oligomers to form an LCP network. Depending on the axial ratio, x, and structural characteristics of the mesogenic group, and on its volume fraction in the polymer network, the network may or may not retain the liquid crystal phase of the parent monomers or linear LCP. A schematic representation of such networks is given in Fig. 4 where the stiff mesogenic groups are described by straight heavy lines and all flexible units by wavy lines. The mesogenic groups in these networks are, generally, rather short with axial ratio of three or thereabouts [106–112]. In some of the networks, mixed main-chain and pendant mesogenic groups are present [106,110]. Typical mesogenic groups are esters such as [112]

$$-O-\bigcirc-\overset{O}{\underset{\|}{C}}-O-\bigcirc-O-$$

and [111]

$$-O-\overset{O}{\underset{\|}{C}}-\bigcirc-O-\overset{O}{\underset{\|}{C}}-\bigcirc-\overset{}{\underset{\underset{O}{\|}}{C}}-O-\bigcirc-\overset{O}{\underset{\|}{C}}-O-$$

or units containing azo or azoxy moieties such as [106, 107, 109]:

$$-O-\bigcirc-N=N-\bigcirc-O-$$

and

$$-O-\bigcirc-N=\overset{\overset{O}{\uparrow}}{N}-\bigcirc-O-$$

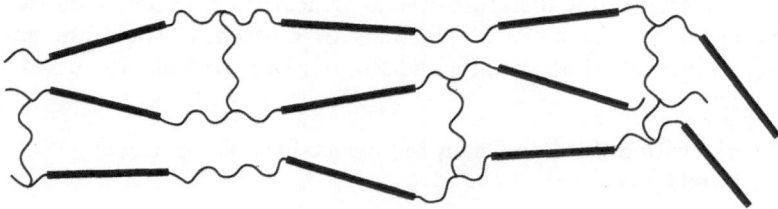

Fig. 4. Two-step network in which all mesogenic groups are in the main-chains and all crosslinks are flexible and effected between flexible main-chain spacers

The flexible crosslinkable units are chemically divergent, ranging from pendant vinyl groups crosslinked with reactive dimethylsiloxane telomer [107, 109]:

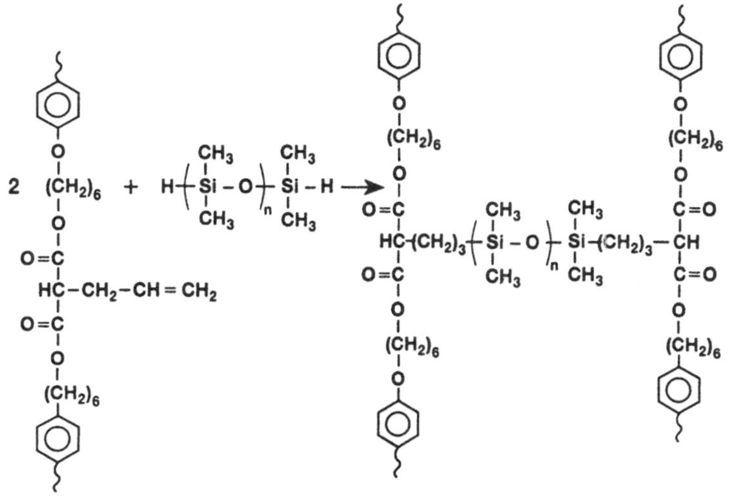

$$2 \; \text{(structure with } (CH_2)_6 \text{ vinyl groups)} + H\!\left[\!Si(CH_3)_2\!-\!O\!\right]_n\! Si(CH_3)_2\!-\!H \longrightarrow \text{(crosslinked product)}$$

to alkene groups positioned in the flexible spacers of the main chain crosslinked by short polystyrene chains [111]:

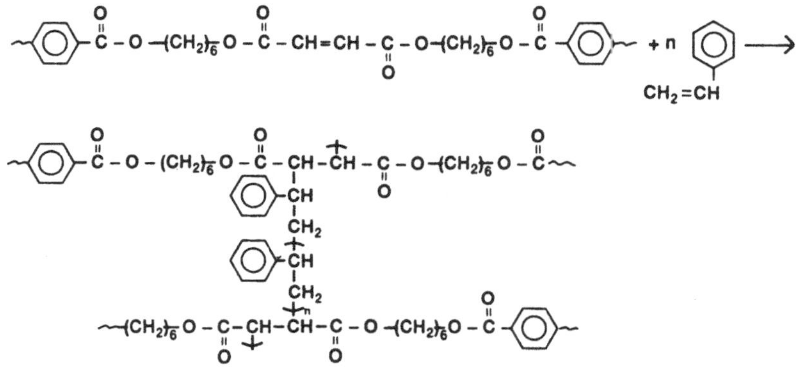

$$\sim\!\!\bigcirc\!\!-\!\!C(=O)\!-\!O\!-\!(CH_2)_6\!-\!O\!-\!C(=O)\!-\!CH\!=\!CH\!-\!C(=O)\!-\!O\!-\!(CH_2)_6\!-\!O\!-\!C(=O)\!-\!\bigcirc\!\!\sim \; + \; n \; \bigcirc\!\!-\!CH\!=\!CH_2 \longrightarrow$$

It is important to recognize that in the last case, linear chains of polystyrene are also being created, as is well known to technologists working in the field of crosslinkable polyester resins [113].

An exception to the two-step synthesis rule is the work of Caruso et al. [112]. Here, long oligomeric units each containing three or five repeat units and terminating with reactive hydroxy groups were first prepared. Then these oligomers were reacted with tricarballylic acid chloride to obtain the crosslinked

LCP network:

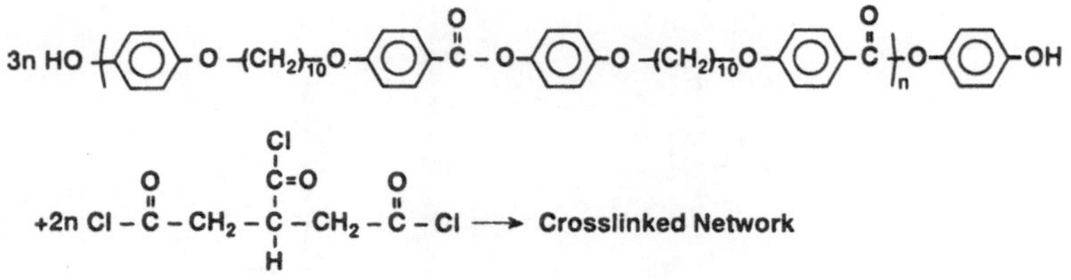

3n HO —[⬡— O —(CH₂)₁₀O —⬡— C(=O) — O —⬡— O —(CH₂)₁₀O —⬡— C(=O)]ₙ— O —⬡— OH

        Cl
        |
    O   C=O   O
    ‖ | ‖
+2n Cl – C – CH₂ – C – CH₂ – C – Cl ⟶ **Crosslinked Network**
        |
        H

forming long chains crosslinked by rather short flexible units at distances corresponding to the lengths of the precursor oligomers. Because the mesogenic group in each repeat unit is very short and comprises a rather small fraction of the total volume of the repeats, it is rather surprising that these networks will be thermotropic in the bulk and lyotropic upon swelling in solvent [112].

1.2.3.2 Networks with Stiff Main-Chain Mesogens, Flexible Spacers and Rigid Branchpoints

We are aware of only two LCP networks falling in this category. In one case, melamine serves as the rigid branchpoint in a network containing main-chain stiff mesogens and a variety of flexible spacers [114, 115]. Typically, hydroxy-terminated oligomers

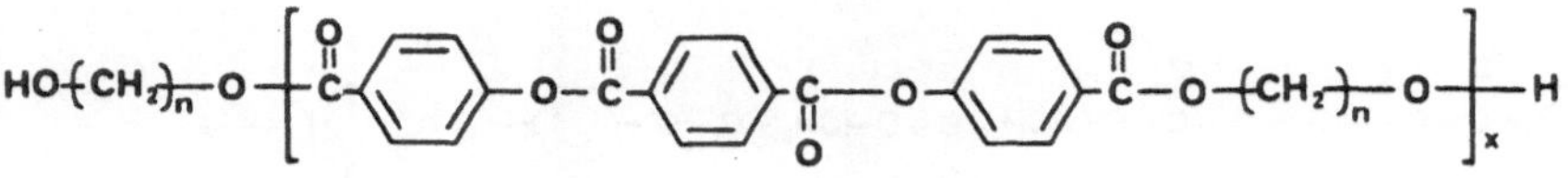

HO–(CH₂)ₙ–O —[C(=O)–⬡–O–C(=O)–⬡–C(=O)–O–⬡–C(=O)–O–(CH₂)ₙ–O]ₓ—H

with x ranging from one to two were first prepared. Then these monomers were condensed together with hexakismethoxymethylmelamine (HMMM) creating branchpoints whose functionality is three but potentially may reach six:

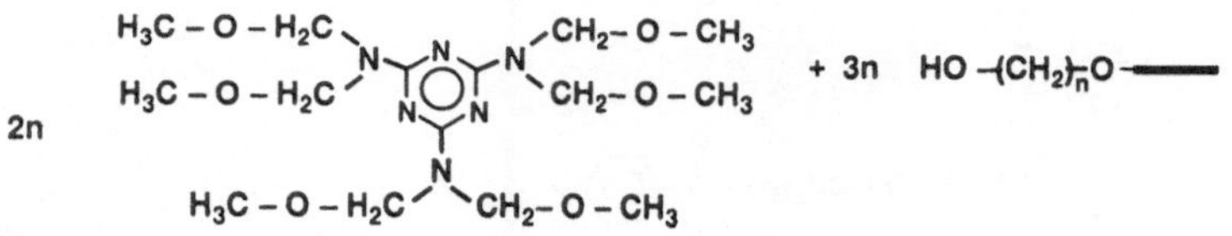

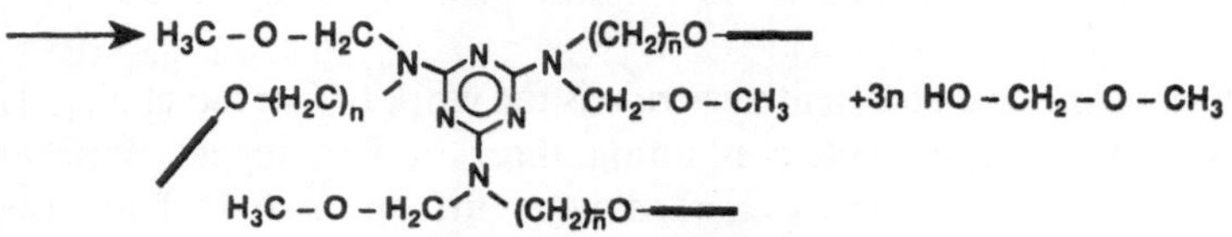

2n

H₃C – O – H₂C \
                N —[triazine ring]— N / CH₂ – O – CH₃
H₃C – O – H₂C /   \ CH₂ – O – CH₃

              |
              N
H₃C – O – H₂C / \ CH₂ – O – CH₃

+ 3n HO –(CH₂)ₙO ——

⟶ H₃C – O – H₂C \
                 N —[triazine ring]— N / (CH₂)ₙO ——
    O–(H₂C)ₙ /   \ CH₂ – O – CH₃

              |
              N
H₃C – O – H₂C / \ (CH₂)ₙO ——

+3n HO – CH₂ – O – CH₃

The branchpoint substitution level on the HMMM nucleus depends on the cure conditions [116, 117]. The presence of unreacted hydroxy groups in the crosslinked networks [114] indicates, however, that not all the functionalities were reacted. Network segments left dangling loose on one end, and segments twice as long as the precursor oligomers and containing a melamine residue in their center, are both present as defects in these LCP networks.

The second member of this category is a two-step network containing linear LCP crosslinked by a flexible unit connecting the centers or two stiff mesogens [118]. As is shown in Fig. 5, the stiff mesogen is converted into a stiff trifunctional group which itself is a part of a four-branch junction point. In this specific case [118], each mesogen is equipped at its center with an aldehyde group

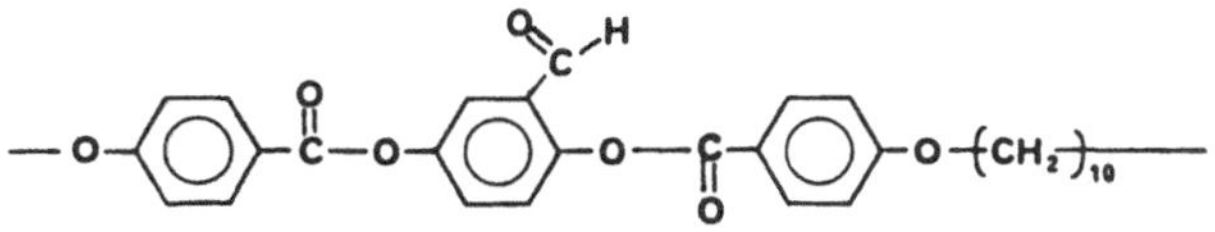

which upon reaction with a flexible diamine forms a flexible tetrafunctional junction containing two Schiff-base moieties:

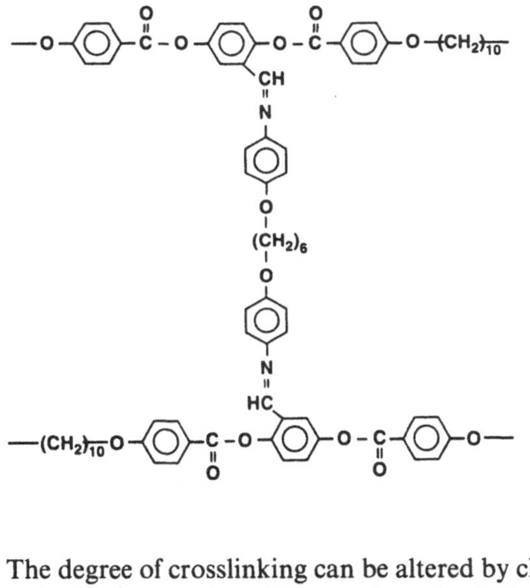

The degree of crosslinking can be altered by changes in the aldehyde to diamine ratio. The network in this case shows both thermotropic and lyotropic behavior over certain temperature intervals.

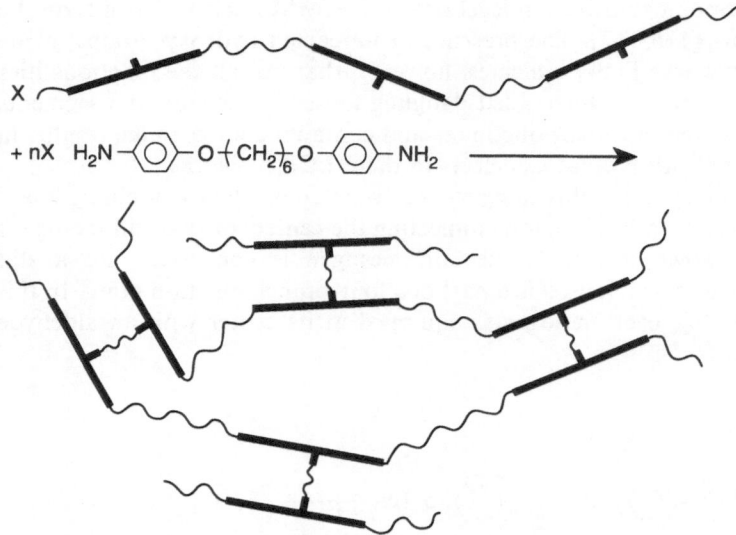

X

$+ nX \quad H_2N-\langle\bigcirc\rangle-O+CH_2\rangle_6 O-\langle\bigcirc\rangle- NH_2 \longrightarrow$

Fig. 5. Two-step network in which flexible junctions connect stiff main-chain mesogens

1.2.3.3 Networks with Stiff Pendant Mesogens, Flexible Main-Chains and Flexible Junctions

A schematic description fitting most LCP networks belonging to this category [119–125] is shown in Fig. 6. An additional interesting photocrosslinkable

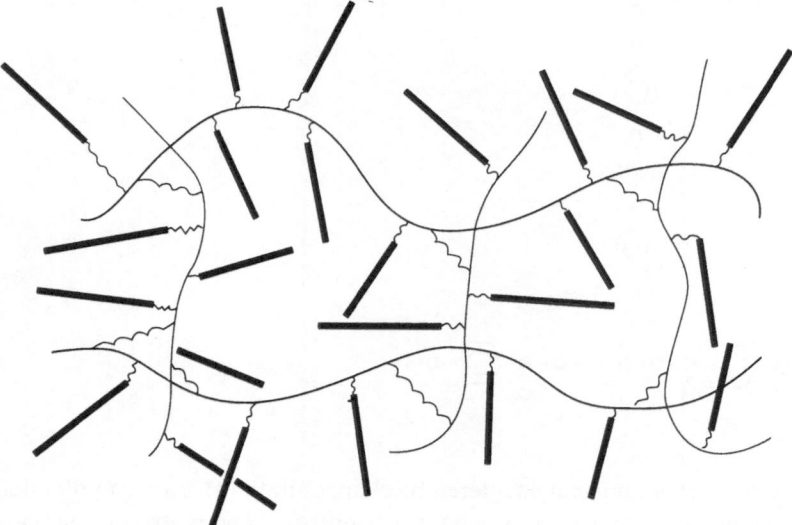

Fig. 6. Network with stiff pendant mesogens connected to flexible main-chains by flexible tethers. The main-chains are connected by flexible crosslinking residues

polymer that should fall in this category but is not claimed as such, contains pendant groups with nonlinear optical activity [126] of axial ratio sufficiently large to exhibit liquid crystallinity. In this case the crosslinking is effected by photoinitiated dimerization of pendant cinnamoyl groups through their double bonds.

In the LCPs of the present category, the flexible main-chains belong to two groups: polysiloxanes [119–122] and acrylates or methacrylates [110, 123, 125, 126]. The polysiloxane polymers were crosslinked by divinyl(dimethylsiloxane) telomers [119] or by groups such as [124]

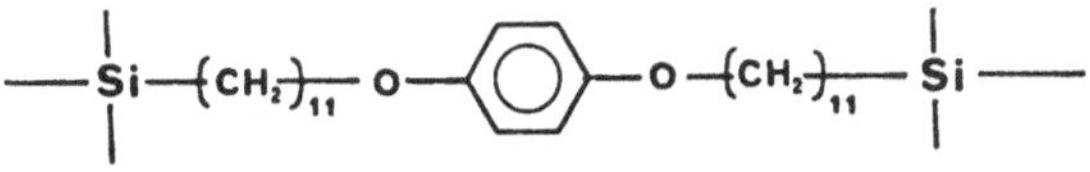

$$-\mathrm{Si}-(\mathrm{CH_2})_{11}-\mathrm{O}-\bigcirc-\mathrm{O}-(\mathrm{CH_2})_{11}-\mathrm{Si}-$$

The acrylates and methacrylates were crosslinked by groups such as [123]

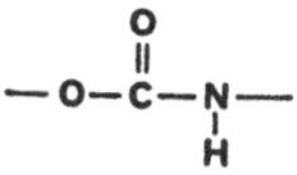

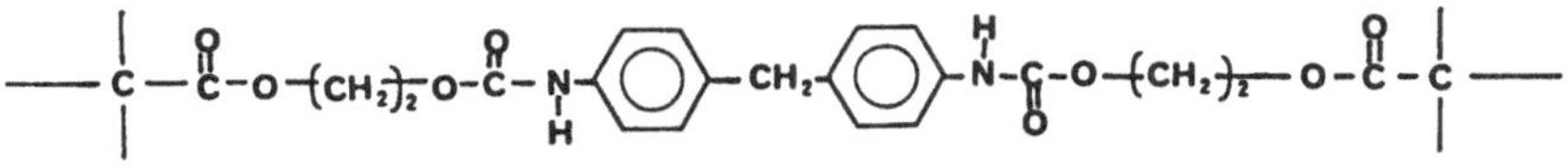

$$-\mathrm{O}-\overset{\overset{\displaystyle O}{\|}}{\mathrm{C}}-\underset{\underset{\displaystyle H}{|}}{\mathrm{N}}-$$

$$-\overset{|}{\mathrm{C}}-\overset{\overset{\displaystyle O}{\|}}{\mathrm{C}}-\mathrm{O}-(\mathrm{CH_2})_2-\mathrm{O}-\overset{\overset{\displaystyle O}{\|}}{\mathrm{C}}-\overset{}{\mathrm{N}}\!-\!\bigcirc\!-\mathrm{CH_2}-\bigcirc\!-\overset{}{\mathrm{N}}-\overset{}{\mathrm{C}}-\mathrm{O}-(\mathrm{CH_2})_2-\mathrm{O}-\overset{\overset{\displaystyle O}{\|}}{\mathrm{C}}-\overset{|}{\mathrm{C}}-$$

or [125]

$$-\overset{|}{\mathrm{C}}-\overset{\overset{\displaystyle O}{\|}}{\mathrm{C}}-\mathrm{O}-(\mathrm{CH_2})_6-\mathrm{O}-\overset{\overset{\displaystyle O}{\|}}{\mathrm{C}}-\overset{}{\mathrm{N}}\!-\!\bigcirc\!-\mathrm{CH_2}-\bigcirc\!-\overset{}{\mathrm{N}}-\overset{}{\mathrm{C}}-\mathrm{O}-(\mathrm{CH_2})_6-\mathrm{O}-\overset{\overset{\displaystyle O}{\|}}{\mathrm{C}}-\overset{|}{\mathrm{C}}-$$

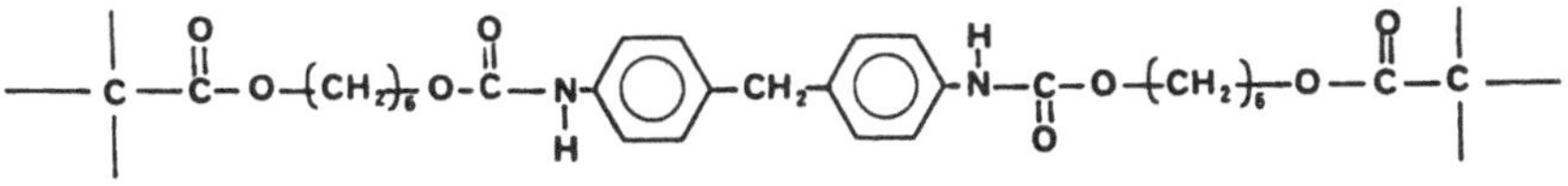

The stiff mesogenic groups in both families are typically [119, 124, 125]

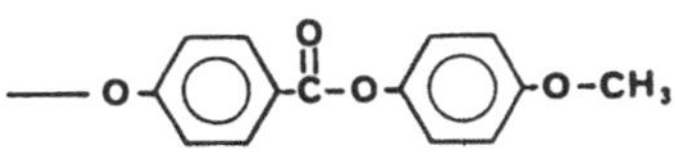

$$-\mathrm{O}-\bigcirc-\overset{\overset{\displaystyle O}{\|}}{\mathrm{C}}-\mathrm{O}-\bigcirc-\mathrm{O}-\mathrm{CH_3}$$

but groups such as

$$\mathrm{N}\!\equiv\!\mathrm{C}-\bigcirc-\overset{\overset{\displaystyle O}{\|}}{\mathrm{C}}-\mathrm{O}-\bigcirc-\mathrm{O}-\overset{\overset{\displaystyle O}{\|}}{\mathrm{C}}-\bigcirc-\mathrm{C}\!\equiv\!\mathrm{N}$$

(with a $(\mathrm{CH_2})_{11}$ tether)

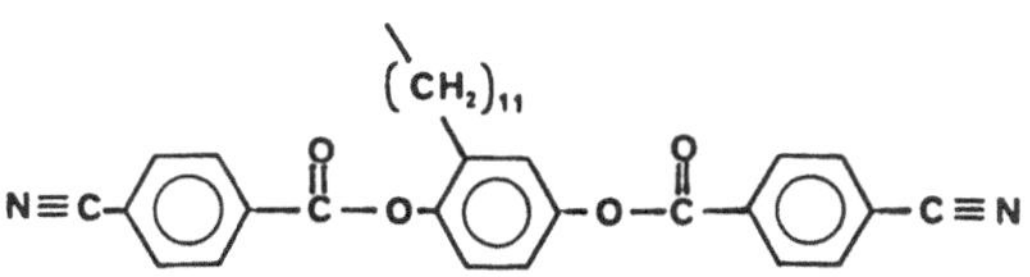

where the center of the mesogen is connected to the flexible main-chain by a flexible tether, were also reported [123]. Cholesteric [119, 124, 127] and chiral [108, 127, 128] mesogenic pendant groups were reported, often mixed with other, non-cholesteric and non-chiral mesogens. In most instances the LCP networks were prepared in one step from an appropriate mixture of mono-functional and difunctional monomers. In at least one case [125] the networks were prepared in two steps: in this first, reactive monomers containing hydroxy groups were polymerized into the polymethacrylate chains, and in a separate second step urethane crosslinks were formed by a reaction of the hydroxy groups with diisocyanate-bearing molecules. Because of the uniform length of the mesogenic pendants in most of the above LCP networks, they are often smectic [110, 119] but many are sufficiently constrained in movement and, hence, exist in the less ordered nematic state [110, 119, 123].

A different LCP network falling in the present category was recently reported [129]. In this case, a mesogenic group serves as both mesogen and crosslinking agent for a substantially flexible diepoxy:

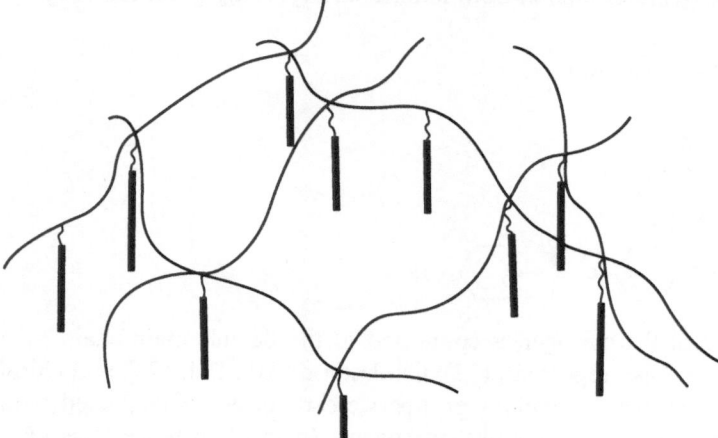

The resulting network is schematically shown in Fig. 7 where the majority of the mesomorphic pendants are attached to the network at the crosslink points. In

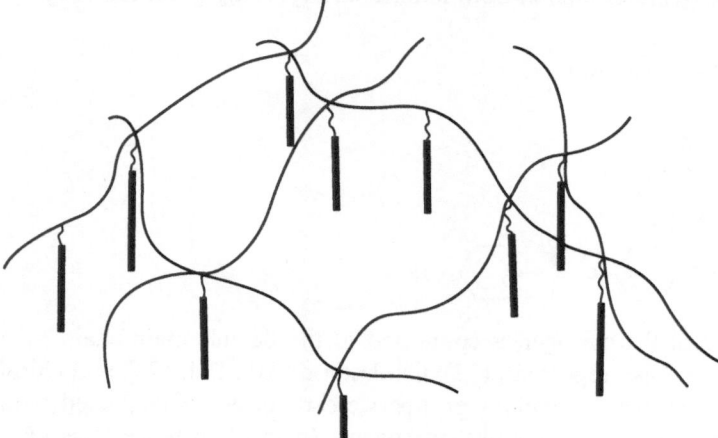

Fig. 7. Network of crosslinked flexible main-chains. The stiff mesogens are attached by flexible tethers to the junction points

instances where only one epoxy group reacts with each of the two amines in each mesogen, then the mesogens remain hanging from the network segments and not from the crosslink junctions.

1.2.3.4 Networks with Stiff Pendant Mesogens Connected at Both Ends to Flexible Main-Chains

Almost all LCP networks in this class were prepared in a single-step addition polymerization from diacrylate esters containing a stiff mesogenic group in between. Most such polymers exclusively comprise the diacrylate residues, but many are copolymers containing monoacrylate pendant mesogens connected to the flexible main-chain at only one end, in addition to the diacrylates. A crude schematic description of such networks is shown in Fig. 8, where the solid bars represent mesogens connected to the flexible chains at both ends, and the dashed bars stand for pendant mesogens attached at only one end. In the sketch the spacers, or tethers, are indicated by zigzag lines. In some instances they are relatively short while in others they may be quite long. The spacers' flexibility or, rather, their ability to decouple the motions of the mesogenic groups from those of the main chain, beyond being a function of temperature, is a function of the number of single bonds along the spacer and the ease with which they may interconvert between one rotational isomeric state and another. In the pioneering work of Liebert and Strzelecki [69, 130–134] and Blumstein and associates [135], the spacers were the rather short carboxylate group and the

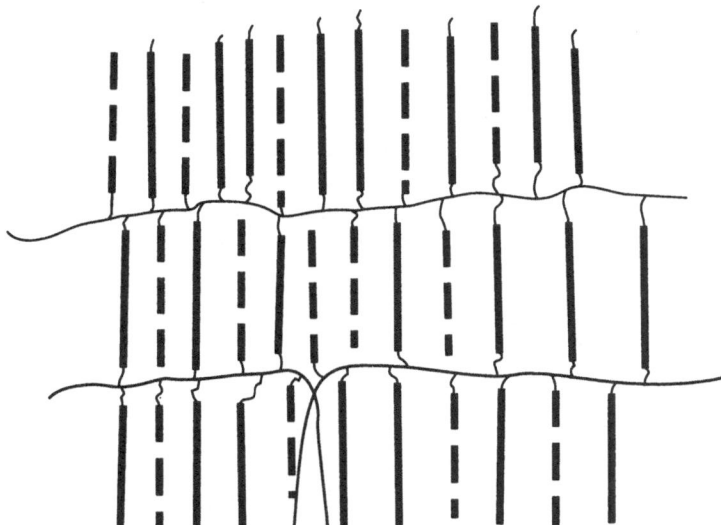

Fig. 8. Network containing flexible main-chains and tethered stiff mesogenic side-groups. The crosslinks are effected exclusively through stiff mesogens

stiff mesogenic unit always contained the substantially stiff Schiff base -CH = N-. The corresponding monomeric species are:

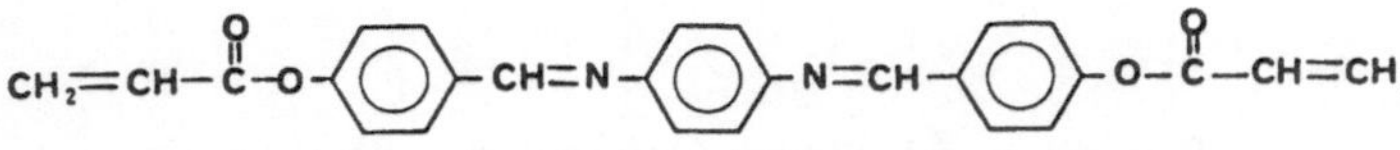

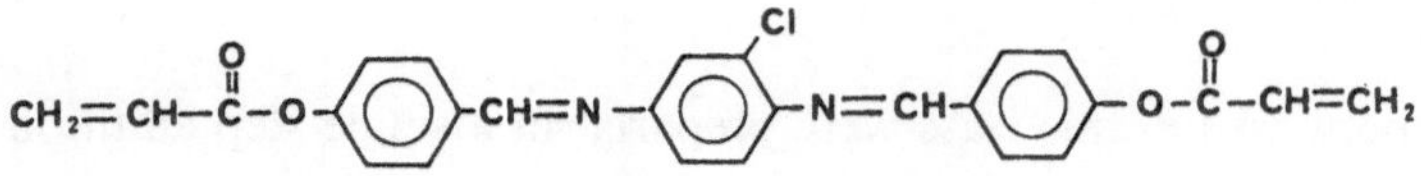

$$CH_2=CH-\overset{O}{\overset{\|}{C}}-O-\text{⟨○⟩}-CH=N-\text{⟨○⟩}-N=CH-\text{⟨○⟩}-O-\overset{O}{\overset{\|}{C}}-CH=CH_2$$

$$CH_2=CH-\overset{O}{\overset{\|}{C}}-O-\text{⟨○⟩}-CH=N-\text{⟨○⟩}\overset{Cl}{}-N=CH-\text{⟨○⟩}-O-\overset{O}{\overset{\|}{C}}-CH=CH_2$$

and

$$CH_2=CH-\overset{O}{\overset{\|}{C}}-O-\text{⟨○⟩}-CH=N-N=CH-\text{⟨○⟩}-O-\overset{O}{\overset{\|}{C}}-CH=CH_2$$

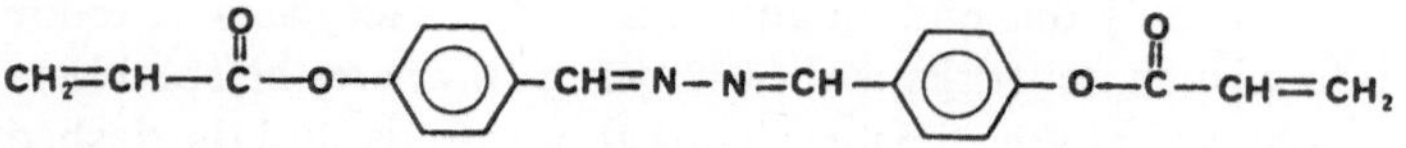

In one case [69] the mesogen contained the ethynylene group instead of Schiff base:

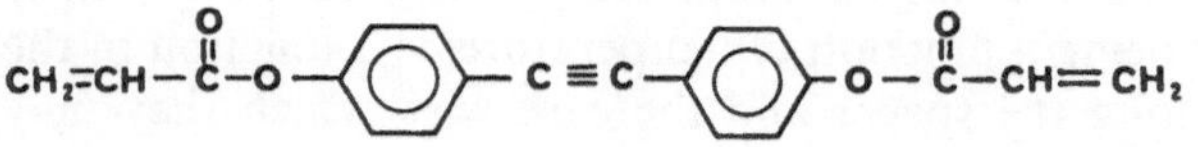

$$CH_2=CH-\overset{O}{\overset{\|}{C}}-O-\text{⟨○⟩}-C\equiv C-\text{⟨○⟩}-O-\overset{O}{\overset{\|}{C}}-CH=CH_2$$

and in another [135], where styrene replaced acrylate as the polymerizable group, no spacer was present between the flexible main-chain and the stiff mesogen:

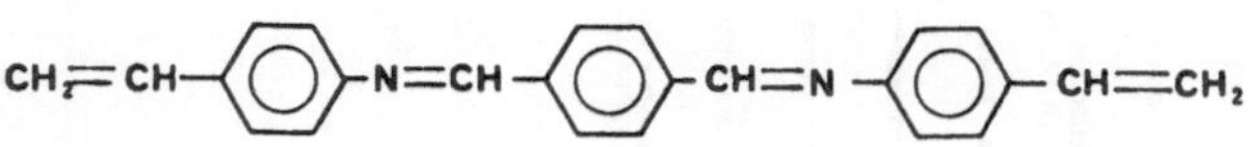

$$CH_2=CH-\text{⟨○⟩}-N=CH-\text{⟨○⟩}-CH=N-\text{⟨○⟩}-CH=CH_2$$

The monomers were generally polymerized in the absence of diluents and catalyst, by heating them above their crystal melting points to produce the corresponding crosslinked liquid crystal homopolymer networks. On occasion a difunctional monomer was mixed with a monofunctional one, such as [131]

$$CH_2=CH-\overset{O}{\overset{\|}{C}}-O-\text{⟨○⟩}-CH=N-\text{⟨○⟩}-R \quad ; \quad R= -(CH_2)_3\; CH_3, -CN, -Br$$

or cholesterol acrylate [132, 136] or a mixture thereof [132] to obtain the corresponding copolymers or terpolymers. When cholesterol-bearing pendants were used, LCP networks with a cholesteric liquid crystal phase were obtained [132].

A group of Celanese Corp. patents [137–139] describes difunctional mesogenic monomers with rather short spacers, as above, but with the aromatic residues in the stiff mesogens being connected by ester groups. A generic structure is

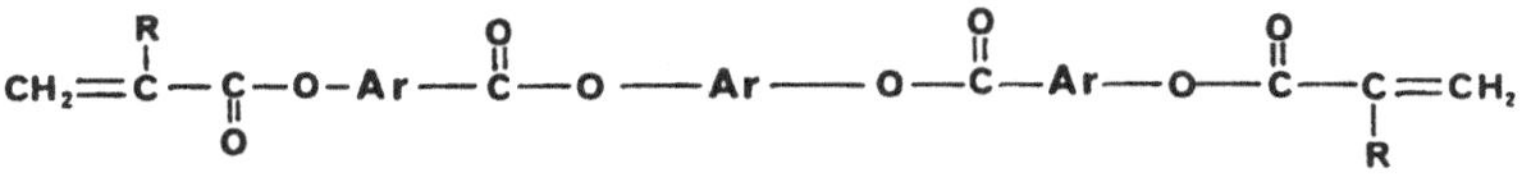

$$CH_2{=}C{-}C{-}O{-}Ar{-}C{-}O{-\!\!-}Ar{-\!\!-}O{-}C{-}Ar{-}O{-}C{-}C{=}CH_2$$

where Ar may stand for

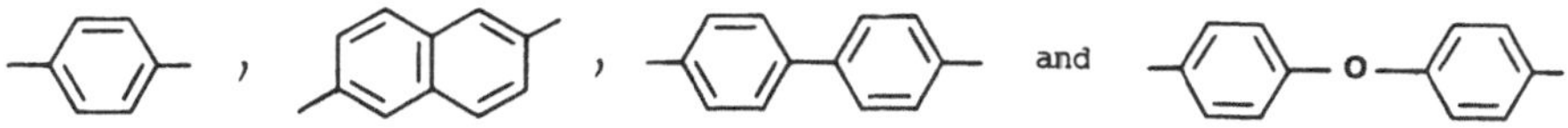

or any combination thereof, and R may stand for any of $-H$, $-O-CH_3$, $-CH_3$ or Cl.

In all the above, the most flexible component of the network was the main chain

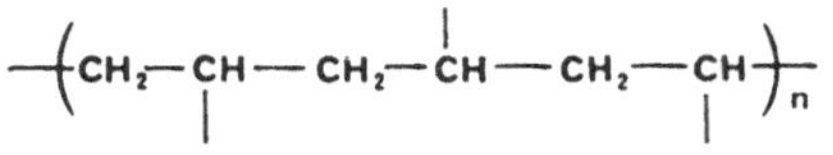

$$-\!\!\left(CH_2{-}CH{-\!\!-}CH_2{-}CH{-\!\!-}CH_2{-}CH\right)_{\!n}$$

With the carboxylate spacers being somewhat less flexible. To increase the overall flexibility of the LCP networks, a group at Philips Research Laboratories synthesized diacrylate monomers containing hexamethylene spacers at both ends [140–143]:

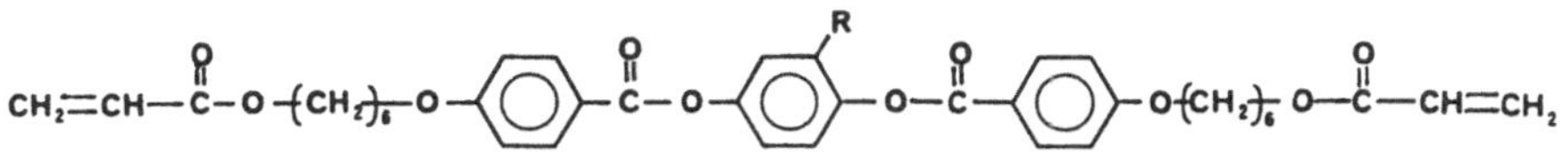

$$CH_2{=}CH{-}C{-}O{-}(CH_2)_6{-}O{-}\!\!\bigcirc\!\!{-}C{-}O{-}\!\!\bigcirc\!\!{-}O{-}C{-}\!\!\bigcirc\!\!{-}O(CH_2)_6\;O{-}C{-}CH{=}CH_2$$

With R $= -H$ or $-CH_3$, and proceeded to polymerize each alone, their mixtures [141, 143], and mixtures of them with a monofunctional monomer containing the same length spacer [142]:

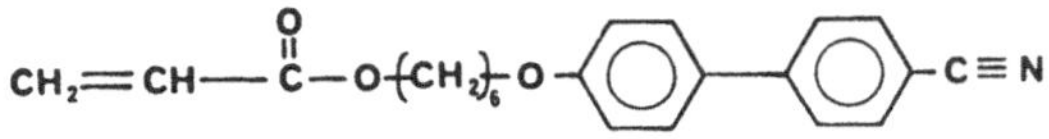

$$CH_2{=}CH{-}C{-}O{-}(CH_2)_6{-}O{-}\!\!\bigcirc\!\!{-}\!\!\bigcirc\!\!{-}C{\equiv}N$$

In this case, the polymerizations were all photoinitiated and were carried separately in the isotropic and in the oriented state.

Depending on the particular difunctional monomer used in systems with either short or long spacers, on the presence or absence of monofunctional monomers, and on the state of orientation of the monomers during the polymerization [144], smectic or nematic liquid-crystal phases were found to exist in the temperature interval between the crystalline and the isotropic melt states. Naturally, when smectic phase was found to be present, it converted to a nematic phase upon heating before reaching the clearing point signaling the isotropic phase.

Recently [145, 146], diepoxy-terminated mesogenic monomers were prepared

$$CH_2\!-\!CH\!-\!CH_2\!-\!O\!-\!\overset{O}{\underset{O}{C}}\!-\!\langle O\rangle\!-\!O\!-\!\overset{O}{C}\!-\!\langle O\rangle\!-\!\overset{O}{\underset{O}{C}}\!-\!O\!-\!\langle O\rangle\!-\!\overset{O}{C}\!-\!O\!-\!CH_2\!-\!CH\!-\!CH_2$$

which were converted to LCP networks, containing flexible chains and very short flexible spacers, by reactions with diamines or triamines [144]. When the monomers were homopolymerized by heating above their isotropization temperature, or when copolymerized with acid anhydride in solution, the polymeric networks turned out to be isotropic [144]. In the diamine-crosslinked liquid crystalline state, a typical junction may look like

$$\langle O\rangle\!-\!\overset{O}{C}\!-\!O\!-\!CH_2\!-\!\overset{OH}{CH}\!-\!CH_2\!-\!\underset{H}{N}\!-\!\langle O\rangle\!-\!\overset{O}{\underset{O}{S}}\!-\!\langle O\rangle\!-\!N\!\left\langle\begin{array}{l}CH_2\!-\!\overset{OH}{CH}\!-\!CH_2\!-\!O\!-\!\overset{O}{C}\!-\!\langle O\rangle\\[4pt]CH_2\!-\!\underset{OH}{CH}\!-\!CH_2\!-\!O\!-\!\underset{O}{C}\!-\!\langle O\rangle\end{array}\right.$$

and the structure of the network may look schematically as in Fig. 9.

A novel variation on the above system, containing flexible spacers between the flexible main-chain and the stiff mesogens, was recently reported [147, 148]. Here, instead of a single mesogen connecting the flexible main-chains, several stiff mesogenic groups are end-linked by flexible spacers in a fashion reminiscent of sausage links. The LCP networks are prepared in two steps. In the first, reactive "sausage link" oligomers are prepared, end-capped by epoxy groups on both ends:

$$CH_2\!-\!CH\!-\!CH_2\!-\!O\!-\!\left[\langle O\rangle\!-\!CH\!=\!\overset{CH_3}{C}\!-\!\langle O\rangle\!-\!O\!-\!(CH_2)_x\!O\right]_n\!\!\langle O\rangle\!-\!CH\!=\!\overset{CH_3}{C}\!-\!\langle O\rangle\!-\!O\!-\!CH_2\!-\!CH\!-\!CH_2$$

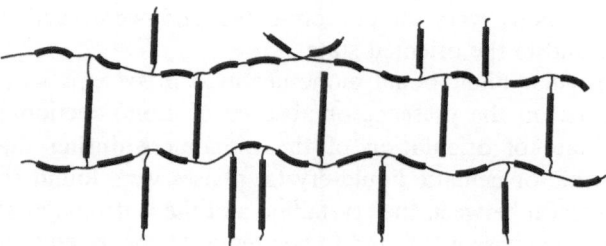

Fig. 9. Network containing both main-chain and side-chain mesogens with crosslinks only between flexible units

With X = 5 or 7 and n ranging from 3 to 10. In the second step, the glycidyl groups are reacted with the diamine methylenedianiline in such a way that flexible main-chains containing repeat units typified by the trifunctional

junction, as well as the tetrafunctional analogue and the difunctional chain-extending diamine unit

are created. Mostly nematic LCP networks were obtained [147]. In this case, the resulting LCP network may appear schematically as in Fig. 10.

1.2.4 Networks Containing No Flexible Spacers

Networks containing no flexible spacers are divisible into two classes: (a) networks with stiff, rigid or wormlike, mesogenic segments connected by either rigid or flexible branchpoints, and (b) networks in which the polymeric

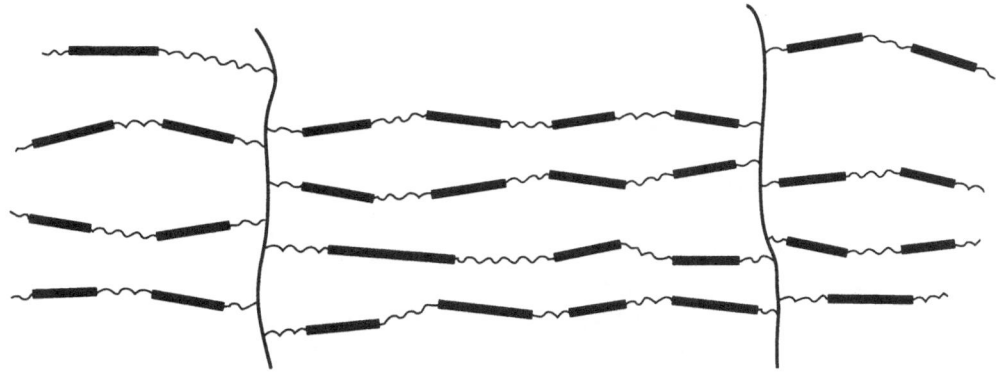

Fig. 10. "Sausage links" network in which reactive end-capped oligomers, each containing several mesogenic units connected by flexible spacers, are reacted to form networks with flexible main-chains

chains are of uniform stiffness, imparting to them liquid crystal behavior in the uncrosslinked state. In the crosslinked networks these polymer chains are usually held together by flexible, often rather long, crosslinking species.

In group (a), rigid branchpoints may be as simple as the 1,3,5–benzenetriamide residue:

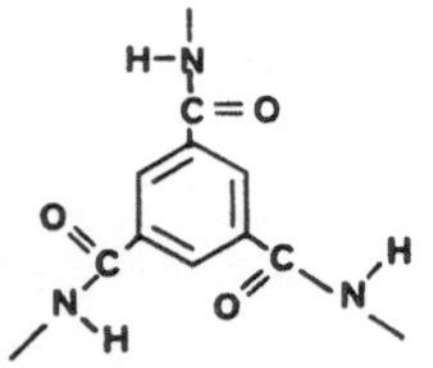

to be discussed in Sect. 3.2.1 below, or the more complex fully reacted poly-silsesquioxane held together by terphenyl mesogenic bridges recently described by Webster and associates [149]:

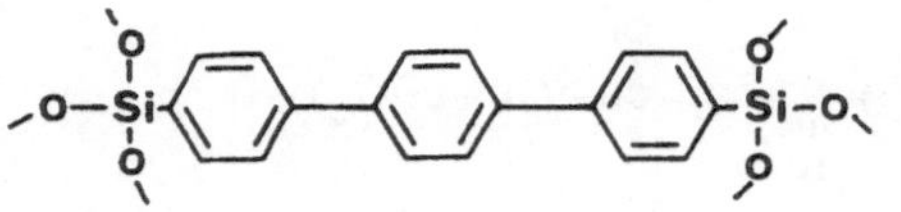

Flexible junctions may be as simple as the tricarballylic amide residue to be discussed in Sect. 3.2.2, or as complex as the cyclic penta(methysiloxane) group recently described by Pachter et al. [150]. Poorly defined flexible junctions are claimed in the patent literature and will be discussed below when appropriate. In group (b), long alkylene chains holding together stiff chains, are typical.

1.2.4.1 Uniform Main-Chain Mesogenicity: Flexible, Wormlike, Rodlike

A group of characteristic values of polymer chains reflects their thermodynamic flexibility or, conversely, their stiffness. In this group one finds the characteristic ratio, the persistence length and the Kuhn segment length. Over the years, much efforts has been invested in attempts to correlate these characteristic values with certain chain structural features and bulk properties of the polymers. Most of the efforts involved attempted correlations of chain stiffness parameters with characteristic numbers describing the chain arc- or contour-length between entanglements [151–167], and through them the entanglement concentration. Several attempts deal with viscoelastic [164, 168, 169] and mechanical [170–173] properties. Building on Flory's thermodynamic model for semi-flexible polymers undergoing isotropic to anisotropic transition [174], one of us [175] derived early on an empirical relationship relating chain persistence length a and chain average diameter d with the onset of anisotropy. Accordingly, for a polymer in the quiescent bulk under equilibrium to exist in an anisotropic state, its rigidity must be such that

$$a \geq 1.58d \tag{7a}$$

Recalling from Eq. (2) that each Kuhn segment equals two persistence lengths [53], we find that for a wormlike chain a minimum of

$$A \geq 3.16d \tag{7b}$$

is required for the onset of liquid crystalline behavior. Such an axial ratio, of about three, is consistent with the axial ratios of the mesogenic groups experimentally observed for stiff mesogenic segments connected in the main-chain by flexible spacers. Very recent computer modeling studies [176] imply that the transition to anisotropic behavior occurs when

$$A \geq 5d, \tag{8}$$

a size in the upper extreme of the experimental observations.

Typical examples of LPCs with uniform main-chain mesogenicity will now be described. The aromatic ones will be presented only briefly here for the purpose of logical presentation. The preparation of linear and network aromatic LCPs will be delayed to sect. 3 and discussed there in substantial detail.

There exists a group of polymer chains that in dilute solution behave as expanded coils. Under such dilutions no liquid crystallinity is evident. However, in bulk or in highly concentrated solutions their chains or long chain-sections prefer to align themselves in more or less parallel arrays, like bunches of grapevine twigs after pruning. Because of their chain length variability they cannot arrange themselves in layers of rather uniform thickness. They therefore exhibit thermotropic nematic liquid crystallinity and in a few instances also lyotropic nematicity. Representative of such polymers are poly(dialkyl siloxane) with alkyl groups such as ethyl, n–propyl or n-butyl [177–182], polyorgano-phosphazenes [183] with side groups such as 1,1-dihydroxyperfluoroalkoxy [102, 175], phenoxy, or substituted phenoxy [175]:

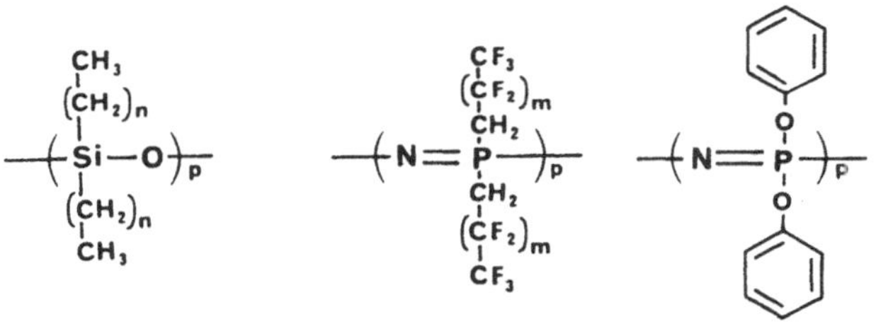

where $n = 1,2,3 \cdots$ and m $= 0,1,2$. Other representative polymers of this class of nematic LCPs are poly(vinyl trialkyl and aryl silane) [102] and poly(n-octyl isocyanide) [184]:

R = methyl, ethyl, propyl, phenyl.

The common denominator of all these polymers is that neither their backbone nor their side-groups contain mesogenic groups, yet all exhibit nematic liquid crystallinity. We believe that the backbone of all these polymers is forced into a rather extended configuration in order to accommodate the packing of the side-groups along the chain. Very small side groups, such as methyl, do not require the backbone to uncoil and extend in order to accommodate them and, thus, they do not induce liquid crystallinity. On the other hand, large groups such as aromatic rings attached directly to the backbone are too large and thick to be accommodated even if the chain were to be fully extended. Therefore, such substituents force the chain into a highly contorted shape, behaving as a stiff coil, to accommodate all substituent groups along the chain. By being highly contorted such chains or long chain sections cannot approach one another to form parallel arrays. Mesomorphicity is, hence, prevented from occurring. Such a behavior was clearly observed in poly-isocyanides [175, 184] and polyisocyanates [185] where an appropriate alkyl or aralkyl group substitution supported mesomorphicity, and a direct substitution of aryls or aralkyls with only one or two carbon atoms connecting the ring to the backbone forced the chains out of their extended configuration into a controted one and by so doing prevented mesomorphic behavior.

The interplay between the energetically preferred torsional angles in the backbone and the size-dependent packing of the side-groups often leads to backbone configurations that are not simple planar zigzag, but are helices of one kind or another. Mesomorphic polyisocyanides prefer to exist in a 4/1 helix [186] while polyisocyanates adopt 8/3 or 8/5 helical configurations [187, 188]. Poly(vinyltrimethylsilane) adopts 15/4 helical conformation [102]. Polytetra-fluoroethylene, which owes its extended configuration to tight packing of the fluorine atoms and is considered by some to exhibit mesomorphic character-istics [189, 190], exists in 13/6 and 15/7 helical conformations [190, 191].

An important group of LCPs, owing its mesomorphicity to the interplay between side-group packing and energetically-favored main-chain valence and torsional angles, is the poly(n-alkylisocyanates) and the copolymers of alkyl and aralykyl isocyanates [185, 192–197]:

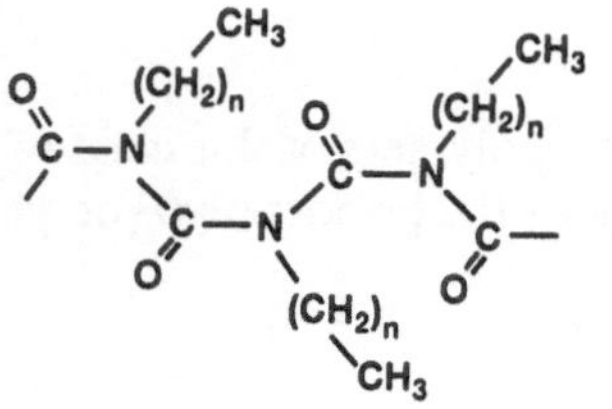

n = 2 to 12
Poly(n-alkylisocyanates)

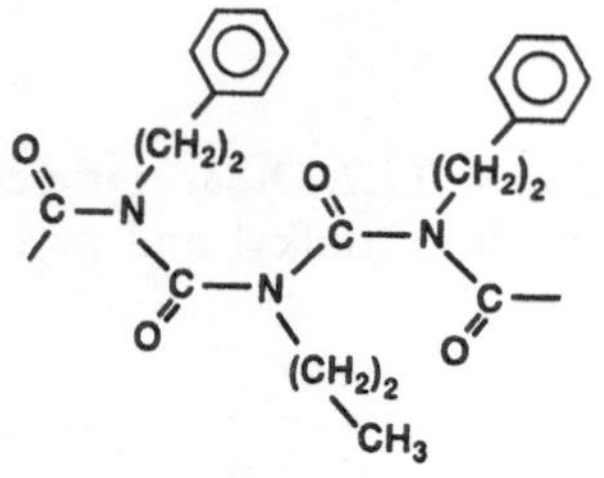

Substituent placement is random
Copoly(alkyl + aralkylisocyanates)

In a fashion similar to many polyphosphazenes [183], the backbone con-formation in mesogenic polyisocyanates is generally *cis-trans*, and the chain adopts a helical configuration in order to accommodate the packing needs of the side-groups. The balancing of *cis-trans* conformation with the helical configuration causes the liquid crystalline polyisocyanate chains to be highly extended, with persistence lengths characteristically about 100 nanometers in size [185, 198–203]. Such persistence lengths are larger than observed for most other LCPs. Computer generated, energy minimized segments of poly-(*n*-butylisocyanate), in Fig. 11, clearly show the rigid helix nature of the chains, with the helix being very close to the one deduced by Shmueli et al. [187]. It should be noted that trans and gauche placements of the side-chains appear not to affect the main-chain rigidity and conformation.

Poly(n-butyl isocyanate)
Energy Minimized

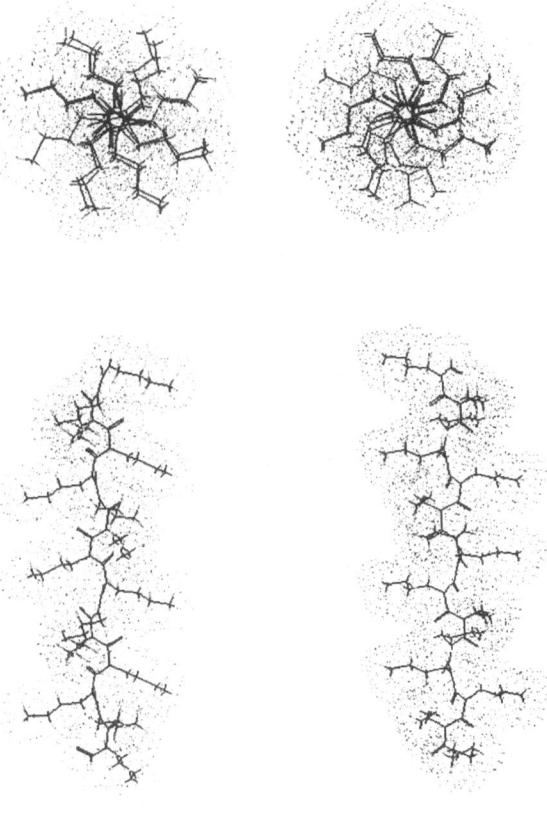

Trans   **Gauche**

Fig. 11. Rigid helical poly(*n*-butylisocyanate). Molecular simulation using POLYGRAPH computer program with energy minimization. *Top*: view down the axis of the helix. *Bottom*: side view. *Left*: the butyl groups are in *trans* conformation. *Right*: the butyls are in *gauche* conformation

The uniform rigidity (or flexibility) of the LCPs described in this section until now is caused by the monomeric repeat distance tightly matching the size of side-groups correlated with one another by characteristic valence and torsion angles energetically accessible along the main-chain and side-groups. A large group of LCPs owe their mesomorphicity to the presence in the main-chain of relatively large and rigid groups, characterized by long virtual bonds, and to the presence between them of either single bonds or short and more or less colinear or coaxial groups having rather high energy barriers between potential rotational isomeric states. Among such polymers one finds aromatic polyesters, polyamides, polyimides, polybenzimidazoles, polybenzoxazoles and polybenzothiazoles. The much shorter chains of unsubstituted and substituted polyphenylenes and the aliphatic staffanes are extreme examples of chain stiffness, approaching rigid rodlike behavior.

Liquid crystalline aromatic polyesters are well known and their structure-property relationships are chronicled in detail in the literature [204–208]. Their generic structure is:

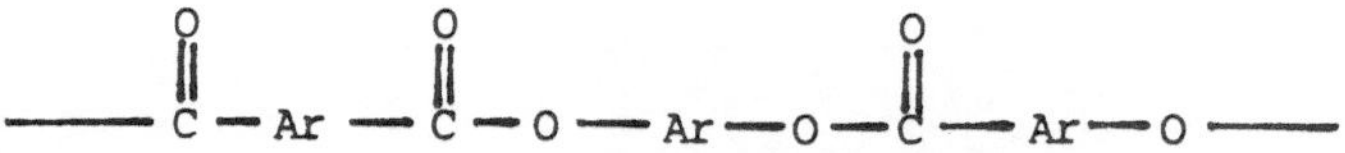

$$ \overset{\displaystyle O}{\underset{\parallel}{C}} - Ar - \overset{\displaystyle O}{\underset{\parallel}{C}} - O - Ar - O - \overset{\displaystyle O}{\underset{\parallel}{C}} - Ar - O $$

where Ar stands for groups as 1,4-phenylene, 4,4′-biphenylene, substituted 4,4′-biphenylenes, 2,6-naphthalene and their various combinations. Because the exclusively para- polyesters tend to be highly crystalline and because such crystals are immune against most solvents they usually dissolve concomitantly with polymer degradation. To prevent this, aromatic groups capable of breaking the symmetry and strict linearity of the chain are frequently added to the polymers. They reduce the melt temperatures to ranges where the polyesters can be processed without excessive degradation. Such groups may be *m*-phenylene, 2,7-napthalene, and various substituents on the backbone rings such as phenyl, *tert*-butyl and a variety of alkoxy groups or halogen atoms. The polymeric products are routinely thermotropic, but can be synthesized to be also lyotropic. For instance, a decade ago one of us synthesized [209] a polyester which is both lyotropic and thermotropic:

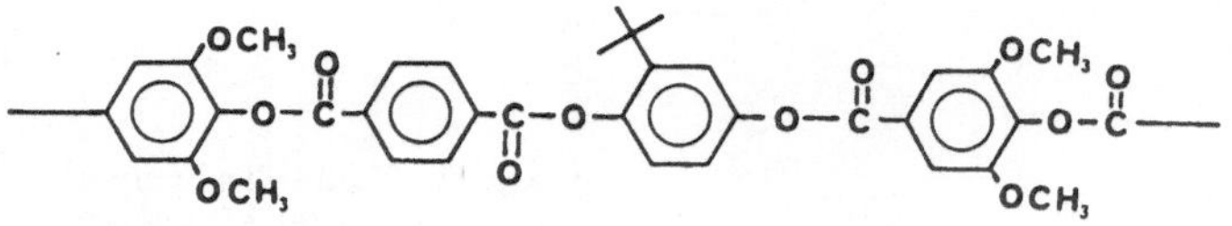

from 2:1:1 molar syringic acid, *tert*-butylhydroquinone and terephthalic acid. This polymer undergoes a crystal to nematic transition at 245 °C and changes to a clear isotropic melt at about 345 °C.

Aromatic polyamide LCPs are very well known [96, 204, 210, 211]. Initially, only poly(*p*-benzamide) (p-BA) and poly(*p*-phenyleneterephthalamide) (*p*-PT) were known:

p-BA   p-PT

More recently, one of us [56, 212, 213] started the routine synthesis of poly(p-benzanilide terephthalamide) (p-BT) and poly(p-benzanilide nitro-terephthalamide) (p-BNT):

p-BT   p-BNT

Later, substituted aromatic polyamides such as

and

were reported [214–216]. Aromatic polyamide LCPs containing occasional methylene groups randomly placed along the chain were also prepared [54]. In this case, the single methylene groups are not flexible enough to serve as flexible spacers, and the chains adopt wormlike curvature with kinks at random locations.

Aromatic poly(ester amide) main-chain LCPs were also described in the literature [61]. A typical one was prepared in two steps from 2:1:1 p-aminobenzoic acid, terephthaloyl chloride and methylhydroquinone:

A group of stiffer wormlike chains includes several fully aromatic polyimides [217, 218] where the Ar stands for para- and occasionally meta- or colinearly-substituted aromatic group, polybenzimidazoles [219–223]:

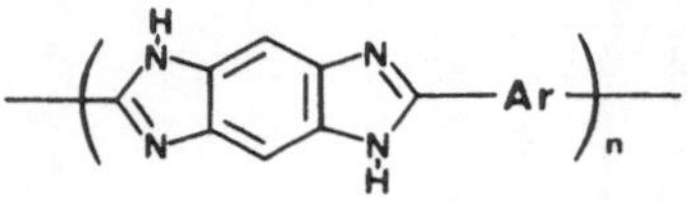

polybenzoxazoles [224–226]:

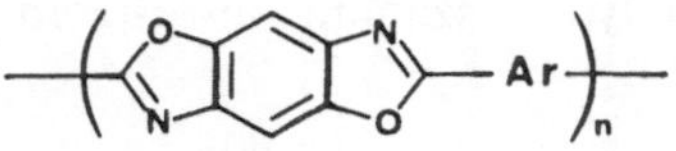

and polybenzothiazoles [226–231]:

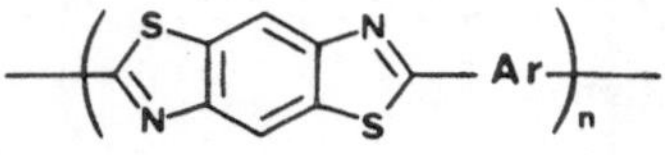

All these polymers are known to be lyotropic liquid crystals [226, 227]. Because their moieties are stiff and can rotate only around a single bond placed in the chain axis direction, a behavior closer to that of rodlike particles ensues [226–230]. These polymers are also interesting because varying lengths of the chain can be twisted, say at 90° relative to the rest of the chain, without substantial chain bending [232]. The strong acid solvents required for the dissolution of most such polymers makes it rather problematic to study their solution properties [223–227], but a variety of substituents attached to the aromatic rings renders some of them sufficiently soluble to facilitate characterization and easier processing [223, 224, 233, 234].

Rodlike hydrocarbon LCPs exist at present only in oligomeric or short-chain lengths. They are either the fully aromatic poly-p-phenylenes [235–237a], the partly aromatic substituted poly-p-phenylenes [238–244], or the fully aliphatic staffanes [245–248]. They are all rendered rodlike because each structural unit is connected to its identical neighbors by a single bond and all these bonds are coaxial. The simplest of these polymers, and also the least tractable, are the poly-p-phenylenes:

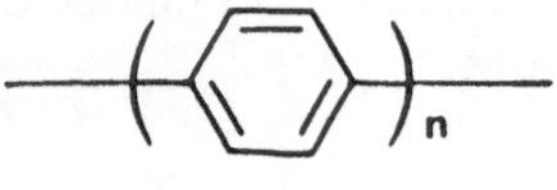

which were obtained in lengths up to $n = 30$. The substituted polyphenylenes are rendered soluble by their side-groups which are most frequently linear alkyls, alkoxy or aromatic groups:

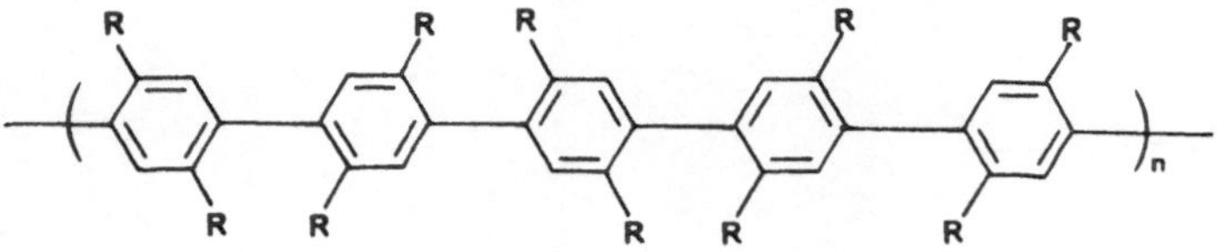

The stiff cage-like repeat units are the reason why the aliphatic rodlike staffanes are so rigid. The most frequently described repeat unit is:

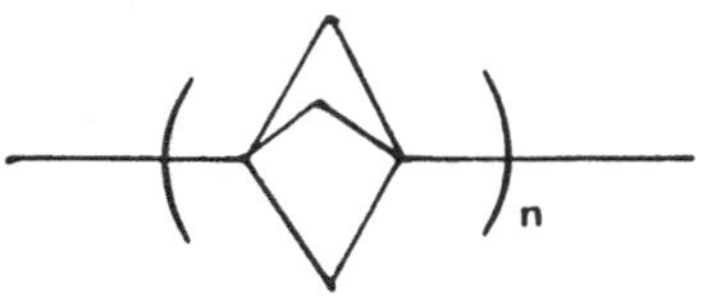

Several such compounds, homopolymerized with $n \geq 3$ or copolymerized with aromatic rings or connected through ester groups [246], were reported to be liquid crystals [246, 248].

The preparation of networks from wormlike polymers such as polyesters and polyamides, and from stiffer polymers such as polyimides and poly-p-phenylenes, will be described in Sect. 3 below. In addition to the chemical nature of the stiff segments, attention will be paid to the rigidity or flexibility of the junction points, and to the sequence of reactions performed in order to obtain various rigid aromatic networks.

1.2.4.2 LCPs from Biological Origin and Their Networks

A large group of biological and biomimetic polymers behave as liquid crystals. They are divided into two major families: polypeptides [249, 250] and polysaccharides [175, 251–262]. Nucleic acids such as DNA [263, 264] and RNA [265, 266] are mesomorphic in concentrated solution. The biomimetic poly(γ-benzyl-L-glutamate) (PBLG) is probably the most investigated of the liquid-crystalline polypeptides, starting in the 1950s [267, 268] and still going strong. In appropriate solvents the glutamate residues form intra-chain hydrogen bonds connecting every third residue and forcing the PBLG chain to adopt a 3/1 helix configuration. This imparts to the whole molecule, or to long sections of it, a rodlike shape. In concentrated solutions ensembles of such rodlike entities cooperatively behave as liquid crystals. Other hydrogen-bonded helical rodlike biological entities may contain double helices with two hydrogen-bonded chains, as in the case of DNA or RNA and the polysaccharide xanthan gum [262], or three hydrogen-bonded chains as in the case of the polysaccharide schizophyllan [260] and the protein tropocollagen [269]. The axial ratio of such polymers may reach into the hundreds [260] with persistence lengths many tens of nanometers in size. Because the helical structures of these polymers are stabilized by intra- or inter-molecular hydrogen bonds, they may undergo a helix to coil transition upon dissolution in hydrogen-bond breaking solvents and lose their ability to form liquid crystalline solutions.

Liquid crystalline derivatives of the polysaccharide cellulose are characterized by wormlike chains with modest persistence lengths in the order of around ten nanometers [175] and modest axial ratios [166, 270]. When the lyotropic liquid crystallinity of hydroxypropyl cellulose was first discovered [251], it was assumed that the many hydroxy groups present in this polymer contribute to the stiffness and stability of its chains by intra- and inter-chain hydrogen bonds. Later, when cellulose derivatives containing no or almost no free hydroxy groups were found to be capable of being both lyotropic and thermotropic LCPs [253, 254, 256, 257, 271], the idea of chain stabilization solely by hydrogen bonding had to be abandoned. We now believe that the stiffness and wormlike configuration of cellulose and its derivatives are caused by a combination of the preferred valence and torsional angles of the bridges connecting the saccharide rings, and the large size of these rings preventing more free chain twisting and folding. Some intra-chain hydrogen bonds may, however, be present and help in stabilizing the wormlike shape of the chains.

The first, and until now apparently the only, cellulose derivative converted into a permanently crosslinked LCP network is hydroxypropyl cellulose (HPC) [272–276]. The HPC chains were either first derivatized with groups crosslinkable in a second step by photoinitiation [272, 274, 276] or were directly reacted with di- or higher-functional crosslinking moiety [273, 275] to produce a crosslinked network. Typical bridging groups connecting saccharide rings in adjacent chains, may be:

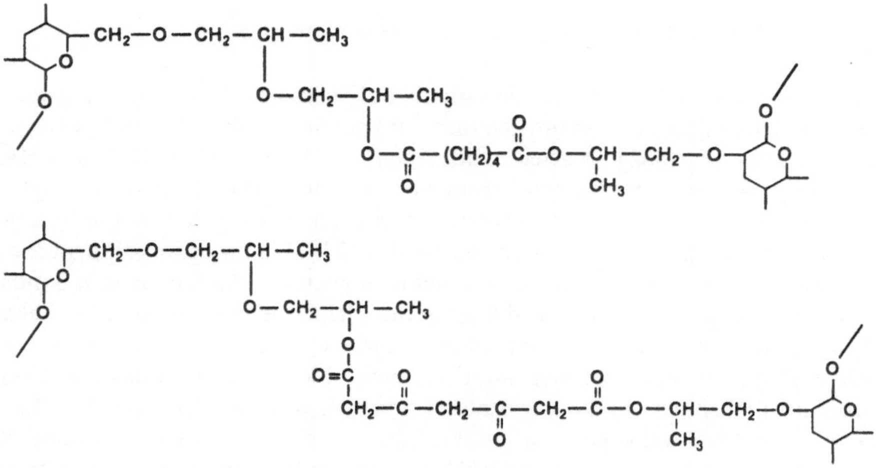

The length and flexibility of the crosslinking residue combines with the overall wormlike nature of the cellulose backbone to allow the main-chains to adopt the most efficient packing, generally parallel, and through it achieve the liquid crystalline state.

1.2.4.3 Networks with Uniform Main-Chain Mesogenicity and Flexible Crosslinking Residues

These networks are prepared in two steps and contain long mesogenic chains whose uniform stiffness is uninterrupted by the crosslinking residues or by rigid branchpoints, flexible junctions or flexible spacers along the chains. The cross-linking entities are flexible and may range in length from about a dozen atoms connected by single bonds to over a hundred such atoms. In practically all cases, the mesomorphicity of the network depends on the conditions during the crosslinking step. When the crosslinking was performed on an unoriented isotropic system, then the network ended up not exhibiting mesomorphicity, but when the crosslinking was conducted on an oriented system or on a system liquid crystalline in the quiescent state, then a highly birefringent liquid crystal-line network was obtained. In most instances falling into the present category, the crosslinking is effected at many sites randomly placed along the long mesogenic chains. There exists, however, one case where long mesogenic chains are crosslinked by end-linking into partly mesomorphic networks. This case, of poly(diethylsiloxane) networks, will be briefly discussed at the end of this section.

Networks with stiff uniform-mesogenicity chains were prepared by Aharoni [277] in two steps from main-chain aromatic polyamide LCPs. Firstly, long liquid crystal chains with amine or carboxyl reactive groups attached to every third up to every tenth aromatic ring were prepared. They were then crosslinked together in a second step by either trimethylene glycol bis(p-aminobenzoate) or dicarboxy-terminated polycaprolactam with a degree of polymerization of around 30. The network fragments including the flexible junctions look like

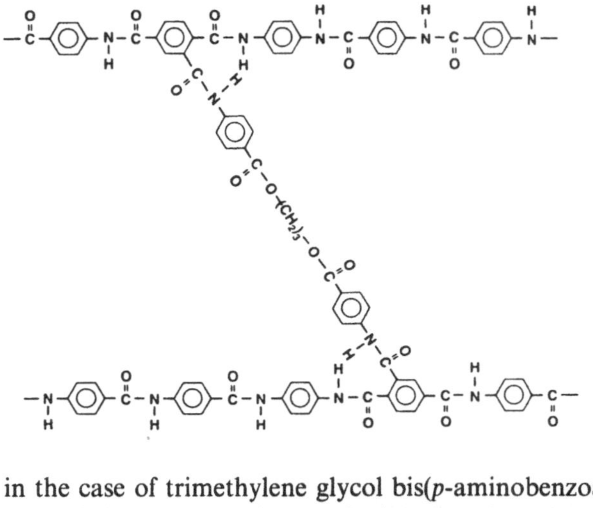

in the case of trimethylene glycol bis(p-aminobenzoate) crosslinker, and in the case of the polycaprolactam flexible junctions (See structure on page no 36). Because the crosslinking was performed in unoriented isotropic solutions at concentrations significantly lower than the phase transition concentration interval, no mesomorphicity of the gelled networks was observed even though the stiff aromatic polyamide chains to exhibit liquid crystal behavior when dissolved at appropriate concentration in solvents such as concentrated H_2SO_4 or DMAc/5% LiCl.

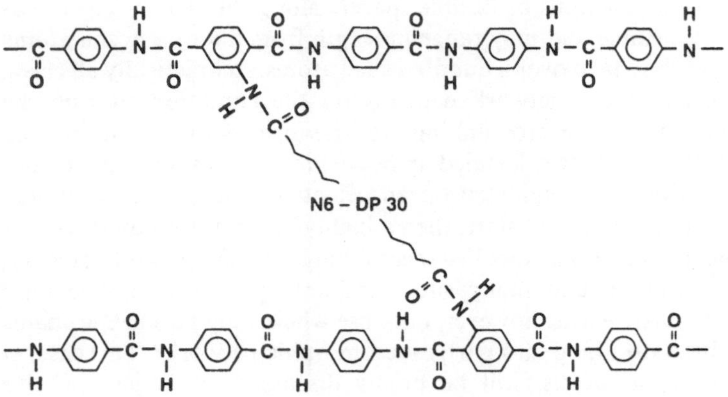

N6 – DP 30

Even stiffer chains than aromatic polyamides were built into networks with flexible crosslinking residues. A specially synthesized polyisocyanate copolymer was prepared by Aharoni [278] containing 25% 1-decenyl- and 75% n-octyl-isocyanate residues. This LCP was then lightly crosslinked by heating in solution in the presence of a free radical initiator (AIBN) to produce a gelled insoluble network. A characteristic structure appears as:

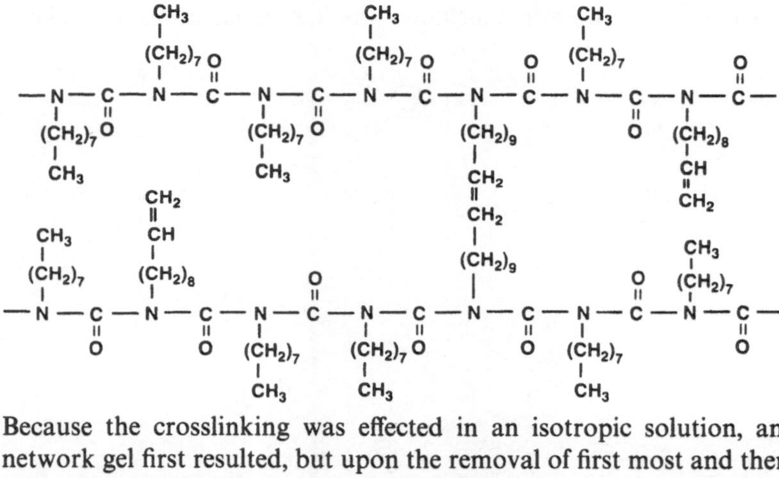

Because the crosslinking was effected in an isotropic solution, an isotropic network gel first resulted, but upon the removal of first most and then all of the solvent, highly birefringent gels and then highly birefringent bulk networks were obtained. It is obvious that the flexible crosslinking residues, each containing 21 single C–C bonds, did not prevent the lightly crosslinked stiff chains in the network from organizing in parallel arrays as the solvent was being removed. Significantly higher crosslink concentration is expected to prevent this from happening.

The stiffest liquid crystal networks falling in this category are those created by crosslinking PBLG [279–281]. Here, the 3/1 helical rods of PBLG are stabilized by hydrogen bonds between every third peptide residue in the direction of the rod's axis, and crosslinking of the rods is effected by partial substitution of the benzyl esters by amide forming diamines such as triethylene tetraamine:

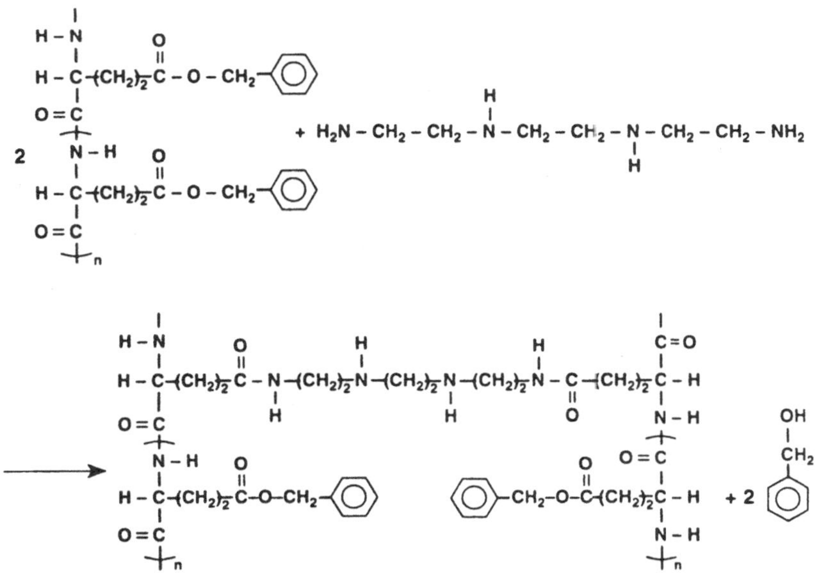

Cholesteric liquid crystalline networks could be prepared when the polymer was present in sufficient concentration in solution during the crosslinking process [279, 281], but when the crosslinking was effected when the solution was oriented under the influence of an applied external field, a nematic structure was obtained [280].

As of now, we know of only one member of the class of networks with uniform main-chain mesogenicity crosslinked by end-linking, the poly(diethylsiloxane) network [282–284]. This network is prepared by first creating hydroxy-terminated long poly(diethylsiloxane) chains and then reacting them with the multi-functional partly-hydrolyzed ethyl silicate. In the quiescent state, the networks exhibit only a low mesophase level, but this level greatly increases when the networks are stretched [283, 284]. Because of the relatively high flexibility of the uncrosslinked precursor polymer, it may be that the apparent liquid crystallinity of the strained network is, in fact, poorly organized strain-induced crystalline phase [285]. Further studies on this polymer and on additional poly(dialkylsiloxanes) will have to be conducted in order to resolve this question.

Several uncrosslinked poly(organophosphazenes) showed liquid crystal behavior despite the fact that their chains are substantially flexible. It is likely that when these will be end-linked as in the case of poly(diethylsiloxane) they, too, will form LCP networks.

1.3 The Fractal Character of Pre-Gel and Post-Gel Highly-Branched Polymers

One important feature of the fractal model is its usefulness in describing real things. It is a mathematical way of representing non-smooth curves, corrugated surfaces and porous media in terms of an exponential dependence of one measure of the object on another. Usually this exponent is a non-integer, fractal dimension of the described object. Imagine a one-dimensional line lying on a two-dimensional surface. Then imagine the line becoming more and more curved and kinky. To do so, more and more length of line will be required and the surface area will be increasingly covered by the line. The fractal model tells us that with the increase in fine details and kinks of the line, its dimensionality increases from its initial one dimension towards the surface's two dimensions. When the line will finally fill the surface completely, its dimensions will be two, identical with the dimensions of the flat surface. When a two-dimensional surface becomes more and more corrugated and the corrugation increasingly detailed, the surface fills more and more space and the fractal model states that the surface dimensionality gradually increases from two to three. When the whole volume is tightly packed with a highly ramified surface the dimensions of the surface reach those of the volume, that is, three. Conversely, when a solid medium becomes more and more porous by the creation of highly interconnected network of fine voids, the fractal dimensionality of the solid gradually decreases from the original three dimensions towards the two dimensions of a flat surface, the dimensions of the initially microscopic and then growing surfaces of the voids. The fractal dimensions, D, of a line will then fall in the interval

$$1 \leq D < 2, \tag{9}$$

those of a surface will be in the range of

$$2 \leq D_s < 3 \tag{10}$$

and those of a solid with increasing intricacy of voids will decrease in the interval

$$3 \geq D_f \geq 2 \tag{11}$$

with D_s and D_f being, respectively, the surface and mass dimensions.

The simplest mass-fractal structure will be one which has a central point and is built up from that position into space (all this discussion will refer to three dimensions). For example, a random flight in space after many steps N, each of length l, settles down to a distribution

$$(3/2\pi Nl^2)^{3/2} \exp(- 3R^2/2Nl^2). \tag{12}$$

If one asks for the probability that this locus goes through the point R and any N, for a large N one has

$$\rho(R) = \int_0^\infty dN(3/2\pi Nl^2)\exp(- 3R^2/2Nl^2) = (3/2\pi)(1/l^2R). \tag{13}$$

For many other structures there is a similar but more complicated function than Eq. (12) and a fractal power law other than R^{-1} replaces Eq. (13) (Fig. 12a). There are, however, other situations where the density of the material cannot exceed a certain value ρ_m. This may have an effect of flattening the distribution of $\rho(R)$ near the origin so that $\rho(k) < \rho_m$ (Fig. 12b) but it can also mean that $\rho(R \sim 0)$ becomes smaller than at intermediate R because of the impossibility of the growing material, a random walk, or whatever is creating the fractal from returning to the small R region because it would somewhere en route violate the condition $\rho(R) < \rho_m$. In the tail of large R a simple power law will apply but in the interior a much more complex situation obtains (Fig. 12c). (For a detailed discussion of this behavior for random walks with a maximum density see the book of des Cloizeaux and Jannink [286]. The mathematics is quite difficult.) This problem has of course a second (or more) length coming into it, and, in general, problems with a single length and a single power law are rare. The relationship between the mass M and the radius R of a highly branched polymeric species conforming with the mass-fractal model is, then,

$$R^{D_r} \propto M, \tag{14}$$

the material density in such fractal polymers decreasing as one moves from their centers or nuclei toward their peripheries or exteriors.

In the one-step process of network formation we are likely to meet the "mixed" situation of a central core with a power law exterior. It is an easily appreciated situation. Suppose relatively flexible molecules agglomerate at

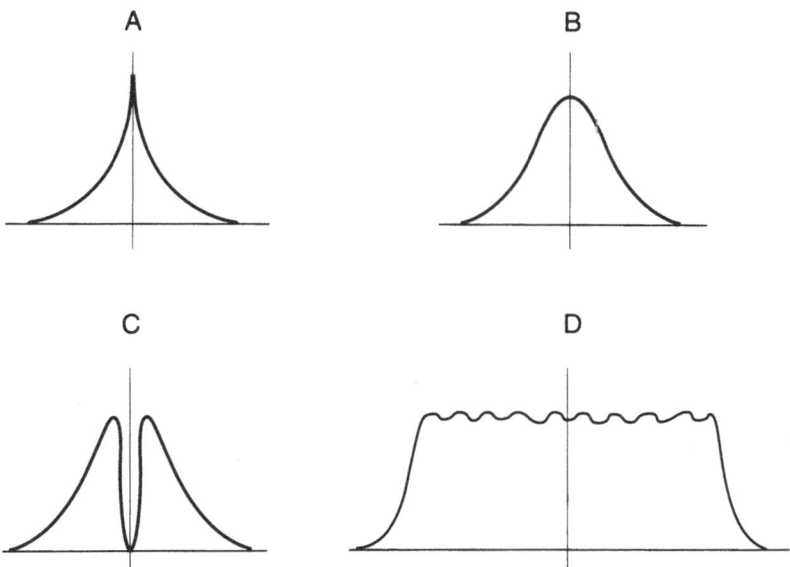

Fig. 12. Potential density distributions in fractal polymers. Details in text

many points in our system, and around each center there grows a simple "flexible" fractal. These then grow until they meet and either interpenetrate, or push up against each other in some way. This means that, although density will peak around each center, the density between centers will not have large regions of being zero, and will have a respectably defined mean value. At the surface of this body the density will start to decline, and that will lead to a power law decay away from the surface (Fig. 12d). This is roughly consistent with the picture of the growth process prior to the gel point of flexible networks drawn by Dušek and Šomvarsky [38] on the basis of extensive experimental results. There are other forms of this phenomenon which are very different physically from that considered here. For example, flexible polymer attached to a colloidal ball will have the same profile.

In the case of branched polymers there exists another kind of possible fractal behavior. Accordingly, the polymer branches or segments fill the volume of each growing branched macromolecule not completely but rather uniformly, leaving the exterior edges of the growing species highly ramified. Such a behavior is usually described in terms of surface fractality. The fundamental difference between mass fractals and surface fractals is, then, that in the former case the density of the material in a particle or cluster decreases exponentially with distance from its center. In the latter case, the density remains rather uniform throughout the volume of the cluster except for a very thin layer at its exterior. The thickness of the layer, where the mass density goes from its uniform value in the interior of the particle to zero out of it, is usually given as the variance from a chosen reference surface usually defined by the radius of the cluster. This variance is an average value obtained from repeated measurements of surface topography [287]. The difference between mass fractal and surface fractal is shown schematically in Fig. 13. It is important to recognize that if the porosity of an object or cluster is uniform throughout its volume, then the object is not fractal, since the fractal model requires its density to decrease exponentially with

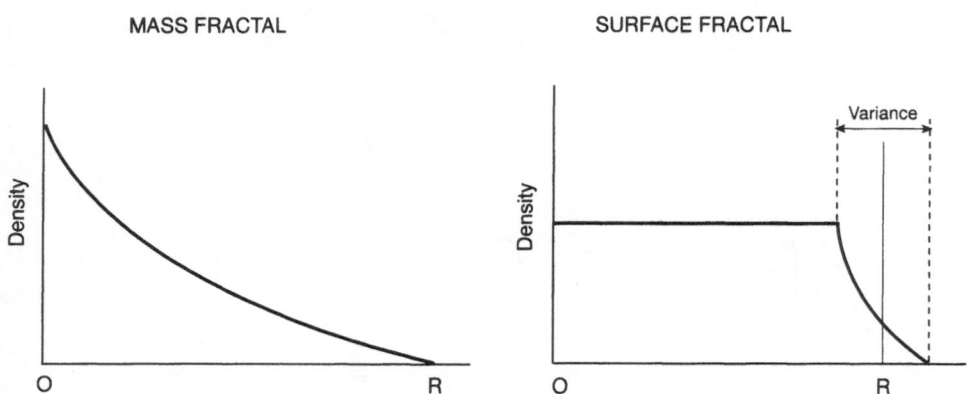

Fig. 13. Schematic of density distribution of mass fractal (*left*) and surface fractal (*right*)

distance from its center [287]. When a highly branched polymeric species is porous on the molecular or segmental scale and these pores pervade its whole volume and reach its exterior, then the resulting exterior is highly porous and ramified on a segmental size scale. It may be that the best mathematical model describing the porous fractal polymers, and the networks formed from them, is a very porous Sierpinski sponge with a very uneven and corrugated exterior.

Our experimental small-angle X-ray scattering (SAXS) and scanning electron microscopy (SEM) results to be shown below, indicate that the rigid fractal polymers (FPs) in the pre-gel and post-gel states conform with the expectations from surface fractality and not mass fractality. Furthermore, the interpenetration of material from one rigid fractal into its neighbors in the evolving network is rather limited, resulting in the retention of the individual character of the FPs at the gel point. The size of the FPs actually visible in their network progeny is about the same as their size in solution right before the gel point, as determined by light scattering experiments. The pre-gel particles behaving as surface fractals may each have grown from a single nucleus with a porous yet rather uniform mass density (Fig. 13), or each have coalesced from several smaller FPs each of which follows the mass fractal model but are sufficiently interpenetrated to produce bodies with reasonably defined mean density (Fig. 12d). Because segmental interpenetration from one rigid fractal to another is expected to be substantially reduced relative to flexible fractals, simply by stiff segments and rigid branchpoints not being able to move out of the way of others, the emerging picture is, then, of less interpenetrated rigid fractal polymers pressed against one another to a level where the dimples in Fig. 12d are deeper, but still far higher than zero density. Because of the nature of the growing species we call them fractal polymers (FPs) but because of their porosity the words "fractal surface" lose their conventional meaning. We prefer, therefore, to use "fractal exteriors" instead of "fractal surfaces" when appropriate. Readers interested in the topic of fractals in general, and their application to the description of polymeric materials, are referred to several broad-based books and articles on the subject [287–295]. One will find in them practically no mention of the kind of polymeric species we are interested in: organic, highly branched rigid aromatic networks.

In radiation scattering experiments, the power dependence of the scattered radiation intensity, I(q), on the scattering vector, q, is directly related to the dimensionality of the investigated sample. Thus, one can obtain information about the dimensionality and fractal nature of polymeric samples from careful measurements and correct plots of the intensity results gathered in low-angle light scattering (LALS), small-angle X-ray scattering (SAXS) or small-angle neutron scattering (SANS) experiments. Defining [296] q by

$$q = 4\pi\sin(\theta/2)/\lambda \tag{15}$$

where θ is the scattering angle and λ is the wavelength of the scattered beam, or, for a small scattering angle approximating [297] q by

$$q \cong 2\pi\theta/\lambda \tag{16}$$

we obtain [296, 298, 299] a power law dependence of the scattered intensity on mass fractals:

$$I(q) \propto q^{-D_f} \tag{17}$$

and on surface fractals:

$$I(q) \propto q^{-(6-D_s)} \tag{18}$$

in which D_f and D_s are mass and surface fractal dimensions, respectively. Logarithmic plots of $I(q)$ against q reveal whether the observed scatterers are mass fractal or surface fractal in nature. Since $2 \leq D_s < 3$ and $D_f \leq 3$, whenever the power law exponent of the scattered intensity falls in the interval between 3 and 4 we know we do not deal with mass fractals but with surface fractals whose $3 < 6 - D_s \leq 4$. When the power dependency is exactly 4, that is

$$I(q) \propto q^{-4} \tag{19}$$

we know the scatterers to be about spherical objects with smooth two-dimensional surfaces. At this point the scatterers are not fractal in character. When the power dependency of $I(q)$ is lower than 3, we know we observe mass fractal behavior consistent with Eq. (17). To be consistent with the proportionality at Eq. (18), D_s will have to be larger than 3 which is an impossibility in a 3-dimensional universe. Most of the LALS, SAXS and SANS work in the literature aimed at elucidating the fractal nature of various scatterers was done on inorganic substances of high porosity [300–304]. Some studies were performed on organic flexible polymers [305–310] and only very few results were obtained from organic, highly crosslinked rigid and aromatic networks [311–314].

One final introductory point. The references in this work are meant to be merely representative and not exhaustive. The presence or absence of a particular reference should be taken in this spirit and with no other implications. Some of the entries in the list of references are identified by a number with an appended lower case letter. These are, generally, very recent references that we only became aware of in the final stages of writing this monograph.

2 Exclusions

Before we describe and discuss the intensely crosslinked rigid aromatic networks, we would like to exclude several classes from consideration. One such
class is entropic networks, some of which contain a substantial fraction of
aromatic residues but these are separated by flexible groups sufficiently long to
decouple each aromatic group from its neighbors. Another is a smaller class
containing highly branched macromolecules, some of substantial aromaticity,
which are constructed in a manner specifically meant to prevent the formation
of "infinite" networks.

In the first class we enumerate all thermoreversible networks and gels where
the junction points or crosslinks are due to non-covalent interactions. By this we
mean ionic and metal-complex interactions [315], inter- and intra-molecular
hydrogen-bond-type interactions [316], polymer-polymer complexation
[317, 318], the formation of helices involving more than a single molecule, and
physical crosslinks created by the formation of microcrystals, paracrystals or
aggregates during the cooling of certain polymer solutions. From the permanent, covalently bonded networks we exclude all flexible-chain ones such as
hydrocarbon elastomers, polysiloxanes [319] and polyurethanes [320]. We
especially mention the exclusion from discussion of two subgroups of flexible-
chain permanent networks. These are LCP networks whose flexible main-chains
contain mesogenic units and LCP networks whose flexible main-chains carry
stiff pendant mesogens. All inorganic or organometallic networks are also
excluded, some of which may be referred to below, together with mixed organic-
inorganic systems and organic interpenetrating networks (IPNs). The families of
epoxy resins and ebonite both fall out of the scope of main-chain rigidity as
defined above and both are excluded. A typical fragment in a typical epoxy
network [321] looks like:

It is characterized by having on the average four single bonds in a row in the
flexible group connecting the more rigid aromatic units. This renders the
network entropically flexible to a substantial degree. The generally observed
ambient temperature rigidity of cured epoxy resins is largely due to the high
crosslink density in the system and the resulting elevation of T_g. Similar
arguments hold in the case of ebonite. By definition, ebonite is natural rubber or
polyisoprene intensely crosslinked with sulfur. The crosslink junction is an
interchain sulfide bridge and the highest amount of sulfur that can be incorporated permanently into the network is 32 wt% [322]. This corresponds to a

network ideally represented by

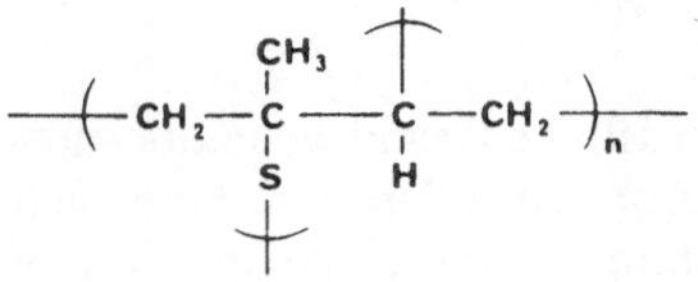

where each repeat unit along the chain is connected by sulfide bridges to two other units which may belong to the same or other chains.

Two additional families of excluded polymers are superficially similar to the rigid aromatic networks of interest, but are substantially different; enough to take them out of consideration. The first such family is one in which high-M flexible chain polymers are crosslinked at all or most repeat units along the chains by relatively short difunctional crosslinking groups which are rigid or claimed to be so. The best known members of this family are the ones prepared by Davankov, Tsyurupa and associates [323–326], consisting of high-M polystyrene chains "hypercrosslinked" by various difunctional and trifunctional groups of varying rigidity typified by:

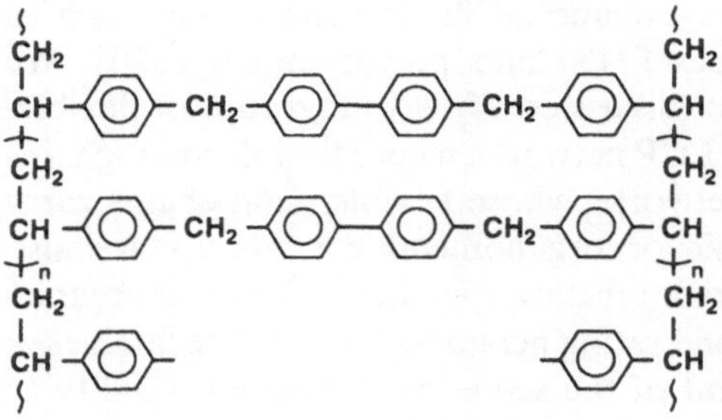

The most rigid of the crosslinks are expected to serve as rigid struts between the polymer chains. The authors [323–326] maintain that the exceptional swelling ability of the hypercrosslinked polystyrene is due to the rigidity of the crosslinking struts. This is contradicted by Negre et al. [327] who demonstrated that intensively crosslinked polystyrenes show similar levels of swelling without the crosslinks necessarily being rigid. Recent reports of networks consisting of flexible-chain polymers crosslinked by rigid struts describe aryl-bridged polysilsesquioxanes [328] and poly-N-(4-vinylphenyl) maleimide [329, 330]. They are typified by

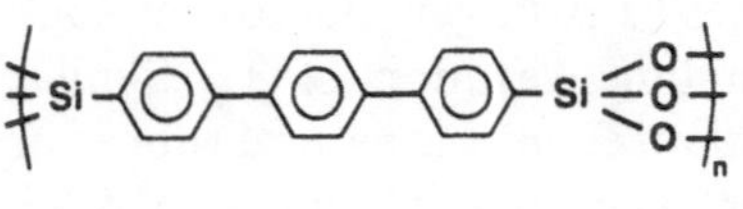

Polysilsesquioxane 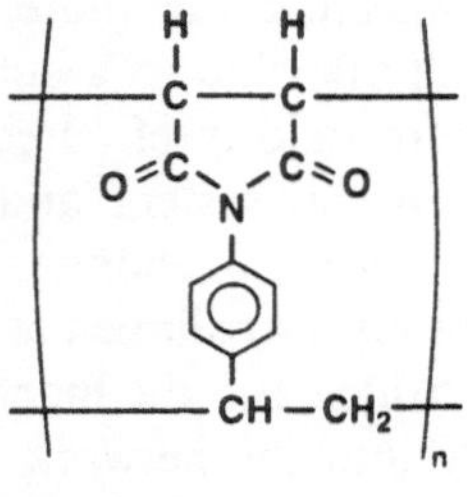   Poly–N–(4–vinylphenyl)maleimide

Even though all these materials contain rigid aromatic struts, they also contain runs of single bonds along the flexible backbones of their primary chains.

The second family contains various dendritic polymers, collectively known as dendrimers [331]. The vast majority of them comprise branches which are in fact not thermodynamically rigid [332–339]. Several of the more recent dendrimers are more rigid, but were synthesized in a fashion insuring that "infinite" networks will not form [340–350a]. Therefore, despite the fact that dendrimers are highly branched and their surface fractality was demonstrated [351], they will not be considered below as network components.

3 Synthetic Highlights

3.1 One-Step and End-Capped Rigid Networks

3.1.1 Networks with Stiff Hydrocarbon Segments

To the theoretician, perfectly regular defect-free networks with stiff hydrocarbon segments are most desirous. To the synthetic chemist the deceptive apparent simplicity of such networks has proved to be, thus far, an insurmountable challenge. When such perfect rigid hydrocarbon networks are synthesized and studied by means of X-ray diffraction techniques, they are expected to behave as large single crystals encompassing the whole volume of the reaction medium. In the case where the branchpoints in such perfect rigid networks are tetrafuctional and tetrahedral in shape, the networks are expected to yield a face centered cubic (FCC) diffraction pattern. When the branchpoints are hexafunctional and the six bonds project at right angles, then the repeating motifs in the network are expected to be octahedral. In fact, perfectly regular defect-free stiff hydrocarbon networks were not synthesized yet. Steps in this direction were taken, some culminating in disordered networks and others are currently in the stage of segment preparation. More often than not, these segments are not of identical length but have a distribution of lengths. The length distribution practically assures that any network prepared from such segments will be irregular and, as a consequence, contain varying levels of defects in the form of incompletely reacted branchpoint functionalities and segments connected to the "infinite" network at only one end. The syntheses of rigid hydrocarbon networks and stiff hydrocarbon segments, and typical results are described below.

3.1.1.1 Polyphenylenes and Poly(Substituted Phenylenes)

The creation and study of fully aromatic rigid networks made exclusively from phenylene groups is a challenge not yet met despite the fact that poly (p-phenylenes) are well known for over a quarter of a century [235, 236, 352–354]. They are prepared from benzene, biphenyl or terphenyl by direct coupling in the presence of Lewis acid-oxidant combination [235, 236, 352–354] or from p-dibromobenzene by Grignard coupling in the presence of nickel chloride dipyridyl [237, 355]. Other methods were also reported [236]. At first glance, the poly(p-phenylenes) appear to be excellent candidates for serving as stiff segments. This was found not to be the case, largely because of the insolubility and intractability of the poly(p-phenylenes) but also because of the uncontrolled nature of the polymerization reaction and the rather low degree of polymerization (DP) achieved [236]. The insolubility of poly(p-phenelyne) in common organic solvents, even of oligomers of very low DP, is due to the rodlike nature of the molecules and the absence of solubilizing groups. Both these facts facilitate the parallel alignment and crystallization of the poly(p-phenylene) and its subsequent precipitation out of solution. The presence of solubilizing groups on all or most phenylene rings increases the solubility of the oligomers by several orders of magnitude [238–240]. The realization of the beneficial effects of alkylation on the solubility of poly(p-phenylene) oligomers and polymers produced three methods to create alkylated poly(p-phenylenes). Kovacic and associates [356] demonstrated that they could alkylate oligomeric poly(p-phenylene) by Friedel-Crafts alkylation and render the product soluble. Changing the sequence of reactions by performing the alkylation first and the oxidative-coupling second, is likely to scramble some chain positions and destroy its rodlike appearance [241]. Such isomerization was not observed in systems containing only aromatic rings [354, 357]. Heitz prepared methyl substituted oligomeric poly(p-phenylene) by stepwise synthesis, creating first aromatic-cycloaliphatic moieties of the desired number of rings and then dehydrogenating the cycloaliphatic moieties to obtain aromatic rings [240, 358]. A third, and thus far the most promising method for obtaining alkyl substituted poly(p-phenylene) oligomers and low DP polymers starts with difunctional benzenes substituted with flexible alkyl side-chains, R. In one case [241], the benzene functionalities are dibromo- and the coupling reaction is a Grignard reaction performed under Yamamoto [237, 237a, 355] conditions:

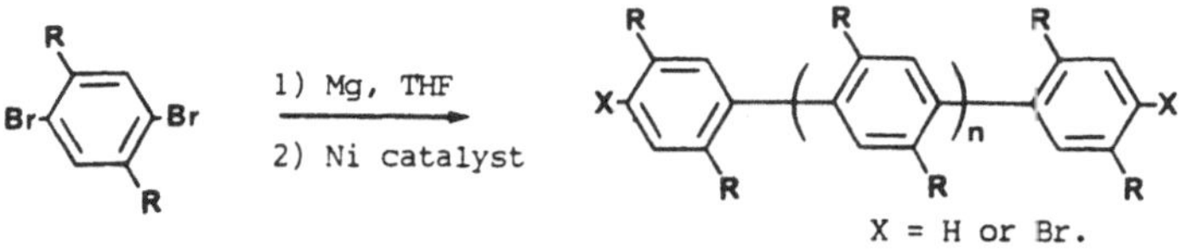

Where the nickel catalyst is Ni(PPh$_3$)$_2$Cl$_2$ with Ph standing for a phenyl ring. This reaction was also conducted with aryl substituted dibromobenzene with similar results [243]:

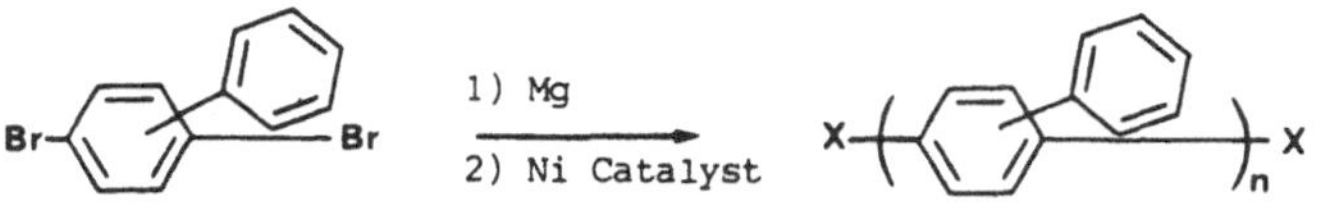

In another case [237a, 242], the starting monomer was dialkyl substituted 4-bromobenzeneboronic acid and the polymerization was conducted in a hetero-geneous mixture of benzene and water containing sodium carbonate in the presence of Pd(PPh$_3$)$_4$ catalyst:

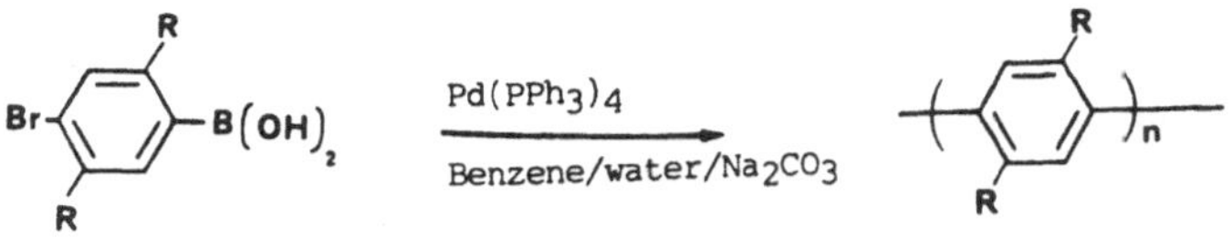

This was followed by the preparation of p-terphenyl and stiff aromatic oligomers terminated by reactive end-groups (telechelic oligomers or telomers in short),

achieved by reacting 2,5-dialkyl-1,4-benzenediboronic acid with a monobromo-
benzene derivative of the desired reactive end-cap Y [244]:

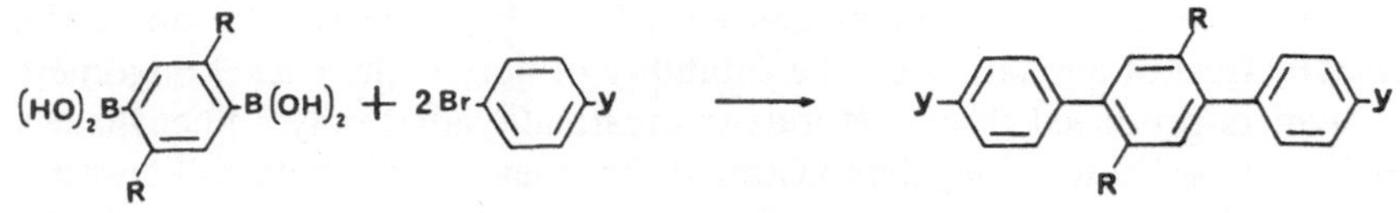

and

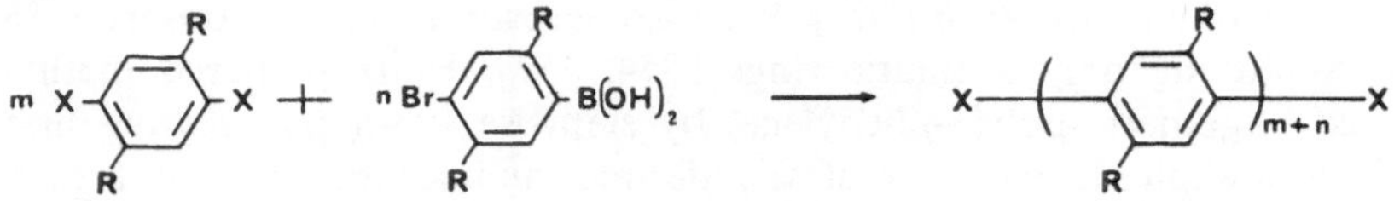

With m being much smaller than n and X being either $B(OH)_2$ or Br. The
oligomers terminated with boronic acid were converted to acetoxy or meth-
oxycarbonyl end-caps by reacting the boronic acid-terminated oligomers with
the respective p-substituted beromobenzene [244]:

Wegner and associates [237a, 244] recognized the usefulness of the alkylated
poly(p-phenylene) oligomers in the preparation of rigid networks with rodlike
segments. When these lines were written, no completed network with long
poly(p-phenylene) segments was reported in the literature.

3.1.1.2 Hydrocarbon Networks with Very Short Aromatic Segments

Polyphenylene networks with only one or two rings between branchpoints were
reported, however, in the literature. Accordingly, reactive oligophenylenes with
terminal ethynyl groups were prepared in a first step from p-diethynylbenzene. It
was claimed that three ethynyl groups cyclotrimerized to produce a phenylene
ring [359–361], but we are not convinced yet that cyclotrimerization is the only
reaction of the ethynyl groups. The molecular weight of the oligomers was
$M = 2300$ and $M_w = 8060$ g/mol [359]. The oligomers were reported to be
soluble in aromatic and halogenated solvents. After a stepwise heat curing, an
insoluble crosslinked, thermally stable and infusible highly aromatic network
was obtained [360]. The ethynyl-terminated oligomer is schematically shown

below:

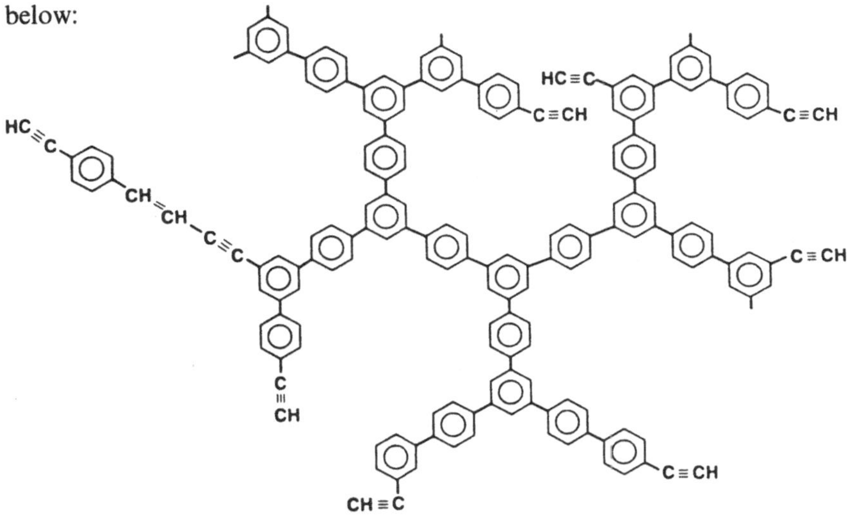

Aromatic networks prepared from long segments end-capped by ethynyl groups will be described in Sect. 3.1.6.3, and in Sect. 3.2.3 we shall describe the two-step preparation of substantially rigid networks using the reactions of ethynyl groups present in long linear chains to form interchain crosslinks.

The poly(p-phenelynes) and their alkylated derivatives have not yet yielded well-defined low defect rigid networks. The synthetic methods used in the attempts to prepare such networks, were successfully utilized in the preparation of networks and network-fragments in which the segments between branch-points are only one or two aromatic rings in length. Starting with tri-substituted phenyl rings, ring coupling and network formation may be achieved under Yamamoto conditions [237, 237a, 355] using a nickel catalyst, or under Suzuki conditions [237a, 362] using a palladium catalyst such as Pd(PPh$_3$)$_4$ in the presence of sodium carbonate. Because the Yamamoto conditions produce substantially less branching than expected [363], the palladium catalyzed Suzuki reaction has ascended in popularity. Accordingly, when monomers such as 3,5-dibromobenzeneboronic acid were reacted with themselves in the presence of a palladium catalyst, hyperbranched polyphenylene macromolecules were obtained [341, 363–365]:

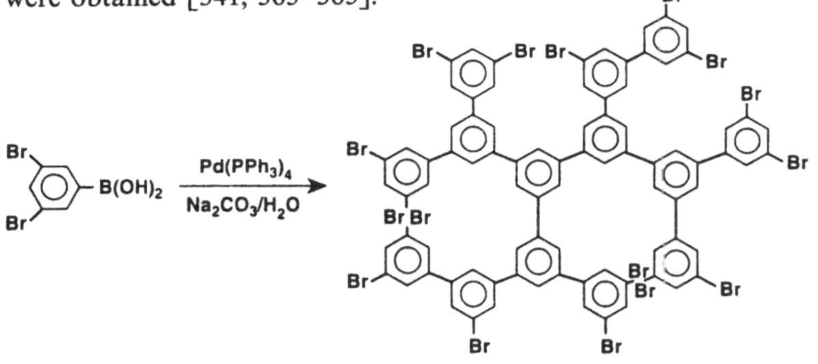

The bromine end-groups were converted to many other functionalities to make the polyphenylene molecules soluble in a broad variety of solvents [363–365]. Until now, the synthetic effort was directed at creating a polyphenylene version of previously described micellar [342, 366–369] or dendrimers and similar highly branched macromolecular structures [331, 332, 336, 338, 343, 344, 370–379]. Because of this, the molar concentrations of the two different reactive groups in the monomer feed were kept unequal. To reach equality between the reactive groups, and by so doing set the conditions for the creation of "infinite" networks, monomers richer in the missing reactive group should be added. Thus, to balance the ratio of bromine to boronic acid in the above reaction, monomers such as 5-bromo-1,3-benzenediboronic acid or 2,5-dialkyl-1,4-benzenediboronic acid [244] should be added. From the stoichiometry and our knowledge of the chemistry, we would expect to obtain an "infinite" polyphenylene network. What is unknown at present is the level of network defects that will be present in such a system, and the level of swelling attributable to the existence of defects in such a network.

In a further synthetic advance, Webster and co-workers recently reported the preparation of "heperbranched" and "hypercrosslinked" rigid polymers [380–384]. (Related to this is the work of Newkome and co-workers on softer branched polymers, e.g. reference 385.) These are characterized by their segments being phenylene rings and their branchpoints being trifunctional carbinol groups. Depending on the monomer structure, the segments between branchpoints can be one or two phenylene rings in length, with or without single-atom bridges in the middle of each segment. The preparation of the "hyperbranched" polymers is conducted in two steps. In the first, a lithio derivative of the aromatic species is prepared by the reaction of its brominated precursor with alkyl lithium [381, 386], and in the second, the lithioaryl is reacted with dimethyl carbonate (Me$_2$CO$_3$) in tetrahydrofurane (THF) at $-80\,°C$ and the temperature then slowly brought to room temperature, followed by neutralization of the reaction mixture. Typical network fragments prepared by this method were reported by Webster et al. [380–384]:

(See Structure on Page no 51)

The polymeric products appear as fine particulate suspension. It will be of interest to determine whether this is due to the relatively low concentration in which the polymerization reaction was carried out, or to an inherent inability of these systems to form "infinite" networks. The significant swelling of the networks in various solvents implies the presence of substantial amount of network defects and inhomogeneity [383].

An interesting family of phenylacetylene rigid macrocyclic compounds, potentially capable of subsequent network formation, was recently reported by Moore and co-workers [386a]. This promises to be an exciting new advance along the road to open rigid networks.

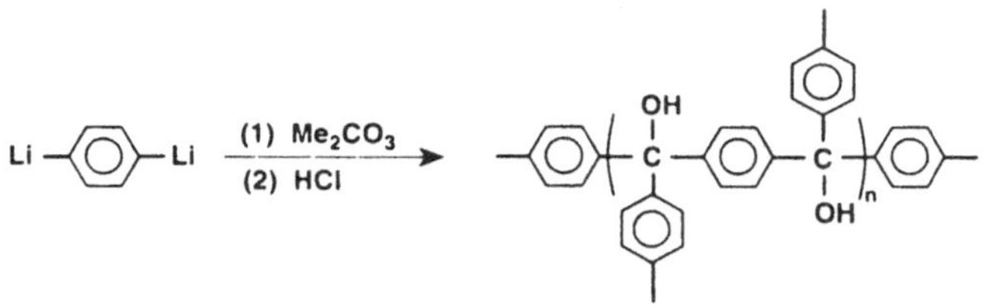
Li
Li
(1) Me2CO3
(2) HCl
OH
OH
n

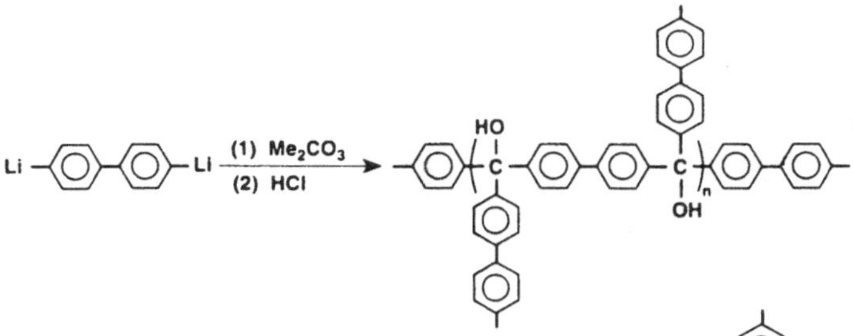
Li
Li
(1) Me2CO3
(2) HCl
HO
OH
n

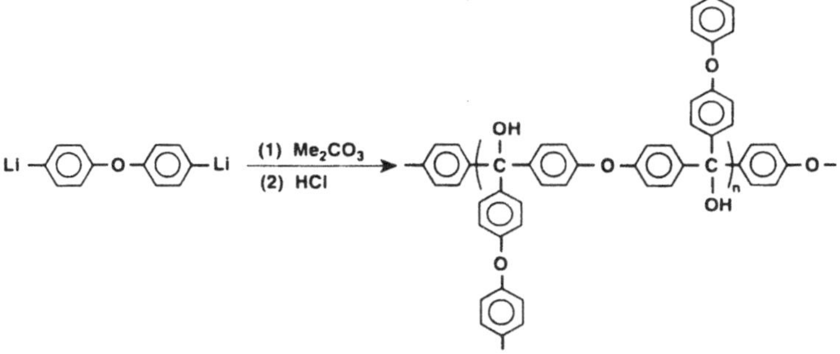
Li
O
Li
(1) Me2CO3
(2) HCl
OH
OH
n

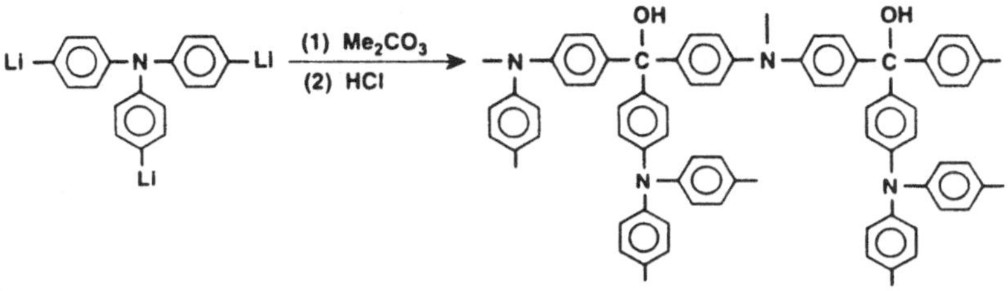
Li
N
Li
Li
(1) Me2CO3
(2) HCl
OH
OH

and

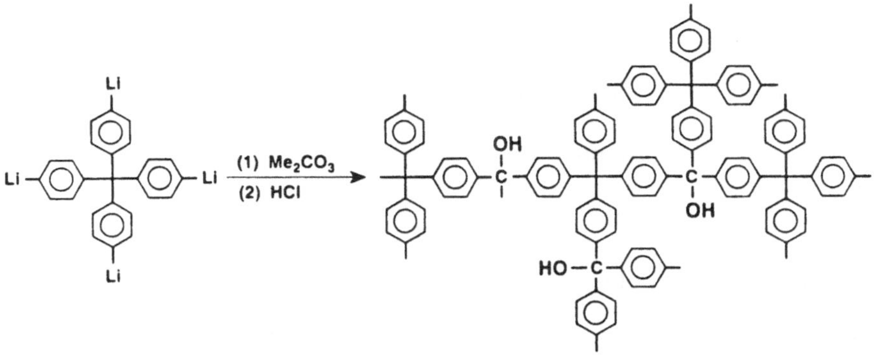
Li
Li
Li
Li
(1) Me2CO3
(2) HCl
OH
OH
HO

3.1.1.3 Stiff Aliphatic Segments

Over the last few years, Professor J. Michl has developed an intriguing approach to obtain intensely crosslinked rigid networks. It consists of preparing terminally functionalized [n] staffanes and then using them as the stiff segments in rigid networks. We are aware of no network prepared by this method and, because they are not aromatic, the [n] staffanes fall out of the scope of this review. They are worth mentioning, however, because of their interesting structure and their implications for network preparation in the future. The use of functional branchpoints such as 1,3,5,7-tetra-substituted adamantane [387] may also be appropriate for this purpose. The synthetic procedures of [n] staffanes were reported extensively by Michl and associates [245, 248, 388–391] and Zimmerman et al. [392]. Typical structures are

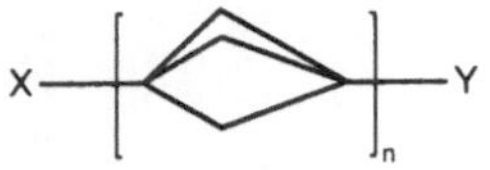

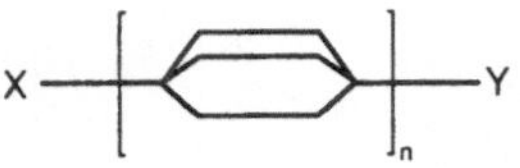

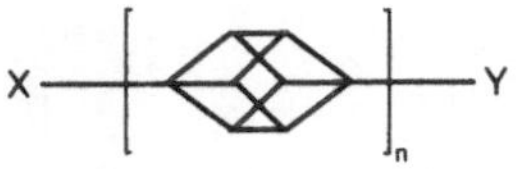

and telomers containing more than a single building block, such as

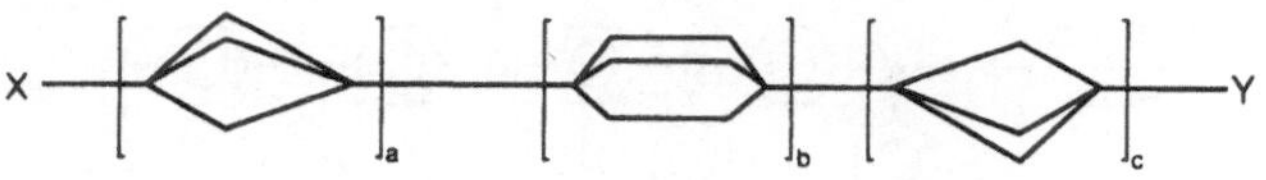

where X and Y are the same or different reactive end-caps. The crystal structure of several [n]staffanes was determined as well as certain features of their liquid crystal behavior [246, 247].

3.1.1.4 Overview

One gathers from the above that, despite substantial effort, we have on hand only the beginning of what promises to be a fascinating field of well-defined rigid networks based on aromatic or cycloaliphatic end-capped stiff hydrocarbon segments.

Only a few properties of rigid aromatic hydrocarbon networks were hitherto described in the literature. They mostly deal with solubility, porosity and surface area, and the relationship between certain polymerization procedures and the nature of the polymeric products. The hyperbranched poyphenylenes and triaryl carbinols are fully amorphous substances [380, 383, 384]. When dry, they are highly porous systems. Unlike their insoluble linear analogues, they are highly soluble in many solvents. Their solubility is further enhanced by functionalization with end-groups compatible with the desired solvent [365]. Branched hydrocarbon macromolecules whose segments and branchpoints are terminated by non-polar groups are soluble in solvents such as chloroform, carbon tetrachloride, methylene chloride and toluene [341, 380]. When such macromolecules were terminated by bromine atoms, they were found to be soluble in THF and O-dichlorobenzene but not in CH_2Cl_2 [363]. When the terminal group was converted to Li-carboxylate, then the highly branched polymer became water soluble. The carboxylic acid terminated polymeric analogues were insoluble in water but soluble in THF [363]. The T_g of the amorphous polyphenylenes is 239 °C and their thermal stability is given as higher than 350 °C in air and 400 °C in nitrogen [382, 384]. The molecular weights of three batches of the branched polyphenylenes were given as up to 32000 [383], $M_n = 3820$ and $M_w = 5750$, and $M_n = 3910$ with $M_w/M_n = 1.81$ [363]. Importantly, the polymerization does not go to completion and the "hyperbranched" molecules cease growing before the monomer supply is depleted [363]. In the case of the convergent growth of "hyperbranched" polyphenylenes [341], the yields of polymeric products were as low as 30%.

An analysis of the branching level in the polyphenylenes revealed that when they were prepared by the Suzuki procedure, the degree of branching was 70% of theory and when prepared by the Yamamoto procedure it dropped to only 40% of expectation [363]. The facts that the actual branching levels in the hyperbranched polyphenylenes are substantially below expectation and the cessation of growth in the presence of available monomer imply that even if "infinite" networks will be made successfully from the hyperbranched polyphenylenes and triaryl carbinols, these networks will be highly defective. An indication that this is indeed the case can be gathered from the very high swelling levels the open networks exhibit in various solvents [380, 383, 384]. The very high surface area measured on several polyphenylenes and poly(triaryl carbinols) [363, 380, 381, 383, 384] is most likely due to a combined effect of the open structure in the interior of the branched macromolecules and the highly corrugated appearance of their exteriors. Such "surface" ramification is expected from the fractal nature of the growth of the macromolecular species. Similar observations by Aharoni on highly branched aromatic polyamide pre-gel particles and post-gel networks will be described below.

Despite the open structure of the hyperbranched stiff macromolecules, they are rather compact with radius of gyration, R_G, far smaller than that for the linear analogues. The compactness is mirrored in the modest change in solution viscosity upon the addition of more than 1 g Li-carboxylate terminated polyphenylene to 1 ml water [363], and in the brittleness and inability to form films [383] exhibited by these polymers. A similar very modest effect on solution viscosity was previously observed by Aharoni et al. [370] on solutions of the highly branched and very compact poly(α,ε-L-lysine) molecules prepared by Denkewalter et al. [332].

3.1.2 Aromatic Polyamide Networks with Stiff Segments and Rigid Branchpoints

A decade ago concepts were set forth of aromatic polyamide dendrimers that are not supposed to coalesce and form "infinite" networks [340]. At the same time the first rigid aromatic polyamide "infinite" networks and gels were prepared by one of us [212, 393] from aromatic amines and carboxylic acids, employing triarylphosphite co-reagents in a polycondensation reaction now known as the Yamazaki procedure [394]. Since then, a broad variety of fully aromatic polyamide networks and fractal polymers (FPs) were prepared by Aharoni, which will be described below. In recent years, three additional polycondensation procedures were described in the literature, creating highly branched aromatic polyamides from monomers containing acid chloride instead of the free acid used in the Yamazaki procedure. Even though the polymeric products of these reactions are not "infinite" networks, the information given is sufficient for the creation of such networks when desired. Therefore the three procedures requiring acid chloride monomers will be described at the end of this section, following the description of rigid aromatic polyamide "infinite" networks.

3.1.2.1 Rigid Polyamide Networks, Gels and Fractal Polymers Prepared in Solution in the Presence of Triarylphosphite

The fact common to all such reactions is the use of triarylphosphite, such as triphenylphosphite (TPP) as condensing agents for the creation of amide groups from aromatic amines and carboxylic acids. The driving force for the reaction is the splitting of one or more aryloxy groups off the phosphite and the conversion of the phosphorus-containing residue from phosphite to phosphate [395]:

A first indication of the potential of triarylphosphites as condensing agents was given by Mitin and Glinskaya [396] over twenty years ago. Here, aliphatic amino acids were dimerized to form peptides by dissolving them in suitable solvents, preferably N,N-dimethylformamide (DMF) or dioxane, and conducting the condensation reaction in the presence of triphenylphosphite (TPP) and imidazole for 18 h at 40 °C. In addition to the peptide, diphenylphosphite (DPP) and phenol byproducts were found in the reaction mixture. This reaction was extended by Ogata and Tanaka [397] to the polycondensation reaction of ω-amino acids and equimolar salts of diamines and dicarboxy acids, to obtain low molecular weight polyamides at room temperature. In this case, it was found that the solvents giving the highest polyamide yield were DMF and N,N-dimethylacetamide (DMAc) and that, in the case of aromatic monomers, the addition of some LiCl increases the solubility of the polymeric products, thus contributing to an enhancement of the molecular weight. Despite the fact that the reaction time was over 40 h, the reduced viscosity of the products surpassed 0.2 only once.

Yamazaki and associates refined the polycondensation by first realizing that pyridine is a far better co-reagent than imidazole [398, 399] and then determining that about 4 wt% of LiCl (or 8 wt% $CaCl_2$) in N-methylpyrrolidinone (NMP) was the best solvent mixture for polycondensations in the presence of TPP or DPP to give high molecular weight (high-M) aromatic polyamides [394]. In this work, it was shown that conducting the polycondensation for about 5-6 h at 80 °C gave the best results. Conducting the polymerization at 60 °C gave only oligomers, while polycondensations executed at 100 or 120 °C gave high-M polyamides whose molecular weight decreased with increasing temperature. Even at 80 °C, 3 h and 10 h reaction times gave polymers with lower-M than 6 h reactions [394].

Over the past ten years we have conducted solution condensations in the presence of aryl phosphites and organic base over a broad range of conditions. From all the solvents tested, only blocked amides promoted the condensation reaction. Among these, the solvent efficiency in promoting the reaction ranked as NMP ≥ DMAc > tetramethylurea > DMF. Except for some initial studies [395], all the polymerizations described herein were conducted in DMAc containing 5 wt% dry LiCl, or DMAc alone when the polymeric product was nicely soluble in it. Unlike Yamazaki et al. [394], we have found that in DMAc/5% LiCl, the best reaction temperature spans the interval 85 < T < 120 °C and, preferably, 90 < T < 115 °C. Within this range, the rate of polymerization greatly increased with temperature [400], while the molecular weight of the polymeric products remains essentially unaffected. In soluble systems, reaction times of about 60 to 90 min are sufficient to effect full monomer conversion. It is interesting to note that, during efficient polycondensation at constant temperature, both the monomer conversion and the molecular weight of the polymeric product appear to be concentration dependent [395]. At rather low concentrations the dilution appears to retard the reaction. At very high concentrations, polymer precipitation or solution microsyneresis greatly slows down the reaction. When the reaction is allowed a long time at temperature, the monomers are finally all consumed even though the molecular weight of the intermediate concentration products (ca. 10 wt/vol.%) remains the highest.

Ester groups remain unaffected under "Yamazaki conditions". This allowed us, over the years, to create zigzag rigid aromatic poly(esteramides) in which all the stiff segments are of identical composition and length [61], flexible and semi-flexible gelled one-step networks [212, 311, 393, 400], gels of semi-flexible two-step networks [277] and a large family of hidrogen-bonded strictly alternating poly(esteramides) many of which exhibit multiple-transition thermotropic meso-morphicity [70]. The presence of pre-formed ester groups in the polymeric backbone combines with their high molecular weights [61, 70] to prove that ester groups do not hydrolize under these conditions. The mesomorphicity of the strictly alternating poly(esteramides) [70] does not appear in the random or semi-random analogues [72]. DSC, X-ray diffractometry and NMR studies show that the alternating poly(esteramides) are highly regular [73–76] and that no transesterification or transamidation occurred under Yamazaki conditions. Therefore amide groups can be created from free caboxylic acids and aromatic amines under Yamazaki conditions in the presence of ester groups without the latter being affected. If, however, the reaction temperature is increased to, say, over 150 °C, then the LiCl in DMAc or NMP attacks the ester groups forming lithium carboxylate salts and chloroalkyl residues. In NMP/5% LiCl at 185 °C, for instance, the reaction is highly efficient, reducing high-M poly(ethyleneterephthalate) (PET) to oligomers in a matter of minutes. In the absence of LiCl this destruction does not take place. The selective attack of LiCl on esters at high temperatures is similar to the selective methyl ester cleavage by LiI reported in the literature [401].

Monomer polymerizability is affected by structure. Experimental results from our laboratory are shown in the next three schemes. In Scheme 1 aromatic and aliphatic diacids and aromatic polyacids easily polymerizable under Yamazaki conditions are shown. In Scheme 2 easily polymerizable aromatic amines and aromatic amino acids are shown. The yields and molecular weights of polymers from monomers of the general structure

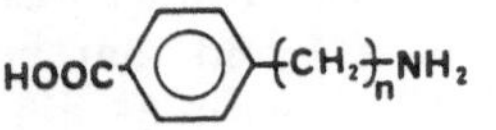

$$\text{HOOC}-\!\!\left\langle\bigcirc\right\rangle\!-\!(CH_2)_n\!-\!NH_2 \qquad n = 1,2,3$$

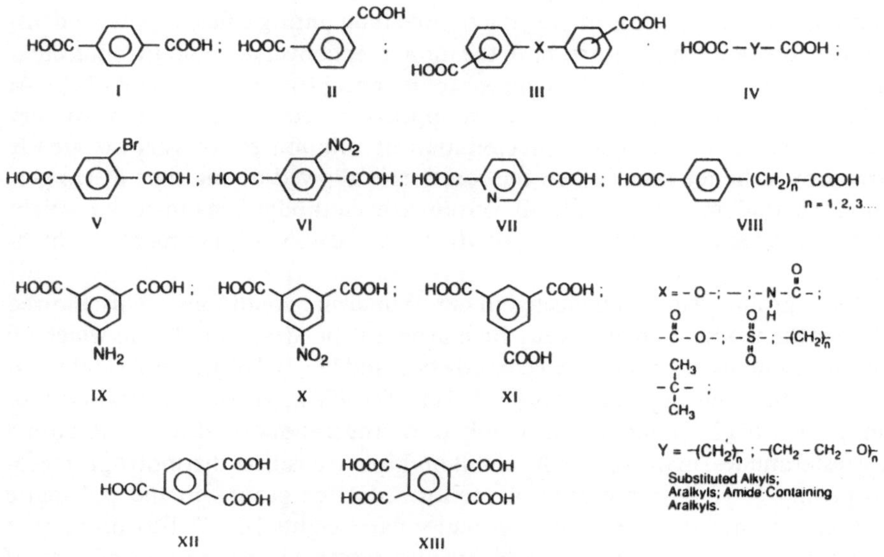

Scheme 1. Easily Polymerizable Monomers under Yamazaki Conditions: Carboxylic Acids

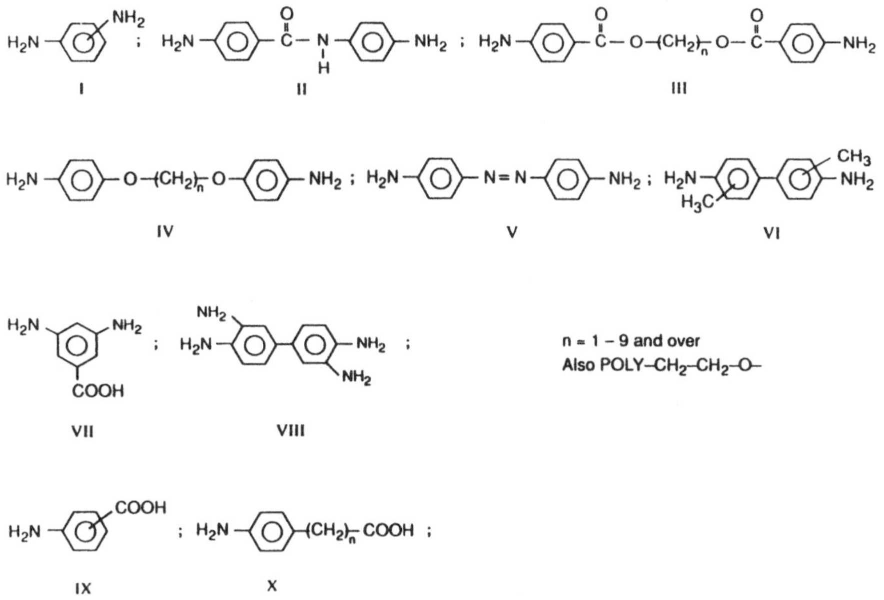

I H_2N—〇—NH_2 ; **II** H_2N—〇—$\overset{O}{\overset{\|}{C}}$—$\underset{H}{N}$—〇—$NH_2$; **III** H_2N—〇—$\overset{O}{\overset{\|}{C}}$—$O$—$(CH_2)_n$—$O$—$\overset{O}{\overset{\|}{C}}$—〇—$NH_2$

IV H_2N—〇—O—$(CH_2)_n$—O—〇—NH_2 ; **V** H_2N—〇—$N=N$—〇—NH_2 ; **VI** H_2N—〇—〇$\overset{CH_3}{\underset{H_3C}{}}$—$NH_2$

VII H_2N—〇$\overset{NH_2}{\underset{COOH}{}}$; **VIII** H_2N—〇—〇$\overset{NH_2}{\underset{NH_2}{}}$—$NH_2$

n = 1 – 9 and over
Also POLY–CH_2–CH_2–O–

IX H_2N—〇$\overset{COOH}{}$; **X** H_2N—〇—$(CH_2)_n$—$COOH$;

Scheme 2. Easily Polymerizable Monomers under Yamazaki Conditions: Aromatic Amines

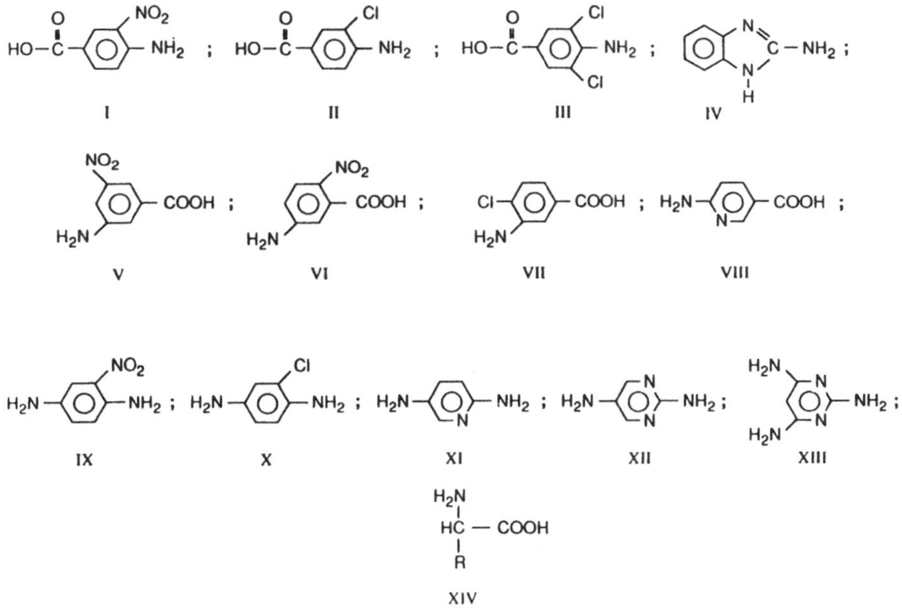

I HO—$\overset{O}{\overset{\|}{C}}$—〇$\overset{NO_2}{\underset{}{}}$—$NH_2$; **II** HO—$\overset{O}{\overset{\|}{C}}$—〇$\overset{Cl}{}$—$NH_2$; **III** HO—$\overset{O}{\overset{\|}{C}}$—〇$\overset{Cl}{\underset{Cl}{}}$—$NH_2$; **IV** 〇$\overset{N}{\underset{\underset{H}{N}}{}}$$C$—$NH_2$;

V $\overset{NO_2}{\underset{H_2N}{}}$〇—$COOH$; **VI** $\overset{NO_2}{\underset{H_2N}{}}$〇—$COOH$; **VII** Cl—$\underset{H_2N}{}$〇—$COOH$; **VIII** H_2N—〇$\overset{}{\underset{N}{}}$—$COOH$;

IX H_2N—〇$\overset{NO_2}{}$—NH_2 ; **X** H_2N—〇$\overset{Cl}{}$—NH_2 ; **XI** H_2N—〇$\underset{N}{}$—NH_2 ; **XII** H_2N—〇$\overset{N}{\underset{N}{}}$—$NH_2$; **XIII** H_2N—〇$\overset{\overset{H_2N}{}N}{\underset{H_2N}{}}$—$NH_2$;

$\overset{H_2N}{\overset{\|}{HC}}$—$COOH$
$\underset{R}{\overset{\|}{}}$

XIV

Scheme 3. Monomers that do not Polymerize under Yamazaki Conditions

greatly decrease with increasing n. Substituted aminophenols

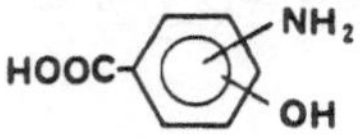

also show poor results.

Monomers that do not polymerize under Yamazaki conditions are shown in Scheme 3. Comparison with Schemes 1 and 2 clearly shows that aromatic amines are much more prone to deactivation than aromatic carboxy acids. The presence of a nitro group anywhere on the ring deactivates the aromatic amine but does not deactivate aromatic caboxylic acids such as nitroterephthalic acid. The presence of chlorine deactivates aromatic amine while bromine does not deactivate bromoterephthalic acid. The presence of nitrogen atom in an aromatic ring, such as pyridine, pyrimidine or benzimidazole, deactivates amine groups attached to the ring but does not deactivate carboxyl groups. Reactions of aliphatic diamines or alpha-amino acids were most often hindered by the formation of insoluble salts when the monomers were mixed in hot or cold DMAc/5% LiCl [395] or by the rapid precipitation of low-M products after the reaction started [395, 399].

A very interesting situation involves the combination of two monomers, nitroterephthalic acid (NTPA) and tricarballylic acid (TCA):

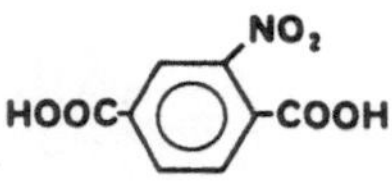

NTPA 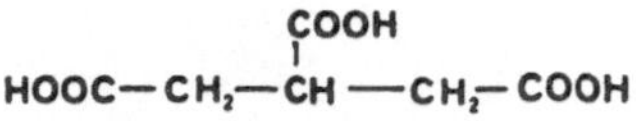   TCA

The two monomers, especially NTPA, polymerize very nicely with aromatic diamines to form linear high-M polymers or well-formed low-defect networks when present in the reaction mixture in the absence of one another. The presence of other carboxy acids, such as terephthalic acid (TPA), 1.3.5-benzenetricarboxylic acid (BTCA) or sebacic acid, does not affect the reactivity of either NTPA or TCA [311, 312, 393, 400, 402]. However, when NTPA and TCA are present together, no networks are formed even when exact stoichiometric amounts of aromatic diamines are present in the reaction mixture. Depending on the solution concentration, varying amounts of gels appear but no "infinite" networks. The reasons for the unexpected behavior of NTPA and TCA in the presence of each other are not understood by us at present.

The fact that polycondensations under the Yamazaki conditions employ free carboxylic acids instead of their chlorides, allows the use of either diacid or amino acid monomers separately or together in the same polymerization. This leads to a remarkable versatility in the creation of rigid networks out of aromatic polyamides. It further allows us to use in a Schotten-Baumann [403–408] type reaction in DMAc mixtures of acid chlorides and free acids with aromatic amines to create certain aromatic branchpoints of particular size of functionality as shown below and some useful components for stiff segments typified by the monomer di(carboxyphenyl terephthalamide) (DCTP):

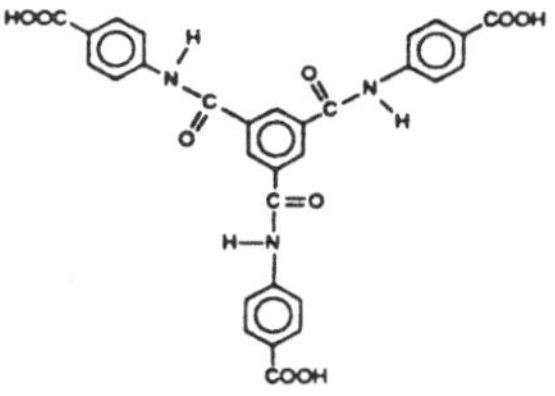

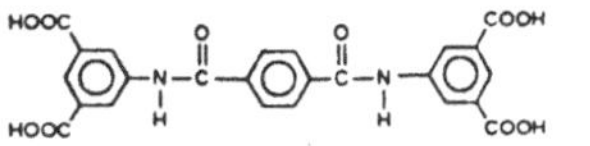

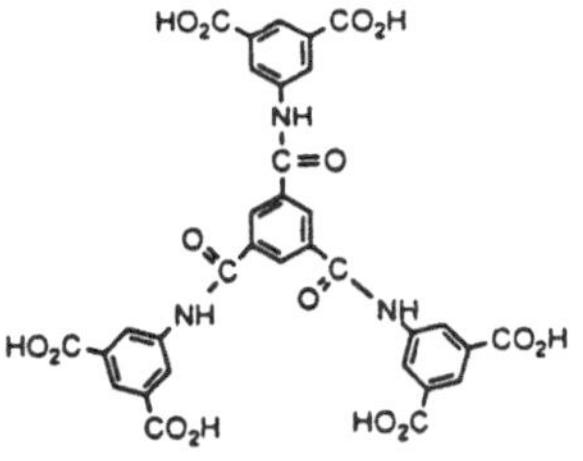

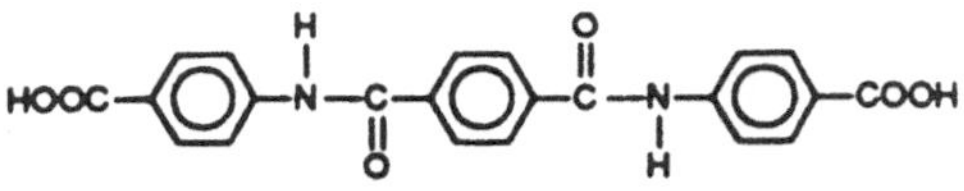

Together with BTCA, TPA and NTPA, these monomers were reacted with amine-bearing monomers, such as 4,4′-diaminobenzanilide (DABA) (number II in Scheme 2) and 1,4-phenylenediamine, to create most of the rigid polyamide networks prepared by Aharoni and associates [212, 311, 312, 393, 400, 409–411]. Networks with stiff segments between branchpoints as long as 208 Å [400] and as short as 6.5 Å [311] were prepared, the latter from equimolar mixture of the trifunctional monomers 5-aminoisophathalic acid and 3,5-diaminobenzoic acid:

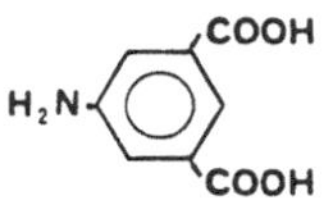

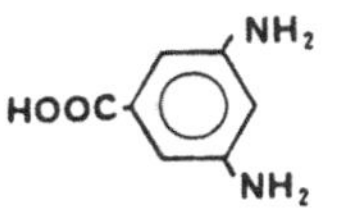

Typical rigid network fragments, each containing one stiff segment between two trifunctional rigid branchpoints, are shown below:

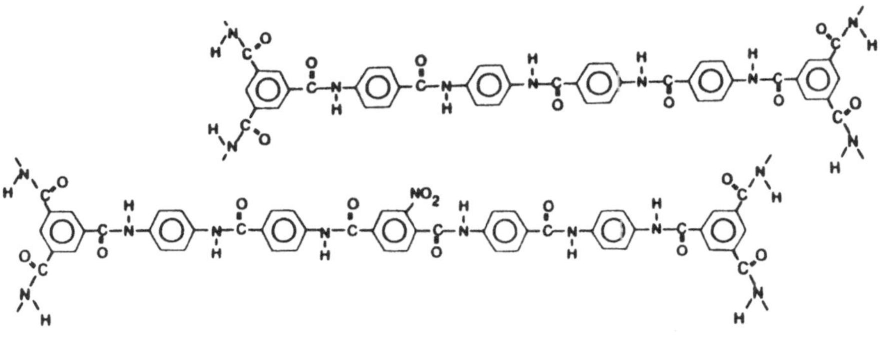

Building on our experience with unbranched stiff zigzag polyamide and poly(esteramide) chains [61, 62], rigid networks and gels of stiff zigzag polyamides were prepared [409]. Here, rigid bends in the segments were inserted by the introduction along them of residues originating from either isophathalic acid or di(carboxyphenylisophthalamide) (DCPIP) [409] monomers:

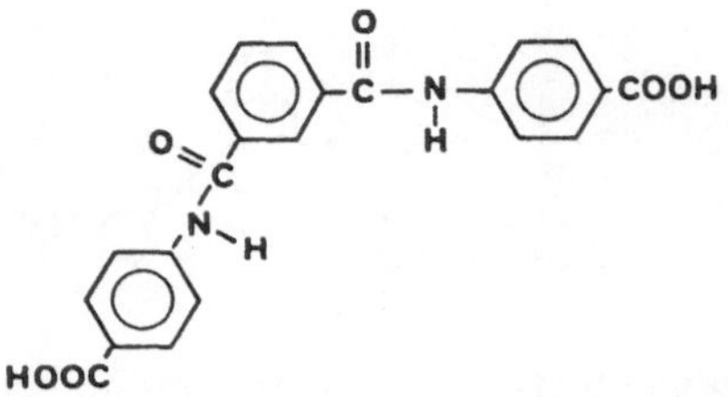

These are expected to serve as swivels whose fixed valence angles combine with a measure of torsional rotation freedom.

Intensely birefringent gels of "liquid crystalline" polyamide networks were prepared by combining stiff aromatic polyamide segments with flexible aliphatic junction points made from tricarballylic acid residues [402]. The flexible junction point, facilitating the parallel alignment of the stiff segments and the consequential birefringence of the quiescent gelled networks, is shown in Fig. 14.

In all the above, we have observed that the use of monosubstituted aromatic rings greatly increases the solubility of the stiff polymeric chains or segments [56, 311] during and after the polymerization, promoting more complete reaction and higher molecular weight of the products. We have used only nitro substitution. Krigbaum et al. [214, 411] used phenyl-substituents and reached the same conclusions. Another important observation is that, in the case of highly branched polyamide structures, pure DMAc is just as good a solvent as DMAc/LiCl. We have found that whenever a substantial amount of interchain or intersegmental hydrogen bonds are present or are likely to exist, solutions of LiCl in DMAc must be used in order to solubilize the polymer. This was observed not only in the case of linear aromatic polyamides [61, 62] but also in the case of networks with very large distances between branchpoints [311, 400]. However, when the segment lengths between branchpoints were sufficiently

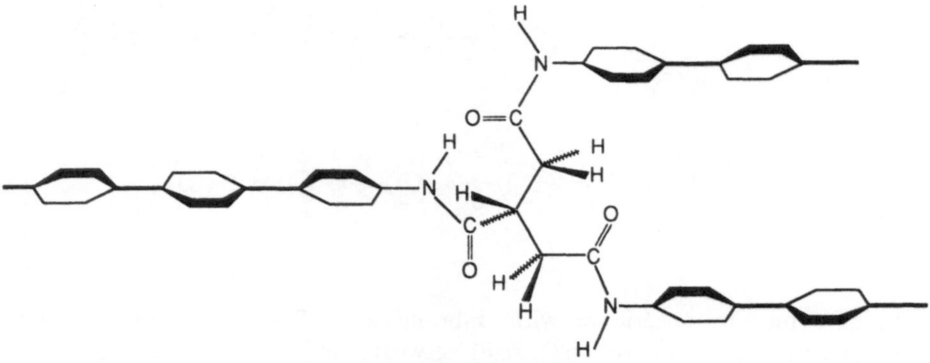

Fig. 14. Flexible junction made from tricarballylic acid, connecting stiff mesogenic polyamide segments in a "liquid-crystalline" network

reduced to eliminate or minimize the amount of intersegmental H-bonds, then DMAc alone served as an excellent swelling solvent [311, 312, 400, 410, 412]. In fact, in studies on the solubility parameters of aromatic polyamides [410], it was found that the swelling levels of the same polyamide network in pure DMAc and in DMAc/LiCl were indistinguishable from one another. We conclude, therefore, that the presence of the LiCl (or $CaCl_2$) in solution help in breaking intersegmental H-bonds when and where they exist. Once these are broken, the pure aprotic amide solvent is equivalent in solubilizing power to its mixture with the soluble metal halides.

In their work on the fractal nature of one-step highly branched polyamides, Aharoni et al. [311] recognized that the critical concentration for the creation of "infinite" networks is inversely proportional to the average segment length. This inverse proportionality leads to a situation in which, for networks created exclusively from trifunctional monomers, C_0^* was as high as 30% and the reaction time to reach the gel point was inordinately long. The poor mechanical properties of these networks indicated them to be highly defective. Increasing the concentration of the networks to $C_0 \gg C_0^*$ was impossible because even at C_0^* we were operating very close to the saturation point of the monomers in the reaction mixture. To operate in a more reasonable range of concentrations and create less defective networks, we opted to make the average segment lengths longer. This was achieved by adding amounts of difunctional aromatic monomers, with the proviso that the total number of amine and carboxyl groups in the reaction bath remained equal. By using monomer mixtures containing diamino- and dicarboxy-aromatic units, and aborting the polycondensation reaction at various points in the pre-gel stage, we were able to obtain fractal polyamides of increasing molecular weights and hydrodynamic radii, R_H. After workup and purification, these FPs were reused in a Yamazaki procedure in the absence of any monomeric or small oligomeric species, and were found to produce "infinite" networks with good mechanical properties [312]. Even more interesting systems have since been obtained by first creating FPs decorated with only one kind of reactive group and then reacting these FPs in various ways to form novel rigid networks [350, 413].

Unlike the previous FPs, highly branched polyamides whose branches are terminated by one kind of reactive group only, are made exclusively from AB and AB_n monomers. For trifurcated branchpoints, AB_2 monomers such as 5-aminoisophthalic acid or 3,5-diaminobenzoic acid are used. Depending on the desired reactable end-groups, a multifunctional nucleus may be used whose reactive groups are the same as the B group in the AB_n monomers. In our work, frequently used nuclei were BTCA and 3,3'-benzidine. Higher functionality nuclei may be used as well. To a solution of the nuclei in DMAc at about 115 °C, are added dropwise a solution of the monomers AB and AB_n in DMAc and pyridine, and from a separate funnel a solution of TPP in DMAc. A Yamazaki reaction is carried for up to 8 h depending on the rate of addition of all ingredients. The molar ratio of AB to AB_n monomers dictates the average segment length between branchpoints, and the number ratio of monomers to

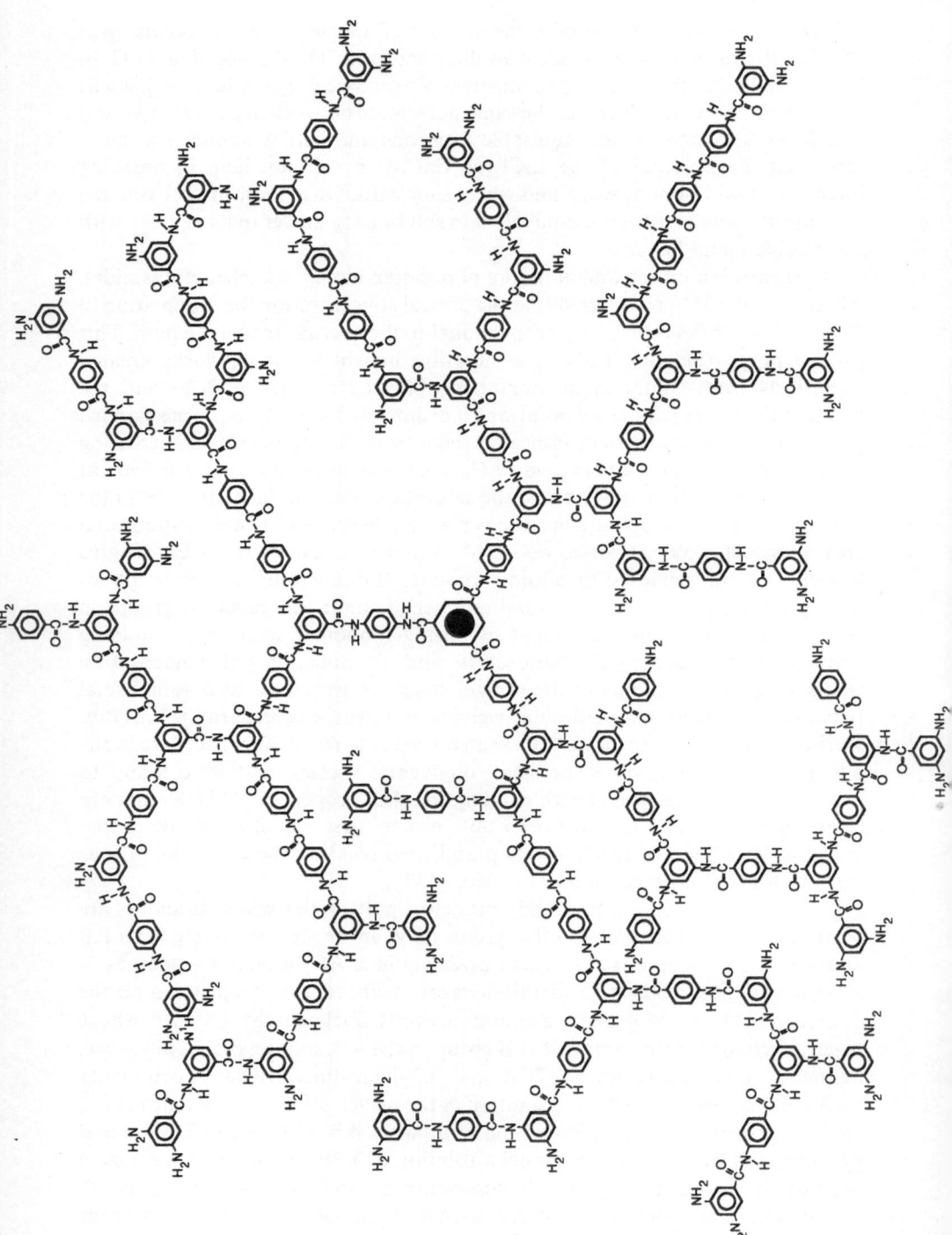

nuclei determines the average size of the resulting FPs. Naturally, the number n determines the branchpoint functionality. In the examples below, the AB monomer was 4-aminobenzoic acid and the AB_2 monomer was either 5-aminoisophthalic acid or 3,5-diaminobenzoic acid. The molar ratio of the AB to AB_2 monomers was $1:1$, keeping the average segment length between branch-points at $l_0 = 13$ Å. Fractal polymers averaging 48 and 64 amine- or carboxy-end-groups were thus prepared. Unlike the FPs made from AA and BB monomers, the FPs made from AB and AB_2 monomers do not react with themselves to form an infinite network when the polymerization is allowed to proceed to completion. A schematic two-dimensional description of a three dimensional polyamide FP decorated with about 48 amine groups is shown in Fig. 15. In this case the nucleus was prepared in a separate reaction from BTCA and a very large excess of 1,4-phenylenediamine.

To test whether the rate of monomer addition affects the final results, two polycondensations were conducted in parallel. Both reactions were identical in all respects but for the rate of monomer mixture addition: in one the monomer mixture was added dropwise over 3 h and in the other the monomer mixture was added to the nuclei solution all at once. Both reactions were allowed to proceed for 3 h from the moment monomer addition was completed. After workup, the products were studied by GPC and solution viscosity. The peak molecular weight in the GPC scans of both batches appeared at identical locations, indicating that the vast majority of polymeric species in both pre-parations had the same molecular weight. The GPC scan of the batch prepared by adding the monomers all at once had two shoulders on the high-M side of the elution curve. Otherwise the curves of both preparations appeared identical. The intrinsic viscosity of the batch prepared all at once was only very slightly higher than that of the batch prepared in the dropwise manner. From the above we concluded that FPs prepared in either fashion end up having about the same molecular weight and molecular weight distribution, except for some aggregate formation that takes place when the monomers are all added at once and does not occur when the monomer mixture is added dropwise.

By adding a different monomer at the end of the polycondensation, or by reacting the FPs in a separate step, FPs with other than amine or carboxyl terminal groups can be prepared. Examples of such are methyl ester, 2-hydroxyethyl ester and various metal carboxylates [350].

An infinite network was prepared at $C_0 = 10\%$ by reacting under Yamazaki conditions 12 g of a 64-amine-terminated AB-type FP with 1.0 g of the short stiff DCPTP [413]. The system gelled and solidified within 15 min from the be-ginning of the reaction. When the monomer charge was reduced to $C_0 = 8.8\%$, no gelation took place. This means that for this system $8.8 < C_0^* < 10\%$. When common one-step rigid networks are prepared from AA and BB-type mono-mers, an increase in concentration of only 1.2% above C_0^* usually produces gels

Fig. 15. Two-dimensional schematic representation of an idealized three-dimensional fully aromatic fractal polymer

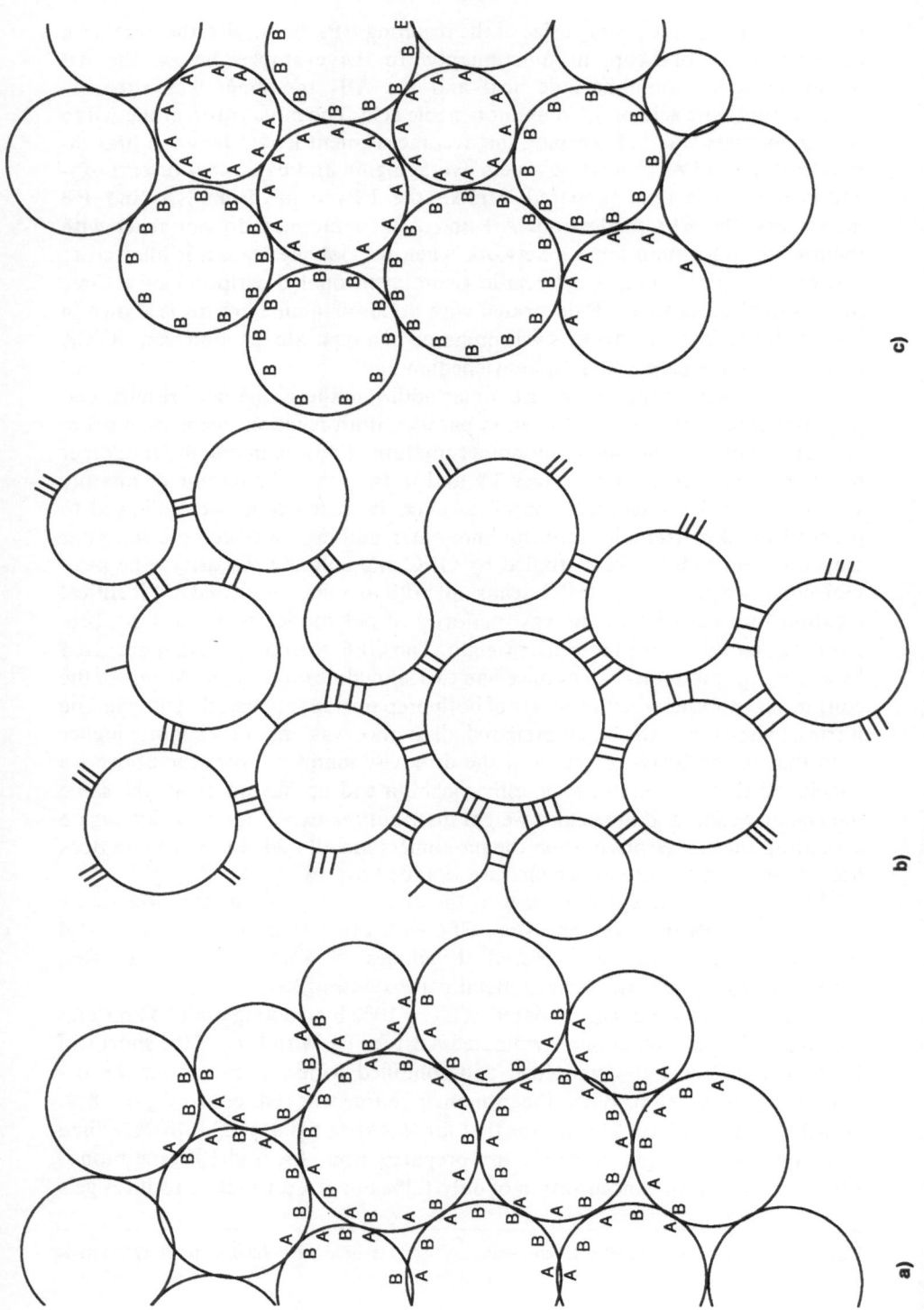

c)

b)

a)

far less rigid than the one obtained from the reaction of FPS and DCPTP, and the time to reach the gel point is far longer. We believe that the differences are due to the fact that, in the "supernetworks" made from FPs and stiff struts, the network is fully swollen at the point of creation and can neither swell nor shrink to accommodate changes in concentration: when the swollen rigid FPs with the attached struts touch one another, a network begins to form.

Another interesting rigid network was recently prepared [413] at $C_0 = 10\%$ by reacting together two kinds of FPs. Here, equal weights of FPs terminated with 48 amines and 64 carboxyl groups each were reacted under Yamazaki conditions. In this case, the reaction proceeded for about 50 min before abruptly reaching the gel point. Even after an additional 4 h of reaction time, the gel product was very weak. Its failure mode, however, was brittle in nature. Equilibration in pure DMAc for over two weeks resulted in a volume increase of 40% of the still weak gel. This indicates a rather defective network, unlike the networks prepared from FPs made from AA and BB-type monomers [312].

In Fig. 16 schematic descriptions of the three rigid polyamide networks prepared from FPs are shown. In Fig. 16a, a network made from AA and BB-type FPs is shown, in Fig. 16b a network made from AB and AB_2 FPs and stiff struts between them is shown, and in Fig. 16c a network made by reacting two kinds of $AB + AB_2$ fractal polymers is shown. In Sect. 3.2.5 we shall return to supernet-works such as the one shown in Fig. 16b. From the Figure, the reason for the high defect level in structure 16c becomes obvious. Unlike the structure depicted in Fig. 16a, where amine and carboxy groups are present at random on the exteriors of all FPs such that any FP may react with any other FP, in the structure depcited in Fig. 16c, when two FPs decorated by amine groups are in contact they do not react and leave a defective region in the network. We may call this an FP-scale fissure. Similar defects are built into the system when fractal polymers containing only carboxy reactive groups touch each other. The statistics of the system are very much in favor of many such FP-scale fissures occurring in networks made from mixtures of exclusively-amine and exclusively-carboxy decorated FPs. When the FPs are terminated at random with both amine and carboxyl groups, a smaller number of smaller size defects are expected.

Some of the properties of rigid aromatic polyamide networks and gels, and of their precursor FPs will be discussed in Sects. 4.1 and 4.2 below.

Fig. 16a–c Three kinds of polyamide networks actually prepared from purified rigid polyamide pre-gel particles in the absence of monomers: a network from AA and BB FPs; b netwrok from AB and AB_2 FPs connected by rigid struts; c network prepared from FPs decorated by A groups and FPs decorated by B groups

3.1.2.2 Rigid Polyamide Networks and Fractal Polymers Prepared in Solution by Other Procedures

In addition to the above described polycondensation of free carboxylic acid with aromatic amine in the presence of aryl phosphites, there are three more amide-forming polycondensations described in the literature. In contradistinction with the phosphite reactions, these require the use of an activated form of the carboxy acid, such as acid chloride. Similar to the phosphite reactions, these reactions are also conducted in aprotic amide solvent such as DMAc or NMP, in the presence or absence of solubilizing salts such as LiCl or $CaCl_2$.

The first of these procedures is the well-known Schotten-Bauman reaction [403–408] in which an AA monomer such as di(acid chloride) is allowed to react, in the presence of an organic base, with a BB monomer such as aromatic diamine to form a polymer:

$$
\underset{\substack{\| \\ \text{Cl-C-Ar-C-Cl}}}{\overset{\substack{O \quad\quad O \\ \| \quad\quad \|}}{}} + H_2N\text{-Ar-}NH_2 \xrightarrow{\text{Base}} \underset{\substack{\mid \quad\quad \mid \\ H \quad\quad H}}{\overset{\substack{O \quad\quad\quad O \\ \| \quad\quad\quad \|}}{(\text{Ar-C-N-Ar-N-C})n}.
$$

$$+ \text{ Base} \cdot HCl$$

When either or both of the comonomers has a functionality higher than 2, a branched system evolves. The major limitation of the Schotten-Baumann reaction is its uses of AA and BB-type monomers, to the exclusion of AB-type monomers. This is simply due to the general unavailability of AB-type monomers in which A is an acid chloride and B is a free amine. Responding to the need, a procedure was developed over the years at duPont by which a reactive AB-type monomer is first prepared and then reacted with itself to create linear [210, 414, 415] or highly branched [416] aromatic polyamide polymers. Accordingly, an amino acid chloride hydrochloride (e.g.,)

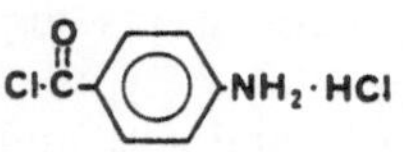

or 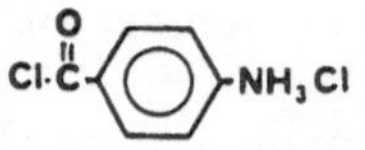) is first prepared by

reacting an aromatic amino acid first with thionyl chloride and then dry HCl [414, 415]. The product is then dissolved in an aprotic amide solvent and allowed to condense with itself. In the case of a trifunctional monomer, a highly branched aromatic polyamide is obtained [416]:

or

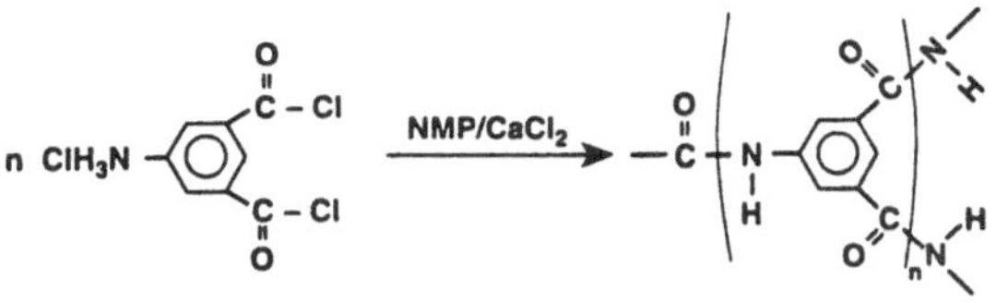

In the reported procedure [416], the desired products were aromatic FPs and not an "infinite" network, leaving the number of acid chloride and amine groups unequal in each preparation. As with the work of Aharoni [350, 413], this led to the FPs being decorated, essentially on their exteriors, with the reactive group present in excess.

In the third polycondensation, the monomers were first prepared from the corresponding aromatic amino acid and thionyl chloride [416] with the HCl treatment being omitted. When the resulting sulfinyl amino acid chlorides

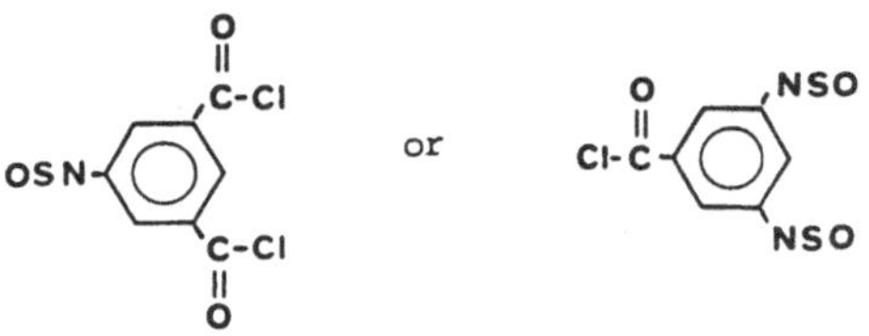

are allowed to polymerize in aprotic amide solvents in the presence of equimolar amounts of water, highly branched polyamides are obtained, containing the molar excess of unreacted reactive groups in their respective free form: car-boxylic acid or aromatic amine [416].

It is obvious that with the appropriate monomer mixture, all the reactions employing activated forms of the carboxylic acid can and will yield gels of "infinite" networks of rigid aromatic polyamides. Here, again, the ratio of difunctional to higher functionality monomers will control the average length of the stiff aromatic polyamide segments.

We would like to differentiate here between an "infinite" network and an "infinite" Cayley tree. An "infinite" Cayley tree is a highly branched macromolecule having a single nucleus from which all branches originate directly or iteratively. The Cayley tree contains no intramolecular closed rings or cycles, especially in instances where the branches are rodlike or stiff. A single branched polymer molecule emanating from a single nucleus thus fills the solvent in the reaction vessel. Dendrimers [331-339] are good examples of Cayley trees with equal length branches whose growth was interrupted before reaching a "infinite" size. Unlike the "infinite" Cayley trees, each "infinite" network originates from many highly branched precursor macromolecules. Each of these precursor molecules starts from its own nucleus. As the precursor macromolecules grow towards one another in the pre-gel, gel-point and post-gel stages, covalent bonds are formed between branches belonging to different highly branched molecules and many rings or cycles are created. When such a network fills all the available reaction medium, it is then called "infinite". It is important to emphasize in this context that, in the case of stiff or rigid polymers, the created cycles generally contain segments originating from different highly branched precursor mole-cules. Both the above "infinite" systems are at variance with flexible polymers where both intramolecular and intermolecular closed cycles frequently exist.

Networks consisting of aromatic polyamide segments connected to one another by junctions created from non-amide segmental end-capping groups, will be discussed in Sect. 3.1.5.

3.1.3 Rigid Aromatic Polyester Networks

To the best of our knowledge, no fully aromatic polyester networks have been reported in the literature until now. Several reports appeared describing the preparation of liquid crystal thermosets from rodlike polyester segments and branchpoints or junctions created by the reaction of reactive end-caps with one another [417-419]. These end-capped polyester networks will be discussed, however, with all other fully or mostly aromatic networks made in two steps by reacting the end-caps of stiff segments with each other.

The recent literature contains, nevertheless, very interesting novel syntheses of highly branched aromatic polyester dendrimers and of unbranched stiff aromatic polyester chains. These synthetic efforts will be described below. They indicate that conditions are presently ripe for the creation of novel aromatic polyester networks and gels.

Prominent among these efforts are ones using the classical Schotten-Bauman [403-408] method, condensing aromatic acid chloride with aromatic hydroxyl in the presence of an organic base to create an ester group:

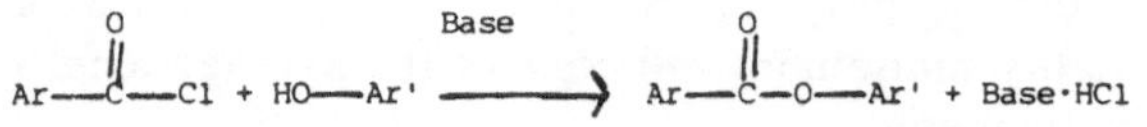

This technique was employed to create liquid crystalline rigid rodlike polyesters decorated with short flexible alkoxy side-chains rendering them soluble [420-422]:

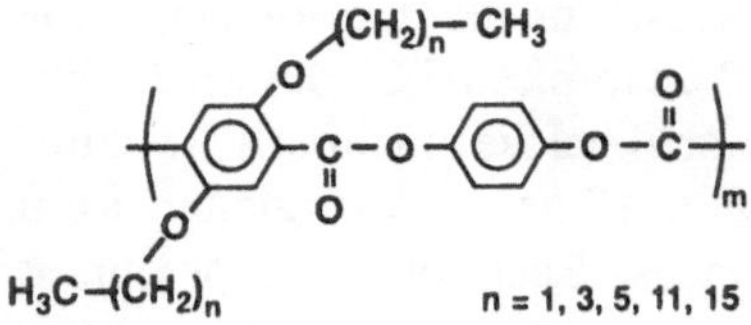

as well as relatively short aromatic ester segments in liquid crystal polymer networks in which the branchpoints are the trifuntional triazine rings [423]:

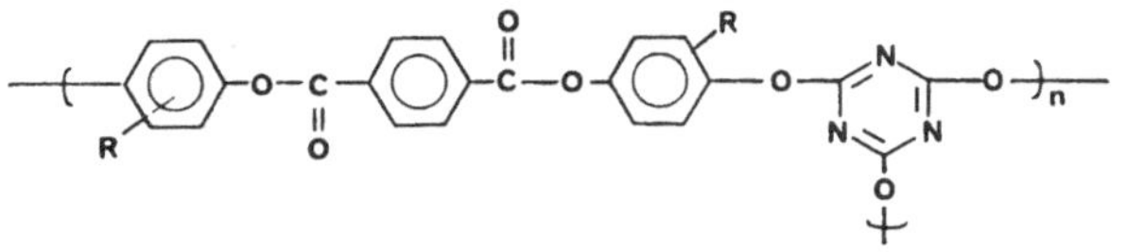

R = methyl or Cl.

Another polyesterification using an aromatic acid chloride was developed by Higashi and associates [424, 425]. Accordingly, by activating in-situ carboxylic acids with diphenyl chlorophosphate, a direct polyesterification of aromatic carboxylic acids with phenolic hydroxyl groups is achieved. This method was used by Aharoni [61] to create relatively high molecular weight aromatic poly(ester amide)

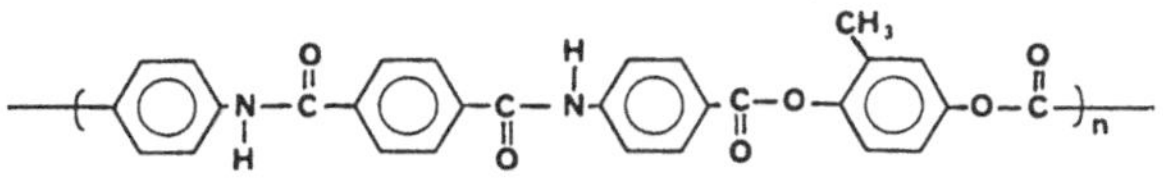

that exhibited both lyotropic and thermotropic mesomorphicity.

A third procedure to activate in situ the carboxylic acids is the use of arylsulfonyl chloride in pyridine as a condensing agent [426-428]. It was shown [429] that this reaction leads to significant amounts of undesirable side products and, as a result, gives low molecular weight polymers.

The use of triphenyl phosphine as a condensing agent for aromatic hydroxy-acids was reported by Ogata, Sanui and associates [430]. Here, hexachloroeth-ane serves as the source of chlorine to generate acid chlorides in situ, and the reaction is carried in a solvent mixture containing pyridine and chloroform:

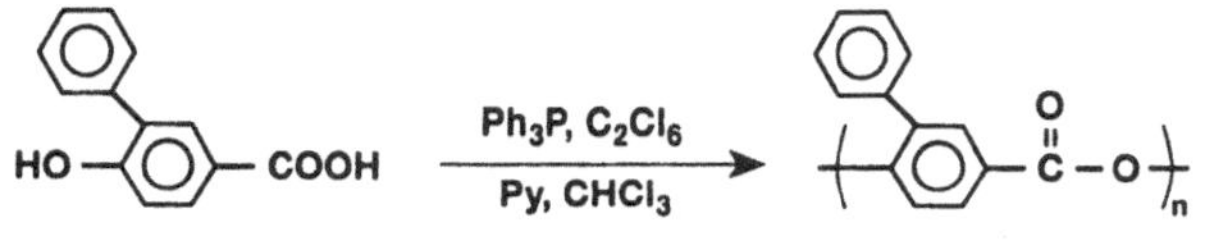

A novel polyesterification procedure avoiding the need for pre-activated carboxylic acid was recently reported by Moore and Stupp [429]. This method reacts phenolic groups directly with carboxylic acids in solution and under mild conditions. It requires the use of carbodiimide dehydrating agent [431, 432] and a novel catalyst which is a 1:1 molecular complex formed by 4-(dimethylamino) pyridine and p-toluenesulfonic acid [429]. Under these conditions, linear polyesters with molecular weights in excess of 15 000 were obtained [429], but the molecular weights of exclusively aromatic polyesters prepared by this method may be lower.

A synthetic procedure gaining popularity, especially for the creation of highly branched aromatic polyesters, is one using trialkyl siloxy derivatives of trifunctional aromatic kernels [433-435]. In this procedure, either 3,5-dihydroxybenzoic acid or 5-hydroxyisophthalic acid is treated with trialkylisilyl chloride in the presence of an organic base. The resulting trialkylisilyoxyaromatic acid is then reacted with thionyl chloride to produce the corresponding acid chloride:

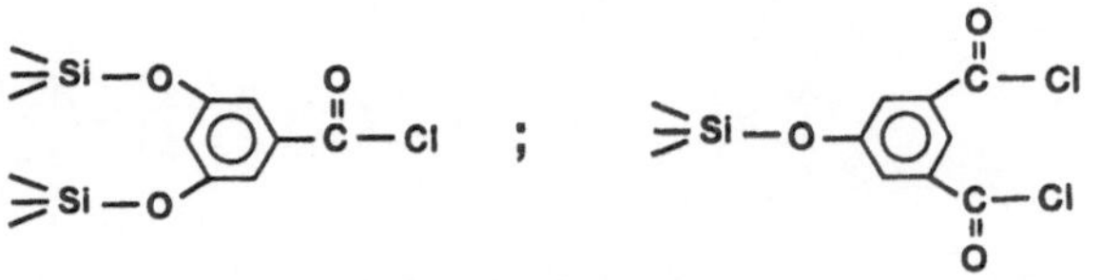

The polymerization is performed either by heating the trialkylsiloxy-protected monomers to effect self-condensation with the removal of trialkylsilyl chloride [433, 435], or by first deprotecting the hydroxyl group and then conducting a solution polycondensation in the presence of an organic base. Highly branched polyesters were achieved, reaching $M_w = 164\,000$ in the work of Frechet and co-workers [433], and one million in the work of Voit and Turner [435]. The molecular weight distribution in the latter case was much broader than in the former. In contradistinction to linear polyesters, the highly branched aromatic polyesters are very soluble in a variety of organic solvents. In this respect they are similar to the highly branched aromatic polyamide FPs.

A report appeared very recently [436] mentioning the preparation of highly branched aromatic polyester by a single-step melt polycondensation of the easily available monomer, 3,5-diacetoxybenzoic acid:

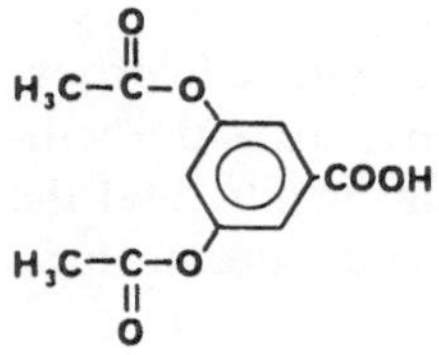

No other information was given in the published abstract.

3.1.4 Rigid Aromatic Networks Containing Single-Atom Bridges

Highly branched networks in which aromatic rings are connected to one another by a single-atom bridge are well-known. By a single-atom bridge we mean that two aromatic groups are linked to one another by two interatomic bonds through a single atom while, at the same time, this atom may participate in more that only these two bonds. Typical single-atom bridges, for example, are the methylene, ketone, amine and ether groups. The aromatic groups linked by such bridges are mostly benzene rings and phenolic moieties, but may include furan derivatives, triazine-containing species and maleimides. In the resin industry, where thermoset networks are prepared in situ in the absence of solvent, such groups are often mixed together in order to obtain a desired property mix, and additional reactive groups are often added to improve the quality of the products such as melamine derivatives and various epoxies. The existing and potential networks comprising aromatic groups connected by single-atom bridges are shown schematically below:

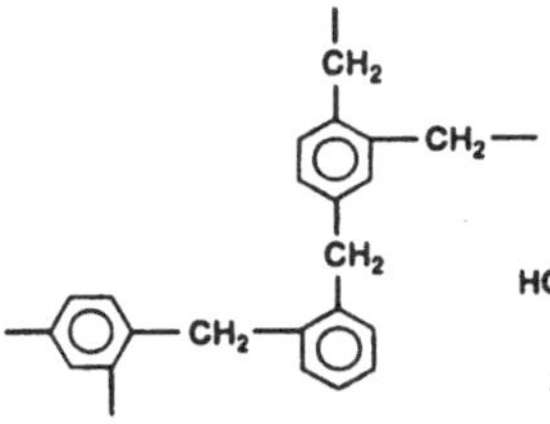

Hydrocarbon 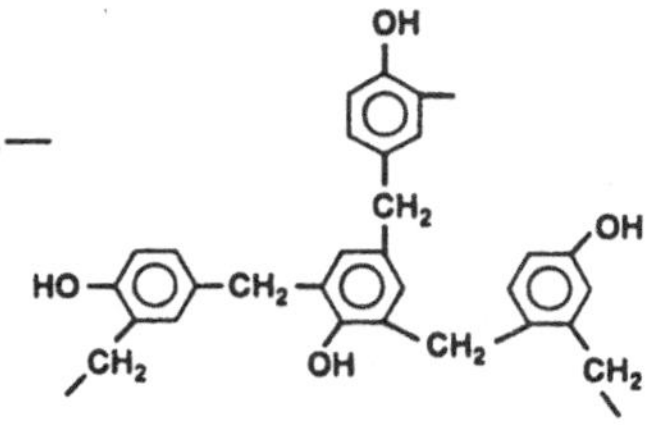   **Phenol-formaldehyde** 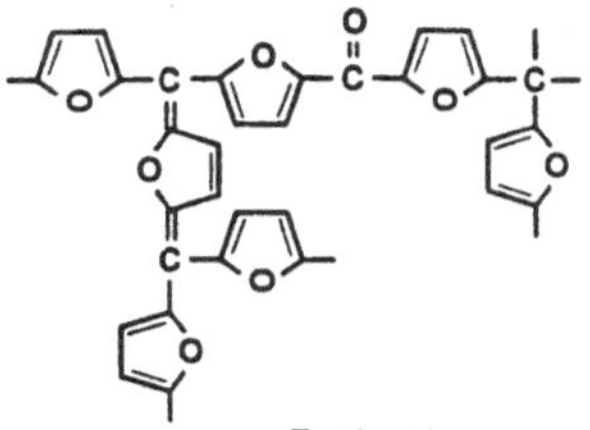   **Furfural**

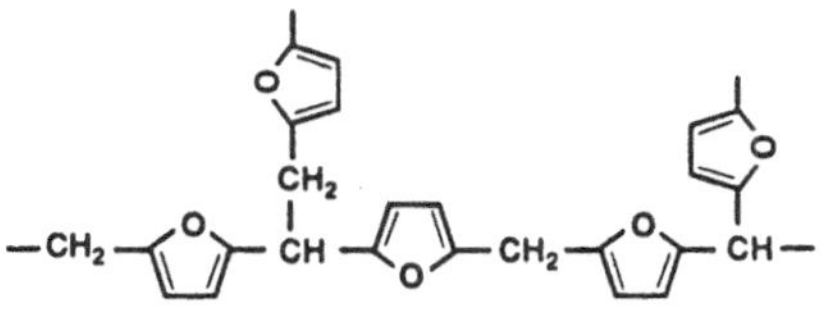

Furfuryl alcohol

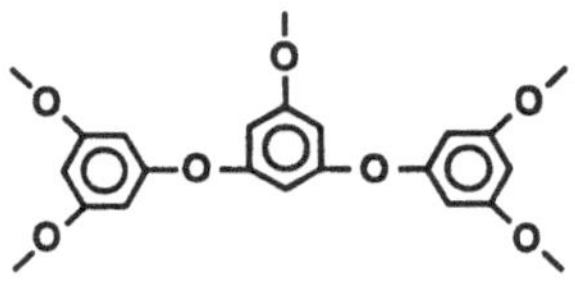

Polyether 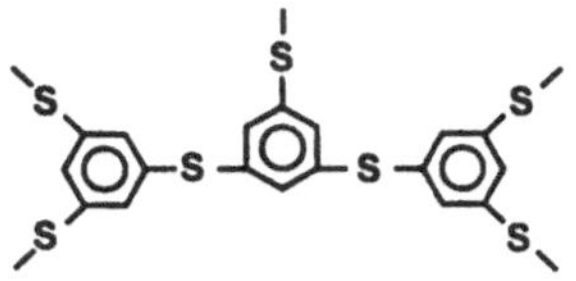   **Polysulfide** 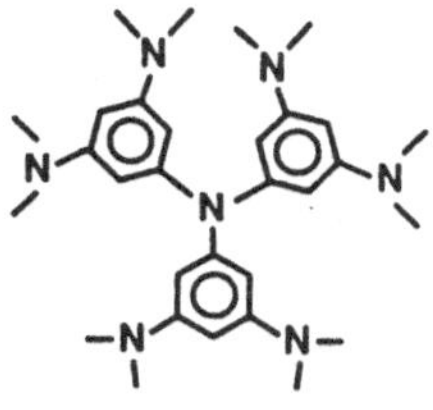   **Polyamine**

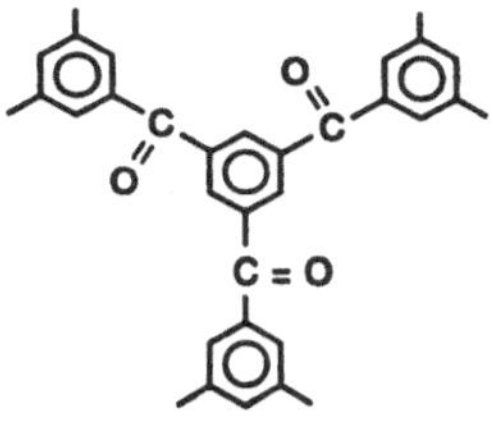

Polyketone 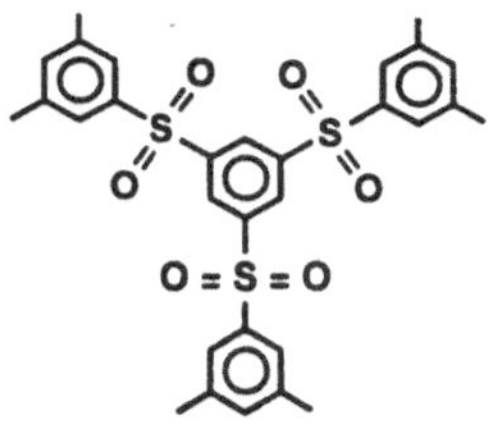   **Polysulfone** 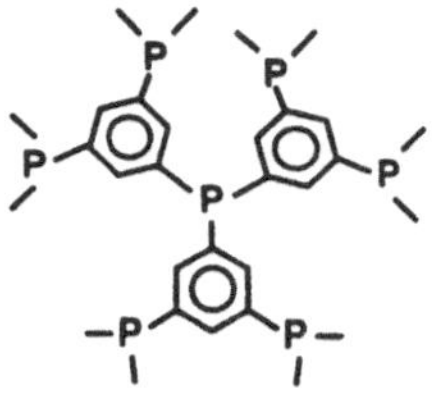   **Polyphosphine**

Not all these networks were prepared and characterized. The synthetic methods for their preparation are already available. Once the need arises, these networks will be created.

3.1.4.1 Aromatic Hydrocarbons with Methylene Bridges

Very highly crosslinked, substantially aromatic networks containing methylene bridges fall into several categories: hydrocarbon networks, phenol-formaldehyde and similar networks, and furan-based networks. The hydrocarbon networks have not yet found broad industrial applications despite the fact that they have been known for over a century. Traditionally, hydrocarbon aromatic networks were prepared [437, 438] under Friedel-Crafts conditions using monomers such as benzyl chloride [439]:

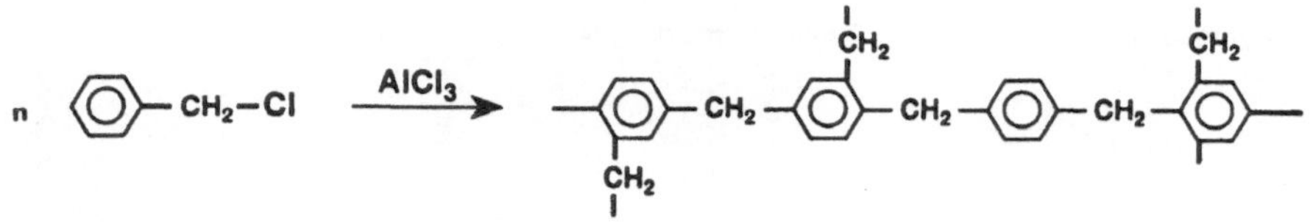

Methyl substituted monomers, as well as one carrying various halogen atoms have also been described in the literature [438]. In fact, hydrocarbon networks prepared under Friedel-Crafts conditions served as a model for Flory's theory for branched polymers containing A-R-B$_{f-1}$ type structural units [440]. Even though the properties of the hudrocarbon networks were expected to be rather similar to those of phenol-formaldehyde resins [437], aromatic hydrocarbon networks never posed a real commercial threat to other resins. This is due to the cost differential between the starting monomers and the fact that the hydrocarbon polymerizations are carried in carbon disulfide solvent using strong Lewis acid reagents, and the purification of the polymeric network products from the metal-containing reaction by-products is not a trivial task.

The Friedel-Crafts reaction was expanded by Jones et al. to prepare networks of aromatic condensed rings connected by methylene bridges [441-444]. The purpose of the work was, in fact, to create linear polymers but, when AlCl$_3$ was used, the products contained substantial fractions of insoluble networks with the rest being soluble oligomeric species. In general, two polymerization methods were used. One was self-condensation of chloromethylated condensed aromatic nuclei, and the second was a polymerization of the unsubstituted aromatic units with a chloromethylating agent [444]. It is interesting to note that when the Lewis acid was SnCl$_4$ instead of AlCl$_3$, only soluble products were obtained. A typical product of the first method contains naphthalene as the aromatic group [444]:

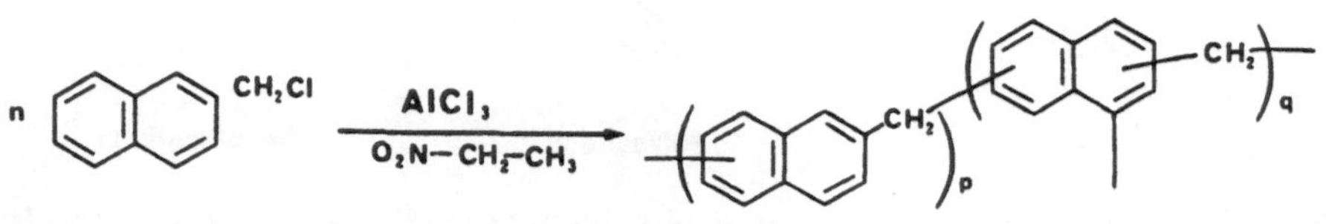

where the number of substitutions and their positions on the naphthalene kernel are not clearly defined. Typical of the second method are polymers containing both hydrocarbon and heteroaromatic groups [441-444]:

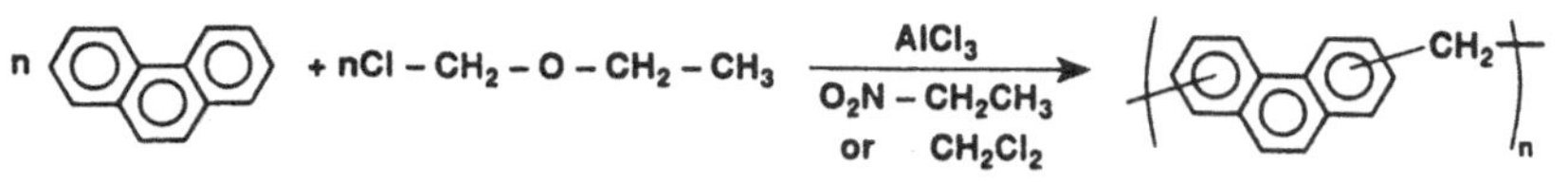

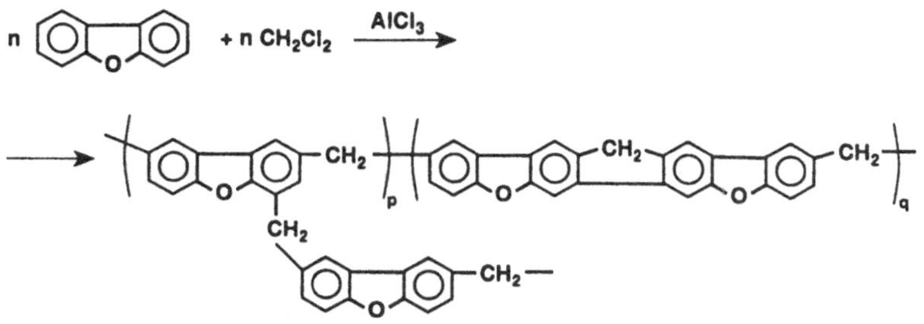

and

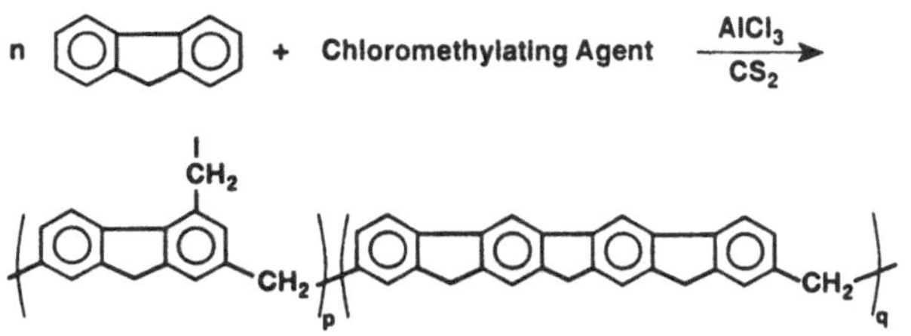

where CH_2Cl_2, chloromethylmethyl ether and methoxyacetyl chloride may serve interchangeably as the chloromethylating agent. In all the above, the degree of substitution on the aromatic kernels is not known exactly, but the fact that all or a substantial fraction of the products were insoluble suggests an average functionality of the aromatic groups higher than 2.0.

Interestingly, when the $SnCl_4$ agent which gave soluble products to Jones et al. was used to polymerize mixtures of condensed aromatic rings and p-xylylene dichloride, crosslinked networks were obtained [445]. Because these polymers contain no conjugated backbones, it is believed that the bridges between the aromatic kernels are exclusively or largely methylenic in nature.

3.1.4.2 Phenol-Formaldehyde-Type Networks

The phenol-formaldehyde resins are used in very large quantities, especially in the lumber and wood-products industry and for thermally and electrically insulating purposes. Smaller applications are found in the laminates, friction and abrasives, and coatings industries [446]. It is not surprising, therefore, that many excellent books and review articles appeared over the years providing updated summaries of phenolic technology and chemical research [446]. The intriguing fact is that despite all the technological advances, including the development by Baekeland [447, 448] of an economical method to prepare moldable phenolic compositions that can be heat and pressure cured to yield hard and heat-resistant products, and the chemical research starting even earlier [449-452], the physics of the phenol-formaldehyde networks remained rather poorly understood. The main reason for this is, of course, the great variability in local structure being built into the network during its evolution. This variability is caused by the presence of a very large number of coexisting kinds of growing species present in the reaction mixture before and after the gel point. Among the factors contributing to this great variability one finds the molar ratio of phenol to formaldehyde, pH, temperature, the nature of the monomeric species prior to oligomerization, the presence or absence of specific catalysts, etc [446]. Thus, for instance, novolak-type resins, in which the phenolic residues are present in numbers equal to or larger than the formaldehyde molecules, are typified by the presence of methylene bridges between the aromatic rings to the almost complete exclusion of other kinds of bridges:

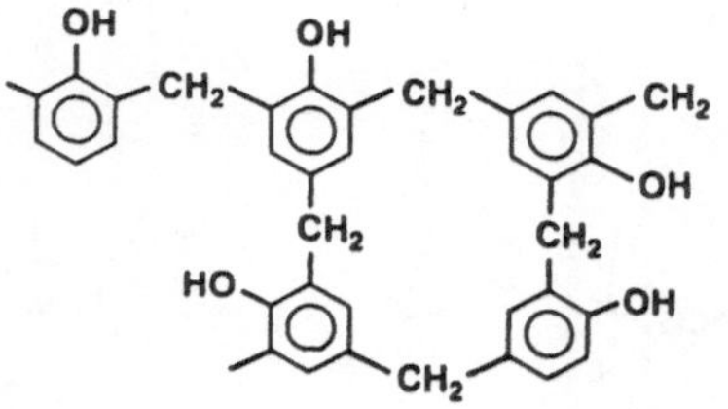

The resol-type resins, richer in formaldehyde, are significantly more complex and may include, in addition to the -CH$_2$-bridges, longer bridges such as -CH$_2$-O-CH$_2$-:

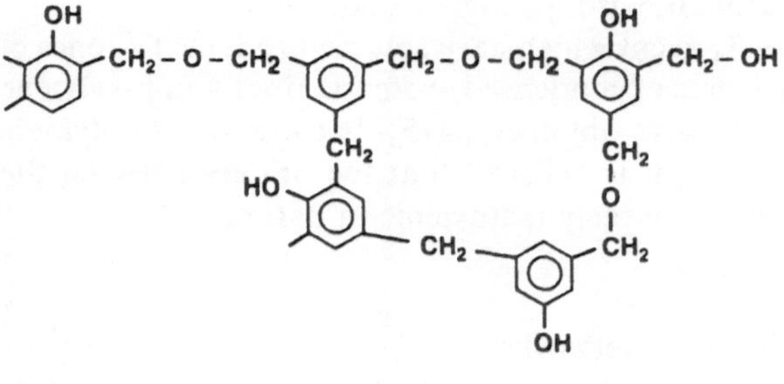

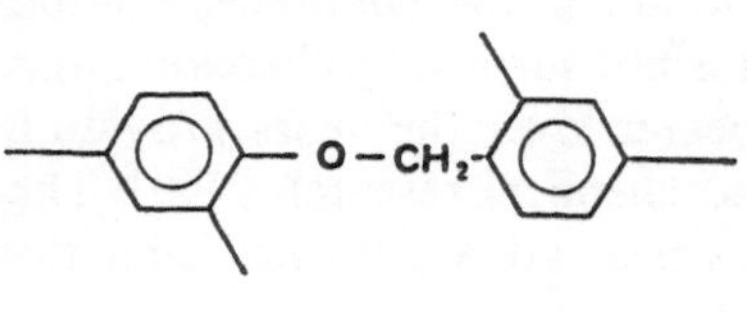

Solid-state carbon-13 NMR studies [453-456] indicated that, in addition to the bridging groups shown in the above schemes, bridges such as may be present at least in resol-type resins, and bridge points such as

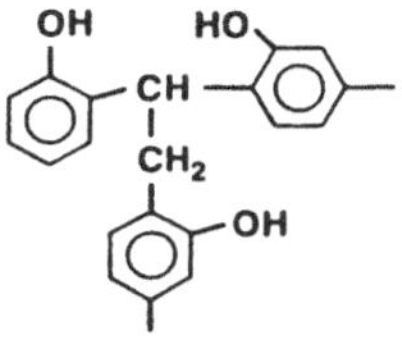

may be present in both resol- and novolak-type resins. More important, these studies [456] clearly demonstrated that the cured solid network is extremely complex and not all the reaction products and mechanisms are fully understood. It seems, therefore, that even where the reactive species in the soluble pre-polymer mixture can be fractionated and spectroscopically identified [446, 457], the fine details of the final crosslinked networks and the nature and conse-quences of their imperfections and flaws are not yet within our grasp.

Despite their complexity and lack of thorough understanding, the phenolic resins find ever expanding new applications in conjuction with other crosslink-able species. A stream of new patents continues to flow along these lines. Among the latest combinations one finds, for example, phenol-aralkyl resins [458] in which some of the aromatic rings are hydroxylated and others are not:

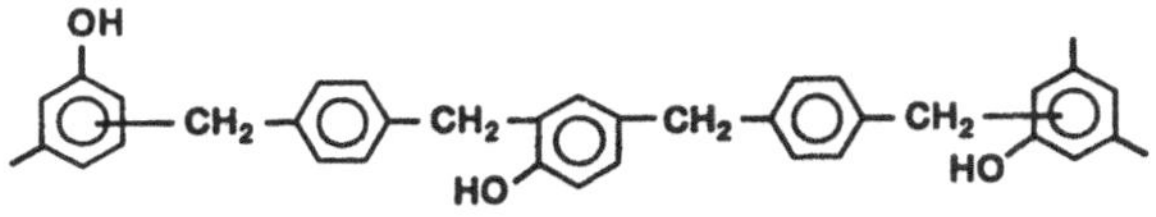

mixed phenolic oligomers [459] such as

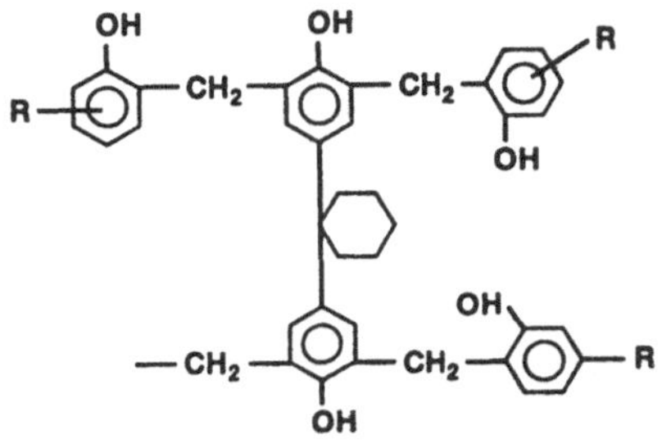

in which R is either an epoxy or an alkyl group, built into substantially epoxy matrix, and phenolic-maleimide intermediates [460] represented by:

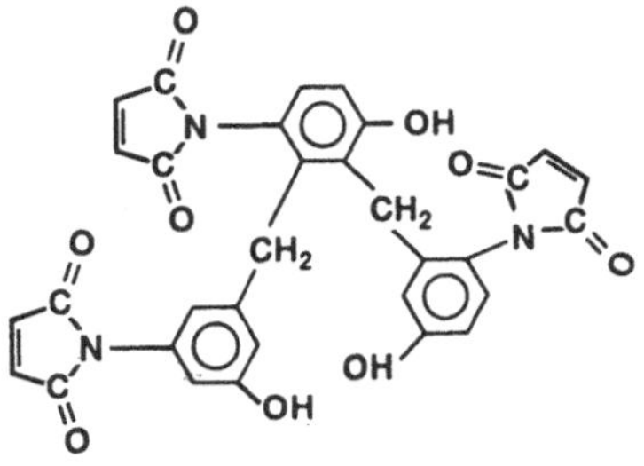

Polyfunctional aromatic crosslinkable compounds containing methylene bridges in which the phenolic hydroxy group is replaced by aniline-type amine group are among the new arrivals in the thermoset resin field. Here, the reactivity and crosslinkability are obtained through the amine groups. Oligomers such as

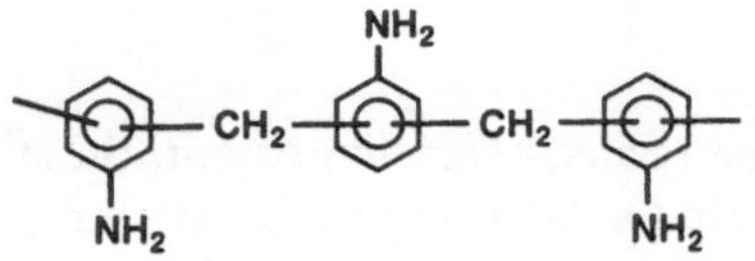

$$NH_2$$

$$—\bigcirc—CH_2—\bigcirc—CH_2—\bigcirc—$$

$$NH_2 \qquad\qquad NH_2$$

may be reacted to form three-dimensional crosslinked polyamide [461]

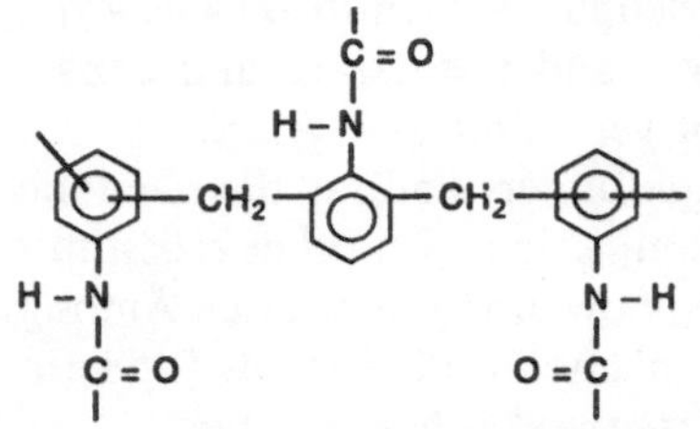

$$C=O$$
$$H-N$$
$$—\bigcirc—CH_2—\bigcirc—CH_2—\bigcirc—$$
$$H-N \qquad\qquad N-H$$
$$C=O \qquad\qquad O=C$$

or polymaleimide [462] networks:

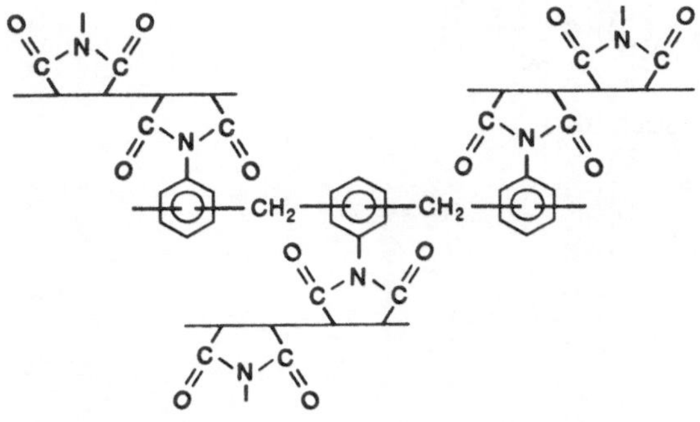

The above oligomeric species are cured in a stepwise fashion alone or with other monomers or oligomers to yield matrices for composites and laminates having improved processing and final properties.

3.1.4.3 Furan-Containing Networks

The fact that furan is only weakly aromatic [463] while, at the same time, substantially dienic [464] leads to its chemistry being rather complex and its polymeric reaction products not fully characterized. The interplay between aromatic and dienic properties leads to a situation where the predominance of either effect depends on the nature and position of the ring substituents, and upon the specificity of the reagents, catalysts and reaction media in each case

[464]. As is the case with phenol-formaldehyde-type resins, the lack of detailed understanding of the reaction mechanisms and complete characterization of the complex polymeric products did not prevent the large scale utilization of furan based network polymers. In both instances, the synthetic chemistry and chemical technology aspects are far better understood than the physical and topological aspects of the solid systems [446, 464-466].

The most commonly used furan resins are those obtained from furfuryl alcohol and/or furfural:

$$\text{Furfuryl alcohol} \qquad \text{Furfural}$$

The furfuryl alcohol is reacted with itself under acidic conditions to produce liquid oligomers.

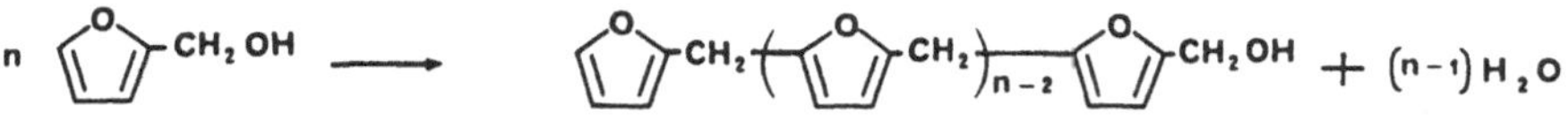

$$n\ \text{(furan)–}CH_2OH \longrightarrow \text{(furan)–}CH_2\text{–[(furan)–}CH_2]_{n-2}\text{–(furan)–}CH_2OH + (n-1)H_2O$$

and in a second, heat curing, stage the oligomers crosslink to form rigid networks typified by the fragment

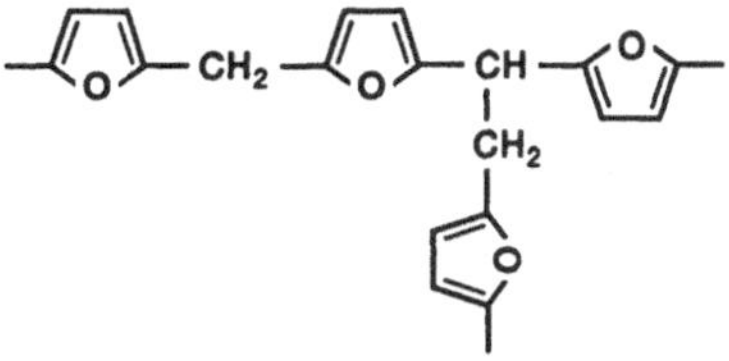

$$\text{—(furan)—}CH_2\text{—(furan)—}CH\text{—(furan)—}$$
$$\qquad\qquad\qquad\qquad\ \ |$$
$$\qquad\qquad\qquad\qquad\ CH_2$$
$$\qquad\qquad\qquad\qquad\ |$$
$$\qquad\qquad\qquad\qquad\ (furan)$$

Solution [467] and solid-state [468] NMR studies revealed that the cured networks contain negligible amounts of methylol and dimethylene ether groups or 3- and 4- positions substitution of the furan rings. It is therefore now accepted that the crosslinking step involves mostly a condensation reaction between methylene groups in the oligomeric chains and methylol groups at the ends of others [465, 468]. The analytical work did not explain, however, the cause for the darkening of the cured resins, suggesting that the chromophore concentration is too low to be detected by NMR or other techniques. It is possible that proton removal from some methylene bridges can produce polyunsaturated sequences involving several furan rings. Such a mechanism requires the appearance of unpaired electrons which were indeed detected in cured furfuryl alcohol resins by ESR techniques [469]. In the case of homopolymers prepared from furfural alone, the cured product is paramagnetic, indicating the presence of stabilized free radicals [464]. This resin inhibits radical reactions [464]. Its proposed typical structure, shown above, suggests that its black color [464] is caused by the presence of either unpaired electrons or extensive conjugation, or both [465]. Various other furan-containing resins were introduced [470] and are commercially available [464-466] but are currently being used in relatively small amounts. Some of them use a single kind of monomer but many employ mixtures of two or more [471] monomeric species.

The phenol-formaldehyde and furan containing networks gained wide acceptance because they possess a combination of useful properties: low cost, easy processibility, sufficiently high T_g for a broad range of applications, and some measure of ductility for the cured networks not to fail in an overtly brittle fashion. Other rigid networks characterized by single-atom bridges did not find broad, or any, acceptance. This is because most of them were hitherto neither prepared nor characterized, and those that were prepared were not characterized or commercialized. One of these families is the highly aromatic rigid hydrocarbons. Other existing and potential rigid aromatic networks containing single-atom bridges will be described below.

3.1.4.4 Aromatic Polyether and Ether-Containing Networks

To the best of our knowledge, no aromatic polyether network has been reported in the literature. In his book, Professor Flory maintains that non-crystalline polymeric substances prepared by Hunter and associates from silver or alkali metal salts of trihalogenophenol appear to be aromatic networks having trifunctional branchpoints [472]. An analysis of the results reported in the original papers [473-475] reveals that, without exception, they correspond to a polymeric structure whose chemical repeat unit is $C_6H_2OX_2$ where X stands for a halogen such as Cl, Br and I. The reaction leading to the formation of the polymeric species involves the abstraction of a halogen atom from positions either para or ortho to the oxygen atom. We believe that the randomness of the reaction leads to a linear polyether with two positional isomeric units randomly placed along rather low molecular weight chains:

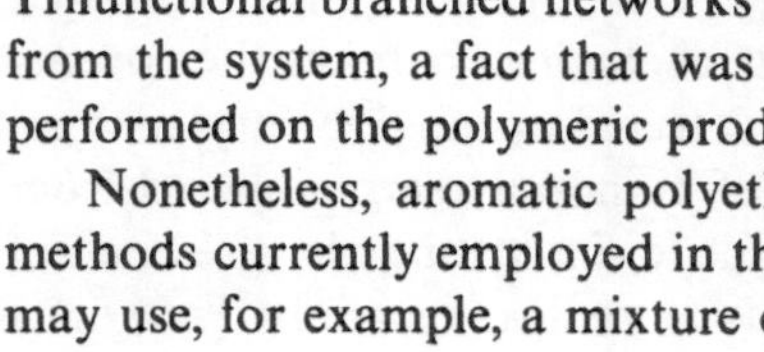

Trifunctional branched networks will require the removal of additional halogen from the system, a fact that was not observed in any of the element analyses performed on the polymeric products [473-475].

Nonetheless, aromatic polyether networks are achievable using synthetic methods currently employed in the preparation of linear polyethers. Thus, one may use, for example, a mixture of the two monomers

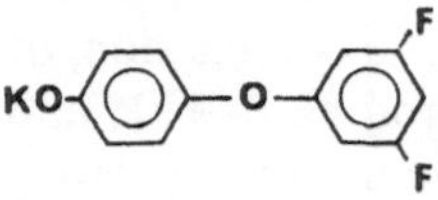

and 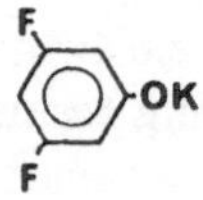 to obtain an exclusively aromatic polyether network. Other potential combinations are

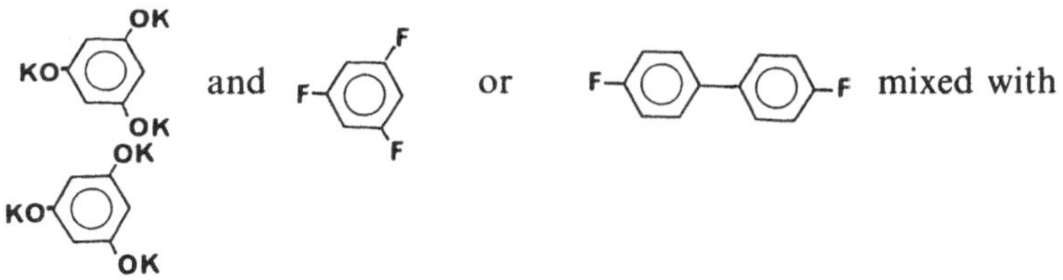

and

or

mixed with

in stoichiometric amounts. As currently used for the creation of linear polyethers [458, 476, 477], the polymerization can be carried out in tetramethylene sulfone (sulfolan) at temperatures around 250 °C or in higher boiling sulfones at higher temperatures [477]. The same polycondensation procedures may be used for the creation of aromatic poly(ether ketone) (PEK) and aromatic poly(ether sulfone) (PES) networks, employing appropriate monomers [476, 477], such as:

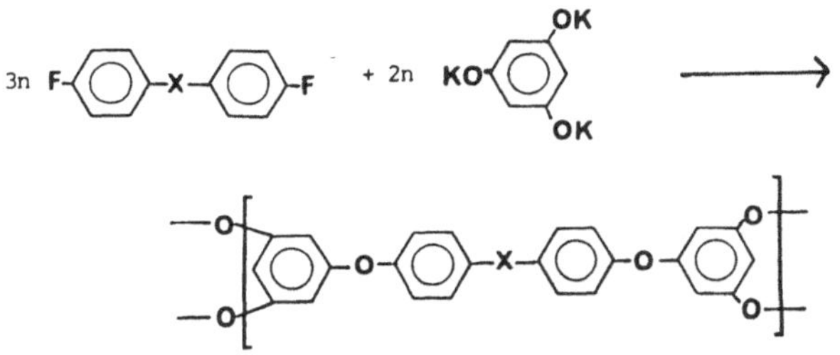

with X being O, CO or SO_2. The inorganic KF salt may be washed with water out of the network during the workup and purification steps.

Poly(ether ketones) may be prepared by three additional polycondensation methods. The first and oldest is a Friedel-Crafts polycondensation in which monomers such as diphenyl ether and 1, 3, 5-benzenetricarboxylic acid chloride are reacted together to form an aromatic poly(ether ketone) network:

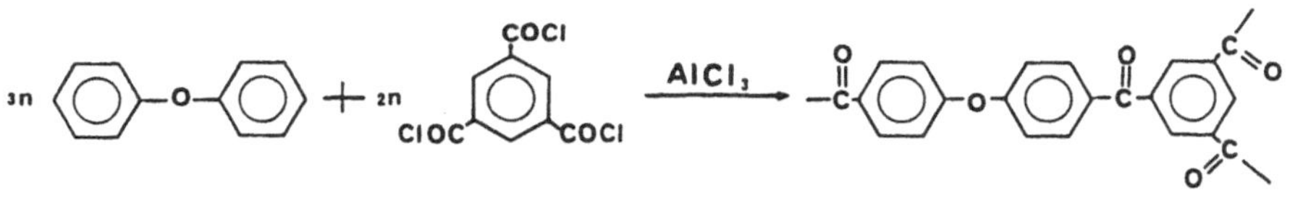

In the preparation of the linear analogues, solvents such as nitrobenzene [478] and methylene chloride [479] were used. A new method capable of creating poly(ether ketone) and poly(thioether ketone) was recently disclosed by Ueda et al. [480-482]. Here, the dehydrating power of a mixture of phosphorus pentoxide and methanesulfonic acid (MSA) is the driving force for the direct polycondensation of aromatic dicarboxylic acids with aryl compounds containing ether or sulfide bridges (X = O, S or a mixture of both):

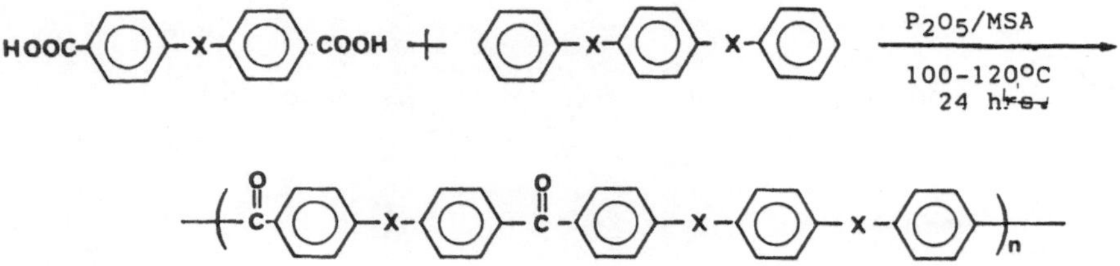

It is obvious that this reaction easily lends itself to higher functionality aromatic species and, therefore, to the creation of aromatic poly(ether ketone) and poly(thioether ketone) networks. Amorphous poly(ether ketones) were recently prepared [483-485] by the complex reaction of the abstraction of fluorine atoms from aromatic kernels and their partial substitution by oxygen using the reaction of potassium carbonate with fluorine-containing monomers:

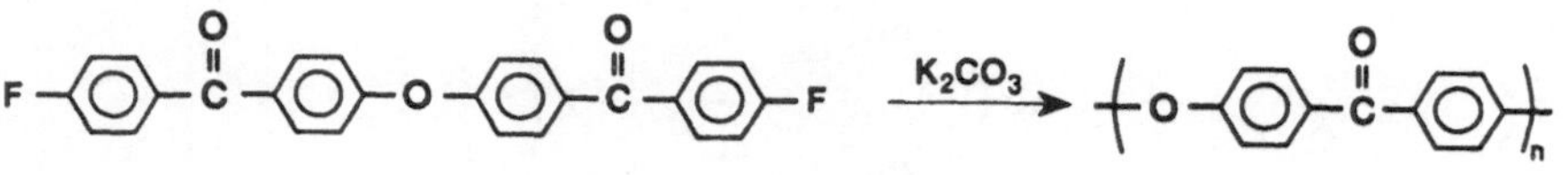

This reaction is carried at temperatures of about 300 °C in the presence of SiO_2 or K_2CO_3 catalysts, using diphenyl sulfone as solvent. In addition to the polymeric products, carbon dioxide and KF by-products are formed. The amorphicity of these poly(ether ketones) is caused by the non-symmetric dibenzofuran species being formed along the aromatic polymer chain [483-485]:

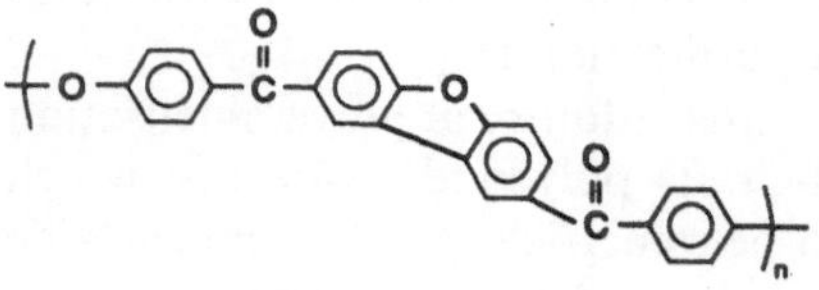

Here, again, the chemistry easily lends itself to the creation of three-dimensional aromatic networks.

3.1.4.5 Bridged Aromatic Networks with Uncommon Electronic Structure

In the quest to obtain polymers with unconventional electrical properties, several highly branched aromatic structures containing single-atom bridges were prepared. These are either carbon atom or nitrogen atom bridges.

Building on the foundation of theoretical predictions [486-490] and experimental results on model compounds [491-494], highly branched organic ferromagnets were prepared [495], containing carbene bridges:

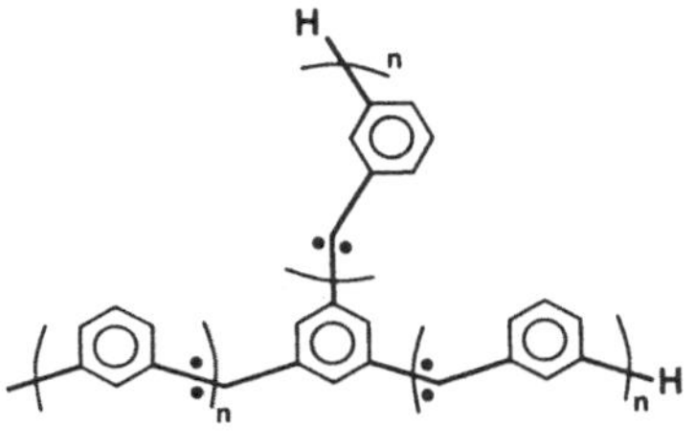

with n being 2 or 3. These compounds, which are stable only at extremely low temperatures, are expected to be steps along the way to an ultimate goal of the following honeycomb structure [495]:

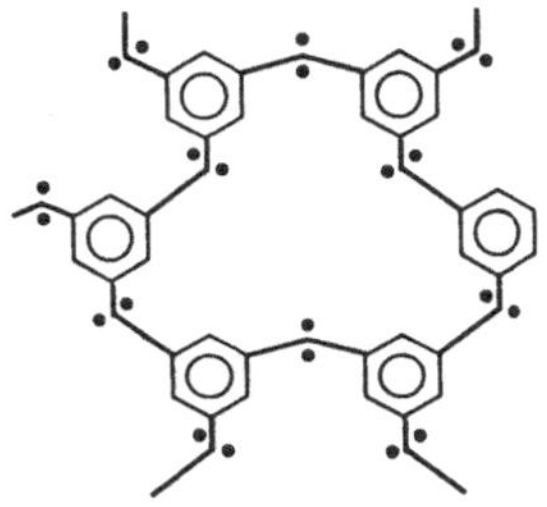

The branched structure shown above was prepared in minute quantities by a multistep synthetic sequence passing through a polyketone and culminating in a polydiazo structure:

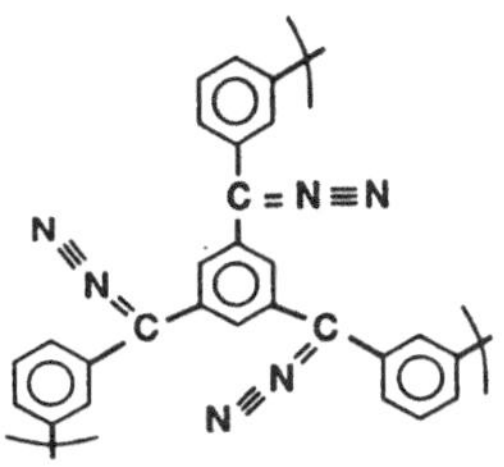

This was followed by the photolysis of the diazo structure in 2-methyltetrahydrofuran glass at 10 K [495]. ESR signals obtained at this low temperature revealed the high-spin state of the branched polycarbene final product. Until now only trifunctional branched polymer was prepared but we believe that, by using difunctional instead of monofunctional monomers, highly crosslinked aromatic networks containing carbene bridges may be prepared.

A different route toward organic ferromagnetic networks was taken by Rajca [496, 497]. Instead of creating an aromatic ring structure with carbene bridges [495], he created an aromatic ring structure with radical bridges. This was done in a multi-step procedure [498], starting with 1,3-dibromobenzene and 4,4' -di-*tert*-butylbenzophenone being treated by butyl lithium in ether and then by ethylchloroformate. In a second step the intermediate product is treated again by butyl lithium in ether, methyl carbonate and ethylchloroformate. These procedures were all carried out at temperatures of -30 to $-25\ °C$. The carbanions were generated from the ether precursors by treating them with lithium metal in deuterated tetrahydrofuran (THF-d_8) [496], and the radicals were created by treating the carbanions with elemental iodine in THF at $-78°c$ [497]:

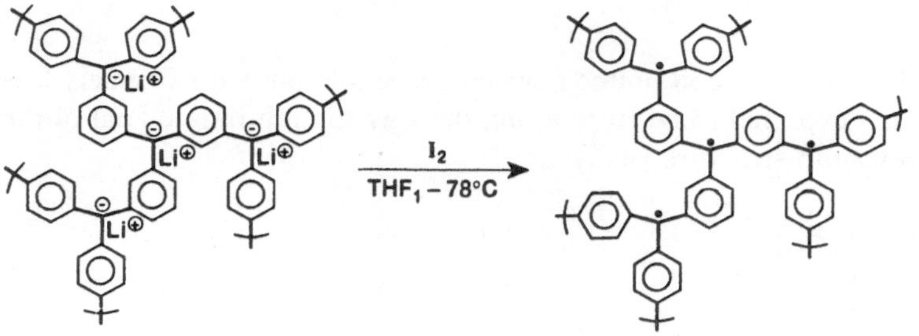

It will be interesting to find whether changes in the monomers, such as the replacement of dibromobenzene by, say, tetrabromobiphenyl, or the dibutylbenzophenone by aromatic polyketone oligomers, will lead to three-dimensional polyradical networks. It also remains to be seen if the envisioned larger ferromagnetic domains will produce ferromagnetic ordered structures at temperatures higher than, say, liquid nitrogen temperature.

Building on the theoretical considerations of Mataga [490] and Ovchinnikov [489], Torrance et al. first failed [499] and then succeeded in synthesizing [500] three-dimensional aromatic networks containing amine bridges. The successful procedure is disarmingly simple and involves the condensation of 1,3,5-triaminobenzene (TAB) in acetic acid. Stronger acids do not produce polymeric products. The rate of polymer formation depends on the reaction temperature and may require over 48 h at $25\ °C$ or a mere 5 min at $110\ °C$ [500]. The polymeric TAB

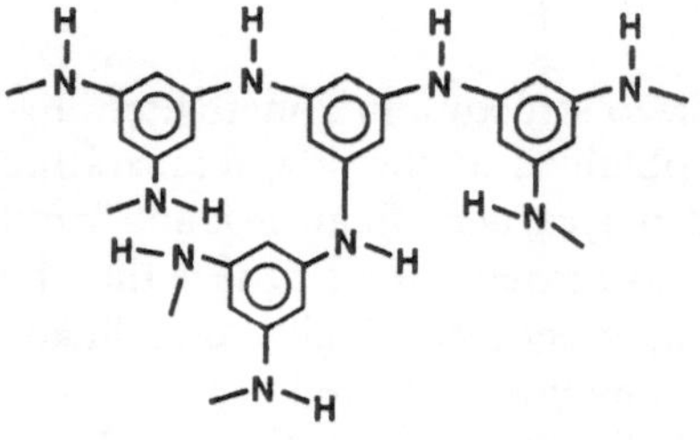

is insoluble in common organic solvents and appears to be amorphous [500]. The polymer can be reversibly protonated and is easily oxidized upon exposure to air. It is interesting to speculate whether it can be deprotonated to the anionic state by treatment with metallic lithium as above [496] or by a solution of sodium hydride in dimethylsulfoxide (DMSO) in a fashion similar to that used to deprotonate aromatic polyamides [501-505]:

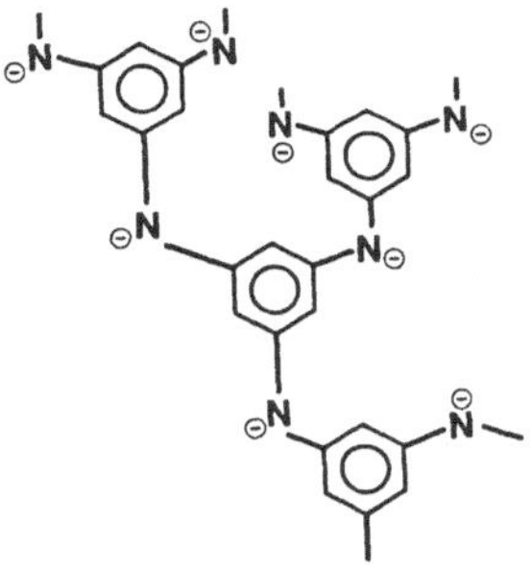

Polyaniline (PANi) is emerging as an electrically conducting polymer with, potentially, a significant commercial future:

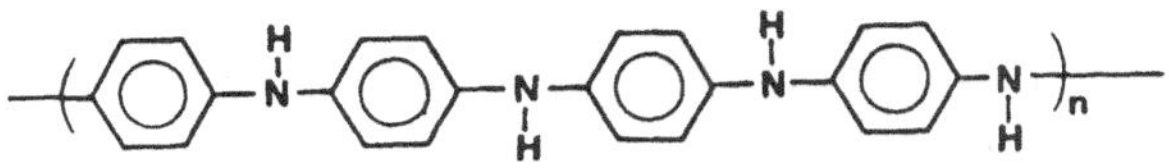

It is currently being test-marketed in its doped and undoped forms by AlliedSignal Inc. under the tradename Versicon. The polymer is generally linear with molecular weights in the range of 10^4 to 10^5, and is soluble in only a few solvents. It has recently been demonstrated that crosslinked PANi can be prepared [506, 507] without diminishing the polymer's electrochemical properties [507]. This is achieved by the use of 1,4-phenylenediamine as a comonomer with aniline during the electrochemical polymerization procedure. The exact nature of the crosslinking site is not yet ascertained, but electropolymerization of p-aminodiphenylamine reveals a propensity towards branching and crosslinking with the apparent creation of phenazine-type branchpoints [508]. Similar structures were obtained by thermal or chemical deprotonation of pre-existing PANi chains [509]. A possible crosslink junction in PANi network may appear as [509]:

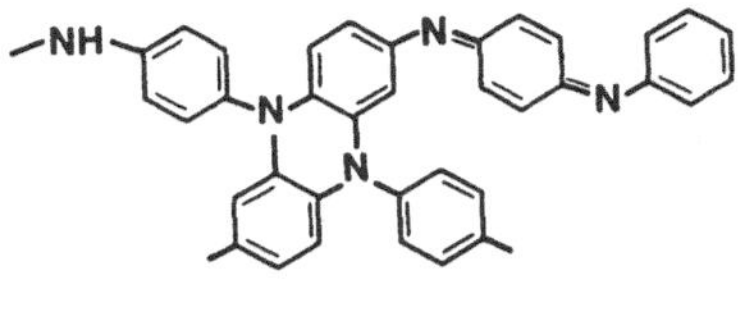

Claims of two-step creation of PANi networks appeared on occasion in the literature. They are based mostly on deviations from expectation of electrochemical properties. In general, no additional characterizations of the materials were reported substantiating the crosslinking claims. In the light of the very complex nature of electrical conductivity in PANi [510], the claims of in-situ crosslinking are not fully validated at present. Because of this, the potential two-step PANi networks will not be further addressed.

Recent disclosures about the electrical conduction properties of various poly(arylamines), many of which contain condensed aromatic kernels [511-513], will undoubtedly increase the interest in this family of aromatic networks with unusual electric and magnetic properties. A brief review describing some of the above and other polymers with potentially interesting magnetic properties recently appeared in the open literature [514].

3.1.4.6 Morphology and Possible Fractality

The morphology of resorcinol-formaldehyde (RF) aerogels [515-517] and of poly(1,3,5-triaminobenzene)(poly-TAB) [500] slowly precipitated from solution was studied by electron microscopy [516, 517], SAXS [516] and dynamic light scattering [518]. The RF-aerogels are morphologically similar to melamine-formaldehyde [519, 520] aerogels with the exception that the former are dark red in color while the latter are almost colorless. All these aerogels are first polymerized in solution at relatively low concentrations. The solvent is then exchanged and finally removed under supercritical conditions [515-522].

As discussed in Sect. 1 above and Sect. 4 below, a disordered material displays fractal behavior when it is self-similar over a range of size scales. This means that over this size interval the material density or its surface will scale with size with an exponent which is non-integer. Furthermore, it should also exhibit scaling behavior in its dynamic properties. An important consequence is that a system exhibiting fractal behavior on one size scale may show non-fractal characteristics on other size scales, either smaller or larger, or both. This situation exists in the RF-aerogels and, we believe, in the poly-TAB system. In both instances primary units of a size up to around 100 Å (10 nm) were observed [500, 516, 517]. These primary units were found by SAXS to be colloid-like particles with smooth surfaces [517]. However, when they are assembled together in an "infinite" network, the network exhibits a fractal behavior [517] over a relatively narrow range of size scales. These size scales correspond to the pore size of the RF-aerogels or, in other words, to the mesh size of the network. The dilute conditions under which aerogels are prepared and, especially, the drying under supercritical conditions create an open structure which is prevented from collapsing upon itself due to capillary contraction. This makes the observation by scattering techniques of network porosity and fractality far easier than for inhomogeneous systems that collapsed upon themselves and densified during solvent removal. Because of the molecular-size porosity observed in the poly-TAB slowly deposited films [500], we believe it is possible that this system may also display fractal behavior on this size scale while not showing it on larger size scale, say, of the order of a micrometer [500].

3.1.5 Rigid Networks with Triazine Branchpoints

From the Six-member nitrogen-containing heterocyclic compounds, pyridine is the most strongly aromatic, diazines are less so, and triazines are the least aromatic. Crystallographic studies, however, reveal that the 1,3,5-traizines in general have intra-ring identical or almost identical bond lengths [117]. Their C-N-C and N-C-N bond angles are not the same [117], resulting in the 1,3,5-triazines having a distorted hexagon shape. The intra-ring bond length similarity provides the aromatic character of the 1,3,5-triazine ring [117]. In this section we are interested in fully or mostly aromatic polymer networks containing 2,4,6-triaryl-1,3,5-triazines,2,4,6-triaryloxy-1,3,5-triazines (called cyanurates) and 2,4,6-triamino-1,3,5-triazines (called melamine):

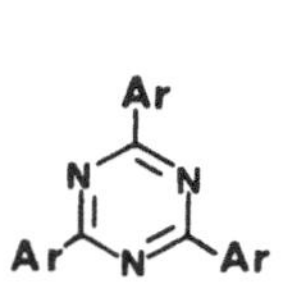

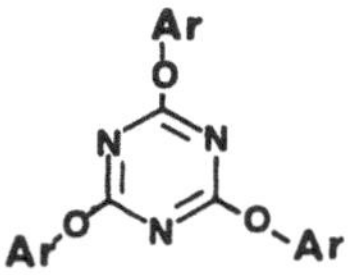

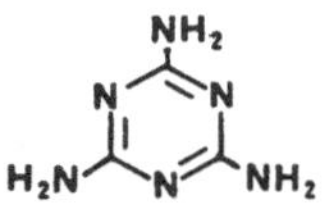

2,4,6-triaryl-1-3-5-triazine   Cyanurate   Melamine

The isomeric 1,3,5-triazine-1,3,5-triazines, called isocyanurates, lost the

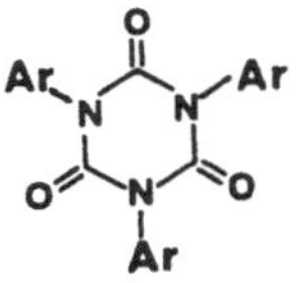

Isocyanurate

aromatic character of the ring and will not be discussed here.

   The 1,3,5-triazine rings may be introduced into a network in their preformed state, as in the case of melamine, or may be formed in place by cyclotrimerization of the appropriate monomers [523]. Thus, by the trimerization of 4,4'-dicyanobiphenyl

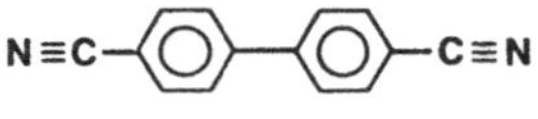

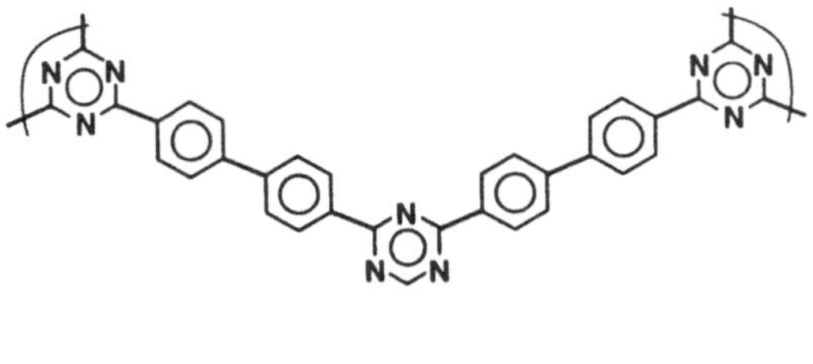

under the catalytic influence of acids such as chlorosulfonic acid, a highly branched fully-aromatic network containing 2,4,6-triphenyl-1,3,5-triazine branchpoints was obtained [524-527]. Similarly, networks with one para-substituted phenyl ring between the triazine branchpoints, and with

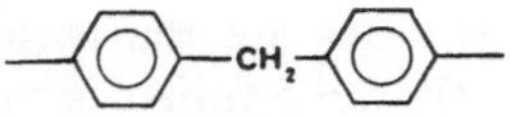

$-\text{C}_6\text{H}_4-\text{CH}_2-\text{C}_6\text{H}_4-$ and $-\text{C}_6\text{H}_4-\text{O}-\text{C}_6\text{H}_4-$ segments were also made

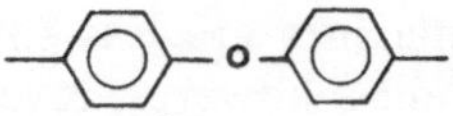

[526]. Networks exhibiting high thermal and oxidative stability, containing segments originating from the monomer

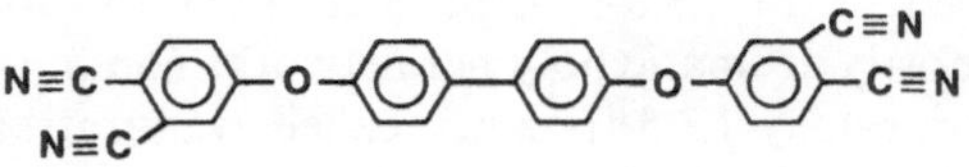

$$N\equiv C-\underset{N\equiv C}{\overset{}{\text{C}_6\text{H}_3}}-O-\text{C}_6\text{H}_4-\text{C}_6\text{H}_4-O-\text{C}_6\text{H}_3}\underset{C\equiv N}{\overset{C\equiv N}{}}$$

that may possibly be end-linked by triazine ring formation, were recently disclosed [528-532].

The cyclotrimerization of aromatic dicyanato monomers to create highly aromatic network with cyanurate branchpoints is far more prevalent. In a typical reaction, monomers such as 2,2-bis(4-cyanatophenyl) propane are heated in the presence of a catalyst to form a network:

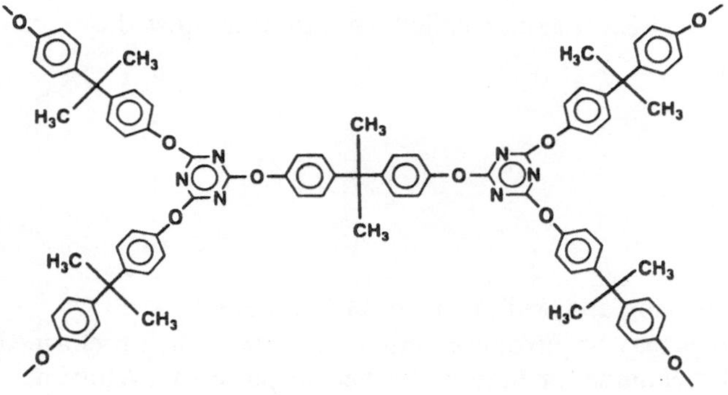

A variety of organic salts [533], organometallic [534] and other compounds [535, 536] may serve as catalysts for the trimerization. Because trimerization was found to be the dominant reaction, and dimerization less than 1% [533, 537], the network fragment shown above may be taken as a true representation of the final network product.

It appears that 1,3,5-triazine branchpoint polymers with the shortest *identical* length segments are those made from bisphenol derivatives. Commercially available monomers that may be polymerized to a network through trimerization brought about by heat curing in the presence of an appropriate catalyst, are 2,2-bis(4-cyanatophenyl) propane, 1,1-bis (4-cyanatophenyl) ethane

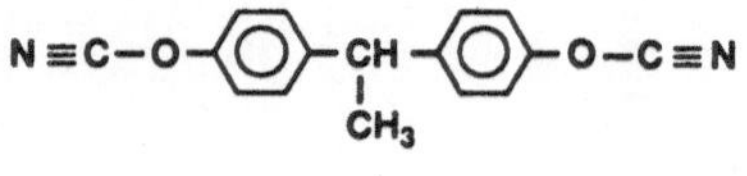

$$N\equiv C-O-\text{C}_6\text{H}_4-\underset{\text{CH}_3}{\overset{}{\text{CH}}}-\text{C}_6\text{H}_4-O-C\equiv N$$

and bis (4-cyanato-3,5-dimethylphenyl) methane

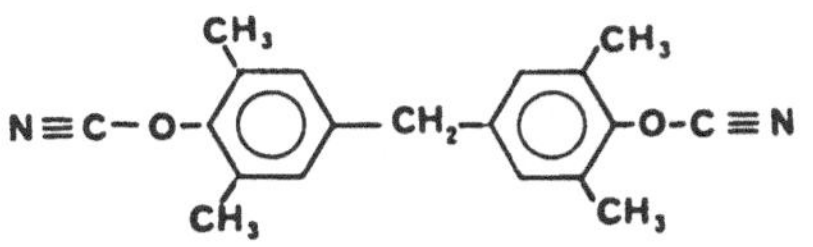

These three monomers, and oligomeric pre-polymers made from them, are being produced and marketed by several companies. Among them one finds Hi-Tek Polymers [538], Rhone-Poulenc [539], Mobay [458] and Mitsubishi [458]. Triazine polymers with shorter *average* distance between the trifunctional branchpoints are the phenolic triazine (PT) resins currently in the process of commercialization by Allied-Signal Inc. [540, 541]. Their generic structure is:

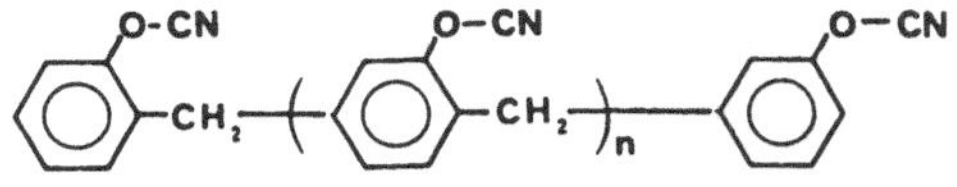

The patents of Das and Prevorsek [542-544] claim additional resins in which not all phenolic rings carry a cyanate group, and the rings may or may not be alkylated. The oligomeric PT-resins are offered in three grades with $M_w = 525$, 1140 and 1420 [545]. Because the average crosslink density of the novolak-based resins, such as PT-resin, is higher than that of crosslinked resins based on bisphenols, the T_g's of the cured novolak-based resins are higher than those of their bisphenol-based analogues.

A polycyanurate with slightly longer segments of *identical* length is one obtained by thermally curing the divalent monomer

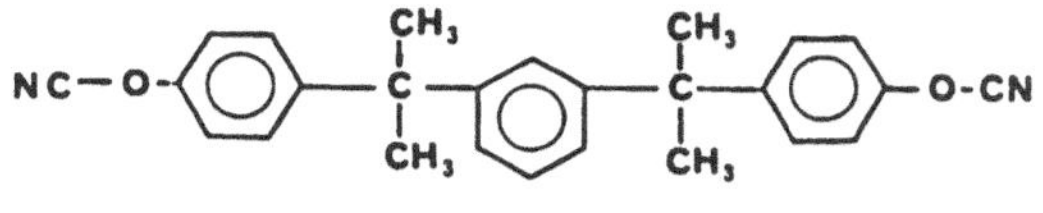

available from Hi-Tek Polymers in monomeric and oligomeric forms [546]. Even longer *identical* length segments between cyanurate branchpoints were recently reported by Barclay et al. [423]:

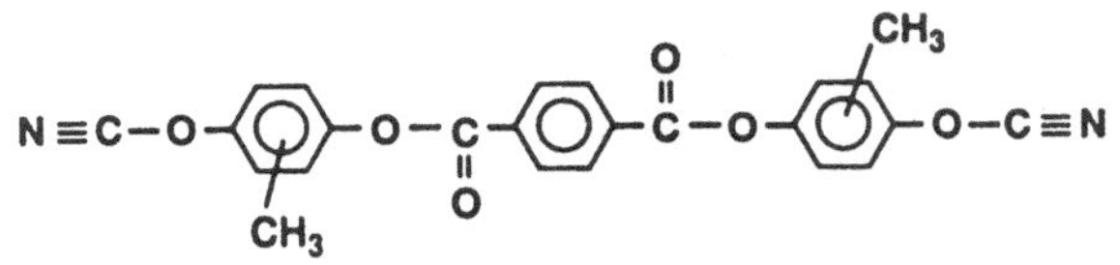

describing novel liquid crystalline aromatic polyester thermosets. Networks with even longer segments between 1,3,5-triazine branchpoints were claimed [547]. These are substantially aromatic yet flexible poly(ester carbonate) segments with molecular weights as high as $M_n \cong 4\,000$:

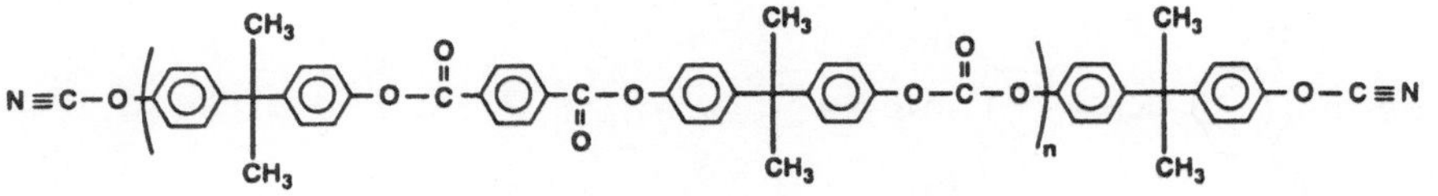

On a higher level of chemical complexity one finds the highly aromatic polymer networks whose branchpoints are based on *both* 1,3,5-triazine and maleimide or similar end-groups containing activated carbon-carbon double bonds. In these, the cyanurate may be built first into a branched oligomer end-capped with a double-bond-containing reactive unit [418, 548-551], or linear segments may be end-capped on one end by maleimide and on other end by cyanate groups. These may later be cyclotrimerized or condensed together to produce the network branchpoints [552-556]. Between the preformed cyanurate branchpoints one finds polyimide and polyamide segments [549, 550]

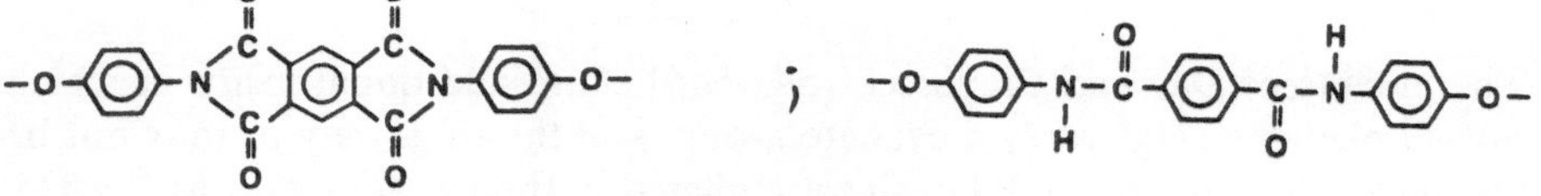

that under advantageous conditions may be capable of exhibiting mesomorphic behavior, and others with a bend in their center or with the less rigid urea moieties that, although highly aromatic, may not be capable of mesomorphicity [549, 550]:

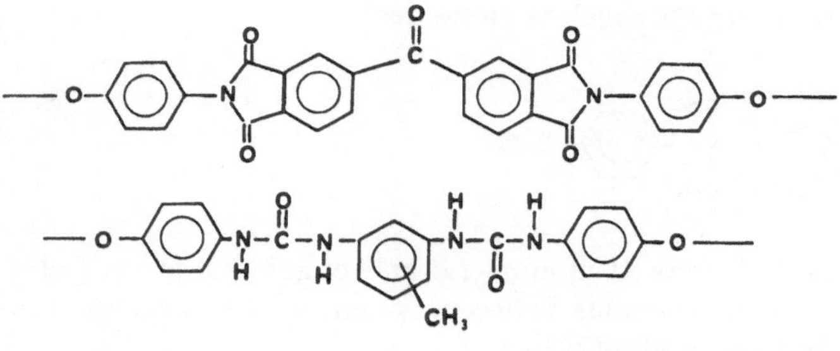

For all these, the starting trifunctional monomer is 3,4,6-tris(4-aminophenoxy)-1,3,5-triazine (TAT) [548]:

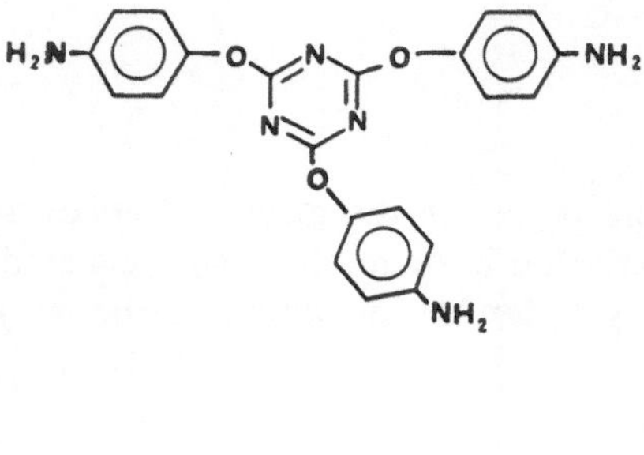

The crosslinking groups containing double bonds were reacted with the amine groups of TAT to form groups such as maleimide, nadimide or itaconimide:

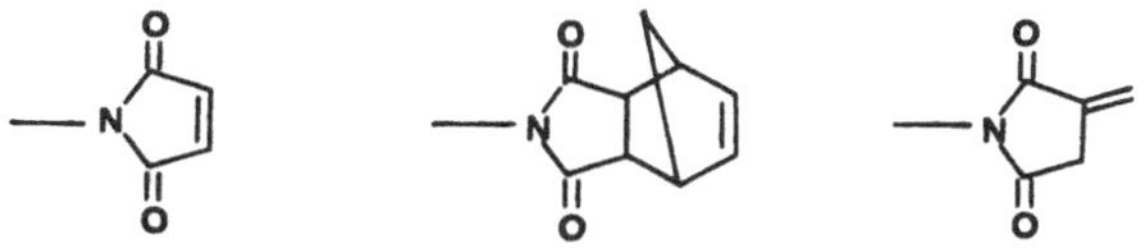

Maleimide   Nadimide   Itaconimide

In the absence of other reactive groups, these react with one another by addition-type reaction to afford crosslinked networks. Thermogravimetric analysis (TGA) showed that the thermal stability of the final networks with respect to the crosslinking groups decreases in the order [550]:

maleimide > nadimide > itaconimide

and, with respect to the nature of the bridging groups between aromatic rings, they were in the order [550]:

polyimide > polyamide > polyurea,

in agreement with expectations.

Aromatic segments terminated with cyanate on one end and maleimide on the other were prepared in two steps from aminophenol species having from one up to a dozen aromatic rings substituted in *para* or *meta* positions. The *para* substitution and the use of relatively stiff bridging groups between the aromatic rings renders the segments mesogenic in nature [556]. The aminophenol entity

$$H_2N—Ar—OH$$

is first reacted in a suitable solvent, such as acetic acid, with maleic anhydride to obtain the respective maleimide [552]:

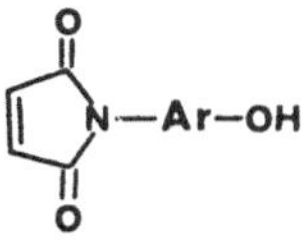

which, after workup, is reacted with cyanogen bromide to obtain the desired reactive network segment [552]

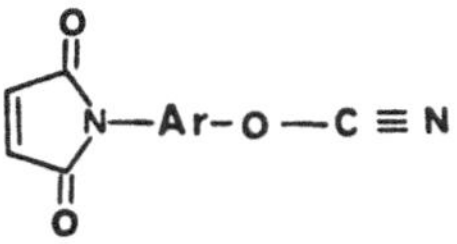

Reactive segments of one kind, or mixtures thereof, are mixed with minute amounts of a catalyst such as cobalt naphthenate, and then heated in a stepwise manner over several hours to produce transparent amorphous crosslinked networks when the segments contain *meta* substitution on the rings and/or when the segments between branchpoints are very short [552, 553]. When the segments are longer and *para*-substituted, the crosslinked networks are substantially birefringent upon examination by cross-polarized light microscopy [556]. By virtue of possessing both cyanate and maleimide end-caps on each segment, the branchpoints may be of several kinds. The cyanate group may trimerize to form the 2,4,6-substituted-1,3,5-triazine ring:

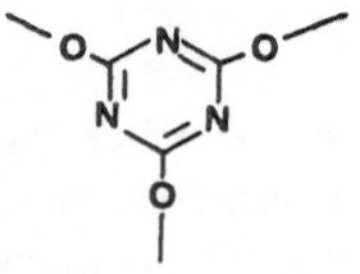

the maleimide may add to itself to form a dimeric structure:

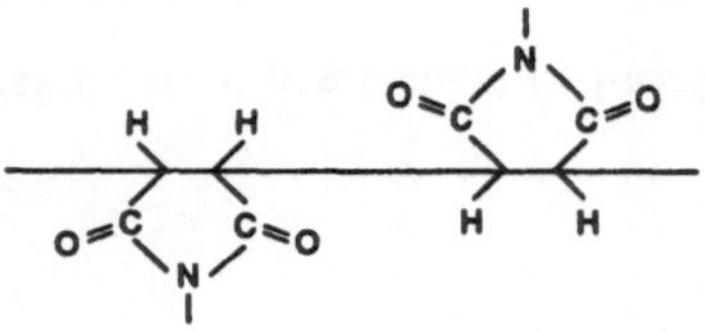

and the cyanate may react with the maleimide to create complex crosslinked structures such as

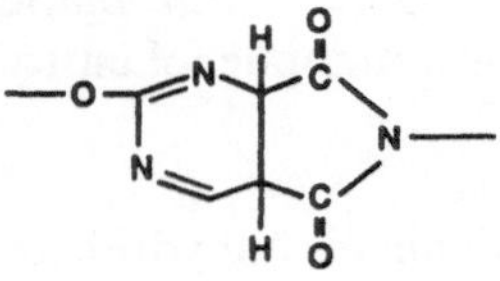

and

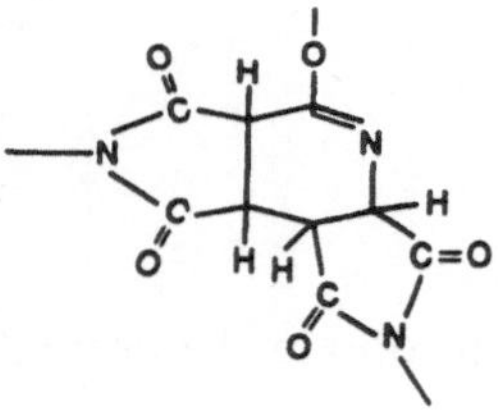

The nature of these structures and their coexistence in varying proportions in the same crosslinked network makes their analysis very demanding. The possible presence of minor amounts of additional crosslinking structures can not be discounted and may further complicate the analysis. In all the above, an additional complication presents itself in networks where the segment lengths are not identical due either to the method of preparation of the networks or to the segments in the network being a mixture of more than one kind of identical length segments.

The third member of the 2,4,6-substituted-1,3,5-triazine trio is melamine. It may be reacted in various propoitions with formaldehyde or with urea and formaldehyde to form a broad range of resins whose properties may gradually change with composition [116]. An idealized structure of a network fragment may be

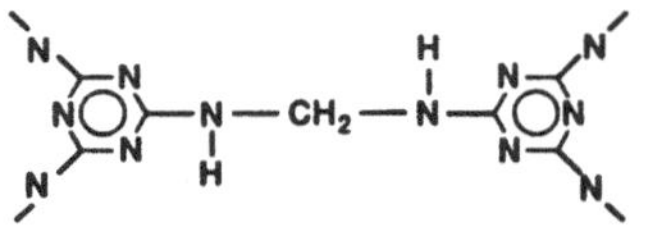

but very often a myriad of other structures [113] coexist with it:

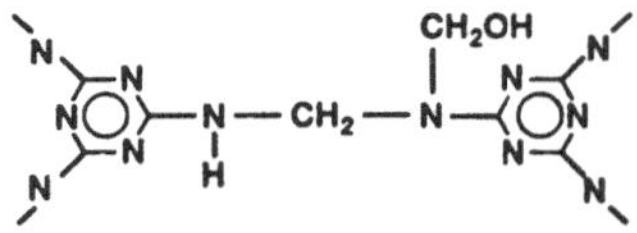

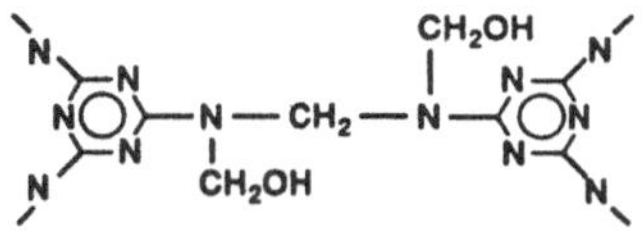

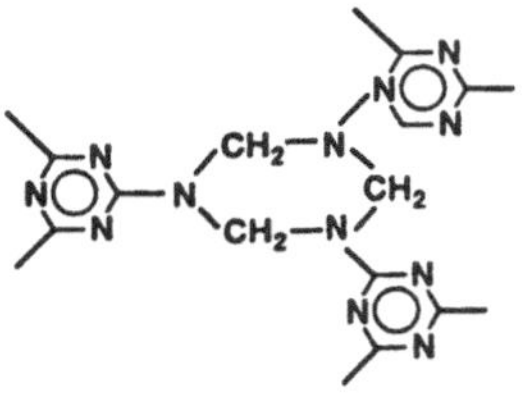

and so on. Even though the number of single bonds between the triazine rings places the melamine-formaldehyde resins out of our definition of aromatic rigid networks, the very short distances between branchpoints render the resins rigid and their growth habit makes them informative for all the triazine containing networks. Two recent patents demonstrate that when melamine-formaldehyde resin is prepared in a manner that prevents its collapse during the removal of water (from the formalin solution), then a highly porous organic aerogel is obtained [519, 520]. The procedure involves a careful replacement of the water by an organic solvent such as acetone, followed by drying the gel by the removal of the acetone with carbon dioxide under supercritical temperature and pressure conditions. The polymeric product is a very porous low density substance, consisting of interconnected minute particles less than 500 Å in diameter. The structure of the aerogel appears to be, hence, of a fractal aggregate character, as is expected from the nature of the polymerization reactions involved.

Studies of the polymerization and gelation process of difunctional monomers forming a network with trifunctional branchpoints consisting of 2,4,6- or 1,3,5-substituted 1,3,5-triazines indicate that the approach to the gel point is inconsistent with the mean field approach to percolation initiated by Flory [39] and Stockmayer [10] and later elaborated by Gordon [557]. This approach calls for a 50% monomer conversion at the gel point, substantially below the 60

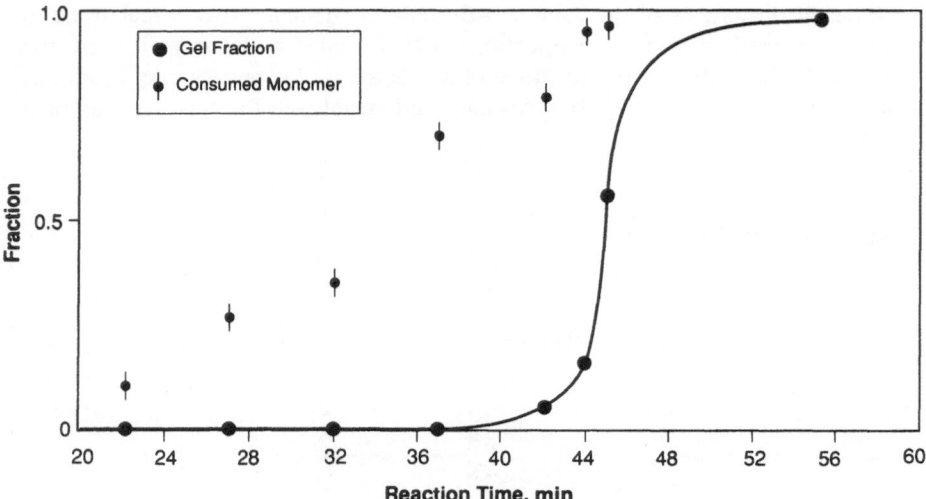

Fig. 17. Changes in gel fraction and monomer consumption as function of reaction time during a one-step solution polycondensation of stiff aromatic polyamide network

to 65% conversion actually observed [533, 537, 558–564]. Attempts to rationalize the difference as due to intramolecular cyclization are based on flexible chain and flexible network models that allow for easy cyclization and random reaction of any end-group with any other, and neglect the fact that the monomers in question, such as 2,2-bis(4-cyanatophenyl)propane, are neither long nor flexible enough to be consistent with these models. The relatively sharp changes in gel fraction that were observed [561, 564] as these systems approach and then pass through the gel point are consistent with the percolation model [30, 565, 566]. They are also similar in character to the results obtained by Aharoni and associates [311, 400] from rigid aromatic polyamides, and to Fig. 17 depicting changes in the gel fraction with reaction time [311]. Taking the above together with the demonstrated fractality of the pre-gel polyamides by Aharoni et al. [311, 400] and the observed fractal morphology of the melamine-formaldehyde resins [519, 520], we conclude that, in general, when rigid networks with trifunctional branchpoints such as the triazines or polyamide rigid branchpoints are constructed in a one-step manner from either monomers, short reactive segments or small network fragments, they pass through a fractal stage at least in the pre-gel state. Some of the fractal characteristics may be carried through to the mature network and are discernible when these networks are prepared under suitable conditions.

3.1.6 Rigid Networks from Stiff End-Capped Segments

Efforts to create highly crosslinked networks of high thermal stability led to the preparation of substantially aromatic, readily processable oligomers with reactive end-caps or with reactive groups built into the chain or its pendants. These reactive oligomers are often called prepolymers. Upon curing, which is usually a sequence of heat treatments at ever increasing temperature, the reactive groups react with each other to form permanently crosslinked junctions between substantially aromatic network segments. Unlike the triazines discussed in Sect. 3.1.5, the crosslinked junctions to be discussed in this section are aliphatic in nature and often contain alkane as well as alkene groups. The aromatic segments in the network may be more flexible or less flexible, depending on the nature of the groups bridging the aromatic rings together, and how stiff they are. In this section, the networks will be grouped in three classes, characterized by the nature of the precursor reactive groups present at the end or along the network segments. These three groups are:

(a) maleimide, nadimide, itaconimide and similar moieties containing unsaturated alkenyl residues;
(b) benzocyclobutene, biphenylene, [2.2] paracyclophane and similar strained-ring species; and
(c) ethynyl-containing groups.

3.1.6.1 Networks with Unsaturated Imide Junction Precursors

Because of the commercial availability and relatively low cost of the starting acid anhydrides, and the ease of preparation of the unsaturated imides, aromatic chain segments and low-M polymers end-capped with these reactive groups are most common. Although exotic unsaturated end-caps appear on occasion in the patent literature, the most common unsaturated reactive groups are maleimide [417–419, 567–570], nadimide [417, 419, 567, 571, 572], methylnadimide [417, 419, 567, 571] and itaconimide [573]. The maleimide end-group is prepared by the condensation of an amine with maleic anhydride, followed by cyclization:

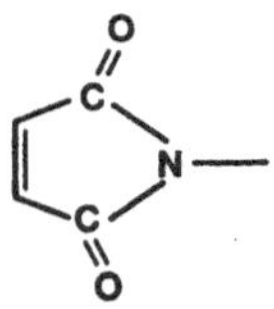

The nadimide end-group

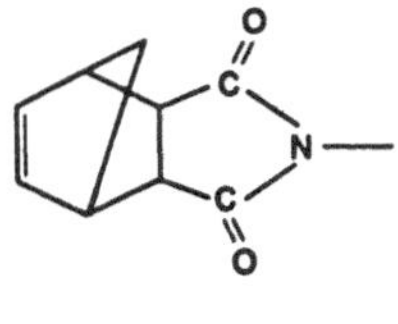

is prepared in a similar manner from 5-norbornene-2,3-dicarboxylic anhydride and an amine end group, the methylnadimide is obtained by condensing an amine with methyl-5-norbornene-2,3-dicarboxylic anhydride

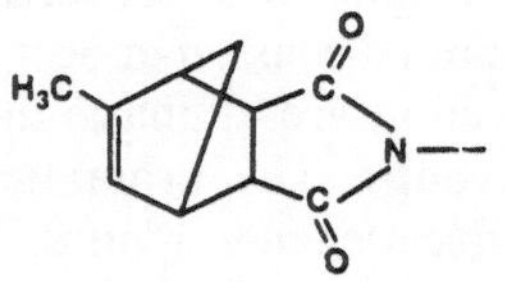

and the itaconimide

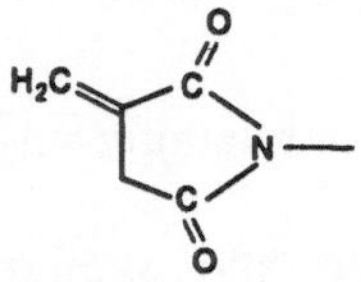

is similarly obtained from itaconic anhydride. During the curing process, the unsaturated end-caps of the oligomers add to one another, creating networks in which wholly or largely aromatic segments are held together by relatively compact and highly branched aliphatic junctions. Thus, maleimide groups combine to form junctions that may be as simple as

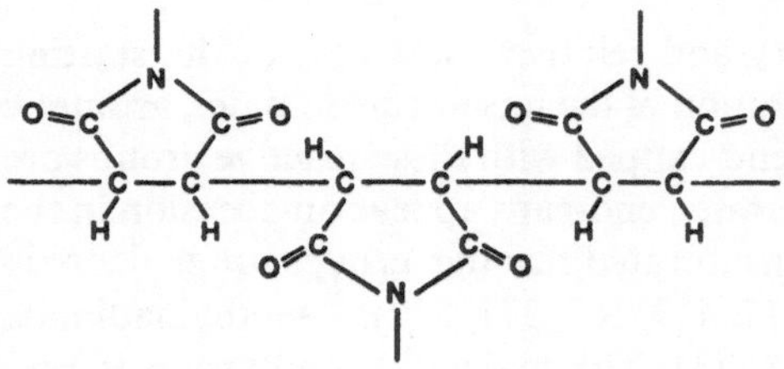

while the nadimide groups may form junctions as simple as

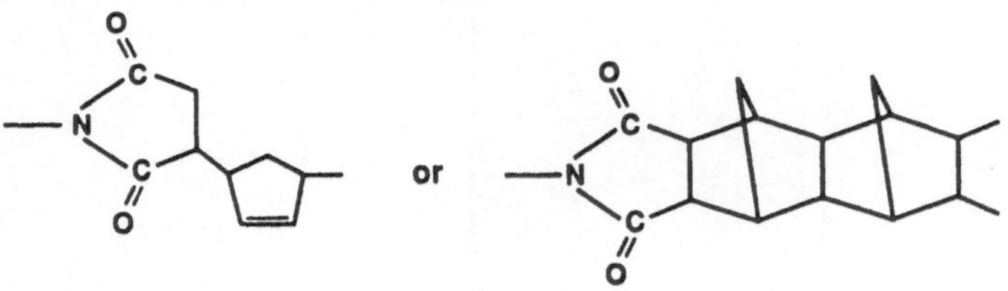

or

and, among many others, as complicated as [419, 572]:

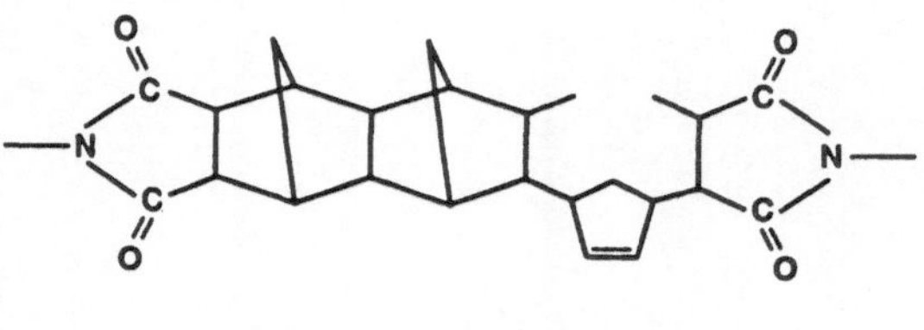

Due to their similar structure, the itaconimides may form crosslinked junctions similar in character to those formed by the maleimides, while the methylnadimide end-groups will add to form the more complex junctions characteristic of the nadimide reaction products.

Historically, all the monomers were reacted together in very high concentration solutions (dopes) in a single step to create, after solvent removal and high temperature curing, highly crosslinked resins [574]. It was felt that the low viscosity monomer solution was conducive to better impregnation of fiber mats to produce better matrix-filler adhesion and better mechanical properties. Later it was recognized that a two-step procedure yields more controllable processes with better final properties. This led to the creation of end-capped oligomers in a first step, followed by their crosslinking in a second, heat curing step.

The aromatic part of the end-capped oligomers may be as short as a few aromatic rings [417–419, 567, 568, 573], resulting in highly crosslinked stiff aromatic networks, or as long as 50 or more aromatic moieties [571]. Depending on their constitution, the aromatic segments between the junctions were designed to have as high as possible T_g and to span the range of flexibilities: from segments with high thermodynamic flexibility to ones that maintain their stiffness practically to the decomposition temperature of the network. Among the more flexible aromatic segments one finds [568]

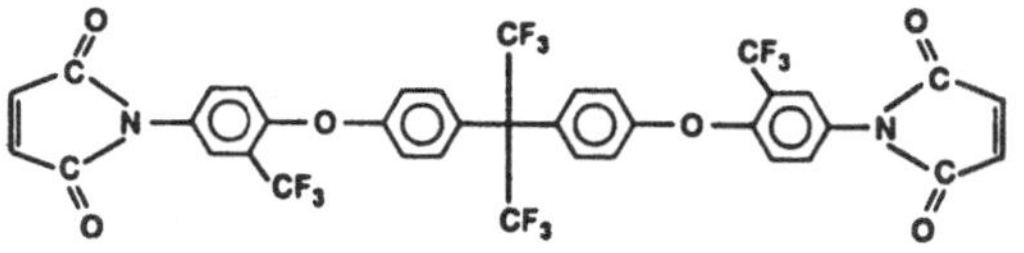

and similar fluorinated ether segments, deriving their flexibility from the presence of the ether oxygen and the angular relationship between the two aromatic rings connected by the $F_3C-\overset{|}{\underset{|}{C}}-CF_3$ bridge. The stiffness and thermal stability of the system originate from the relatively short distance between junctions and the aromaticity of the segments. Somewhat stiffer segments of significantly higher thermal stability are those containing both rigid imide groups and relatively flexible methylene and ketone bridges [572]:

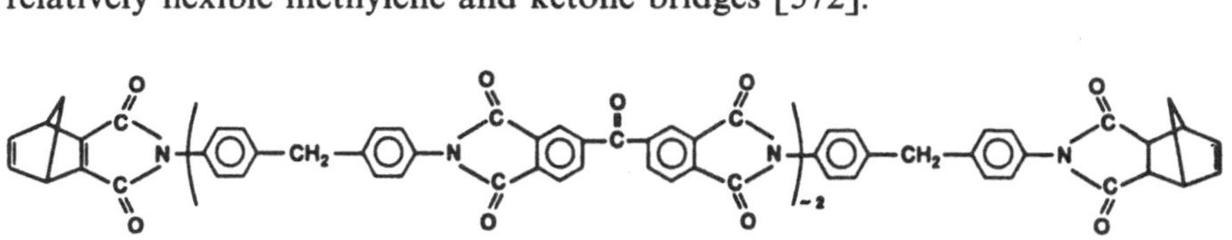

These reactive oligomers were developed by NASA under the name PMR-15 for the purpose of impregnating carbon-fiber mats and polymerizing in-situ to obtain highly aormatic crosslinked polyimide networks. Even more stiff are the systems containing shorter polyimide segments that do not contain ketone bridges [571, 573]. They are typified by [573]:

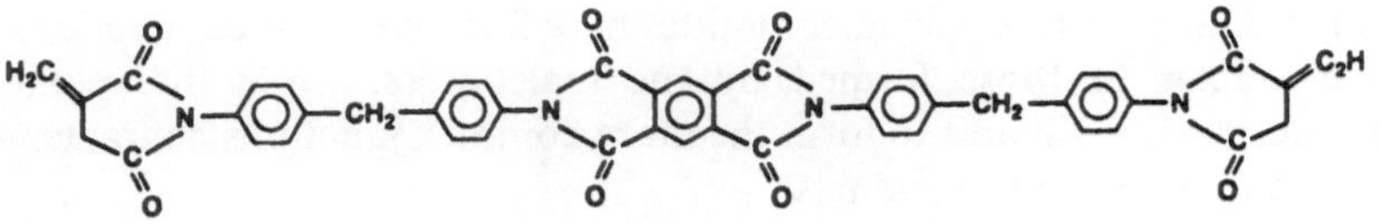

Segments that do not contain entropically flexible moieties are the aromatic esters and aromatic amides [417, 419, 567], exemplified by the mesogenic segments

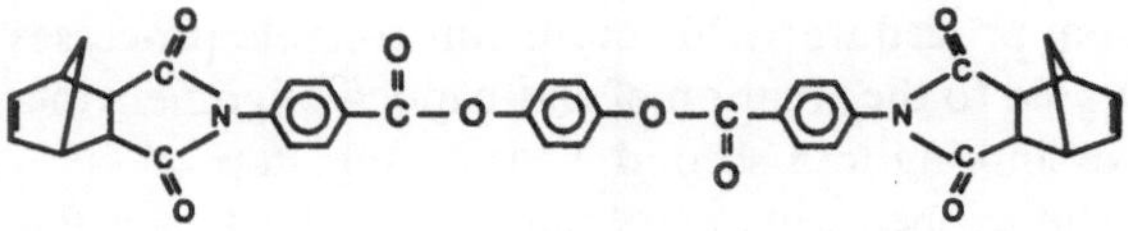

and

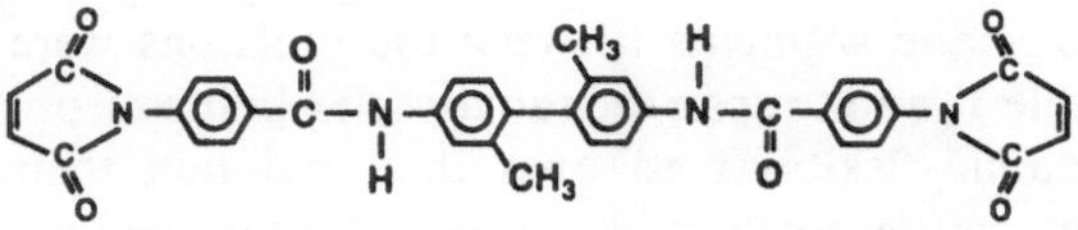

as well as by the stiffer yet not colinear poly(amide imide) [418]

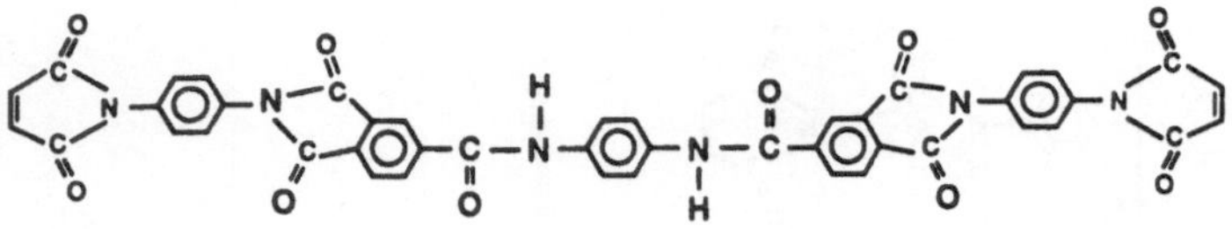

and poly(ester imide) [418]

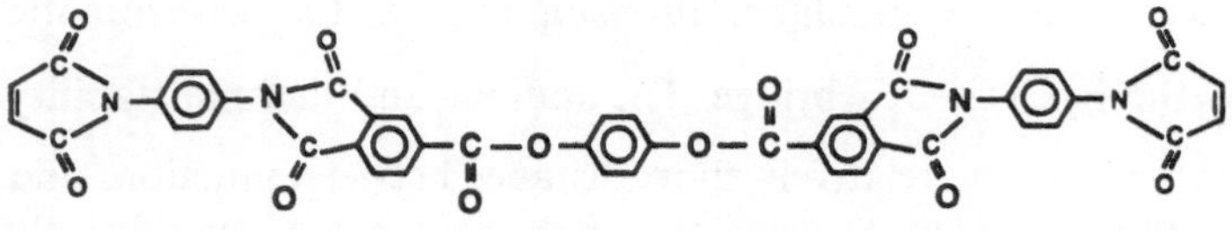

Because of their bent shape, these poly(amide imide) and poly(ester imide) are not expected to manifest liquid crystal behavior despite of their segmental stiffness, high T_g and high thermal stability.

Finally, high temperature bismaleimide networks were recently [575, 576] reported, comprising 4,4'-bismaleimidodiphenylmethane and O,O'-diallyl bisphenol A. Because of the large non-aromatic fraction of moieties in this system, it will not be further discussed.

3.1.6.2 Networks Crosslinked by Strained-Ring Precursors

Another approach to the creation of rigid networks containing alkylene-type junctions is by first preparing more or less stiff aromatic segments terminated by end-groups in which small cyclic moieties are present and the small size of these moieties makes them highly strained. Then, in a second step, the material is heated and the strained cyclic groups open up and new covalent bonds form between the residues of one another. The reaction may stop after joining only two segments together to create a linear broken-rod type chain, or it may continue to create rather massive junctions from which three or more segments emanate. In either case, the strains present in the precursor small cyclic moieties are relieved, and branched junction points are created which may contain unstrained residues, of larger size than was present in the starting material.

The three strained-ring end-cap groups most commonly used for forming junctions between stiff aromatic segments are the benzocyclobutene (BCB) [577–586], biphenylene (BP) [586–591] and [2.2] paracyclophane (PCP) [586, 592] groups:

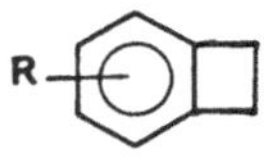

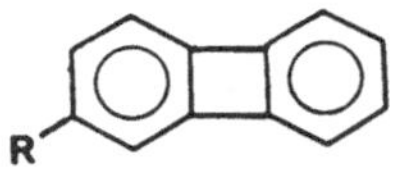

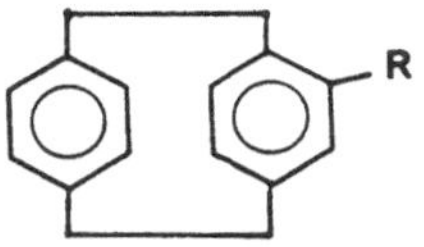

Benzocyclobutene (BCB) Biphenylene (BP) [2.2]Paracyclophane (PCP)
More recent additions to the list of strained-ring reactive crosslinking groups are [586] cyclobutaquinoline (CBQ) and benzocyclobutaacenaphthylene (BCBAN):

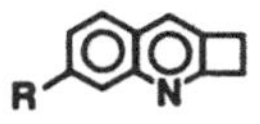

Cyclobutaquinoline (CBQ) 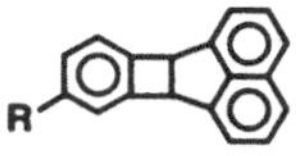   Benzocyclobutaacenaphthylene (BCBAN)

The benzocyclobutene end-group was used with a broad variety of segments ranging from wholly aliphatic to wholly aromatic and from very short to relatively long. Segments of interest in the context of this review are, for instance [582, 583]:

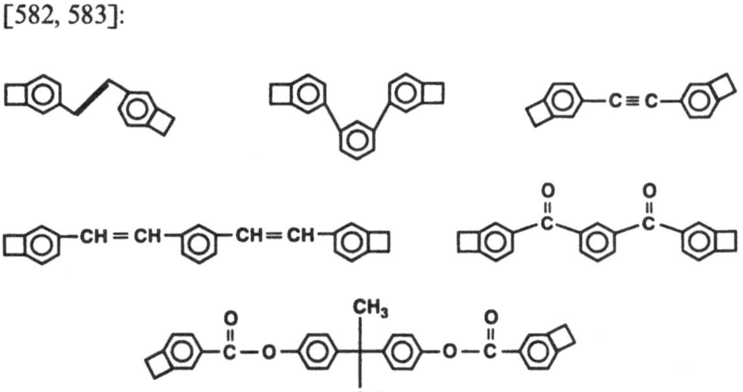

and the polymeric BCB-terminated polycarbonate [583–585]:

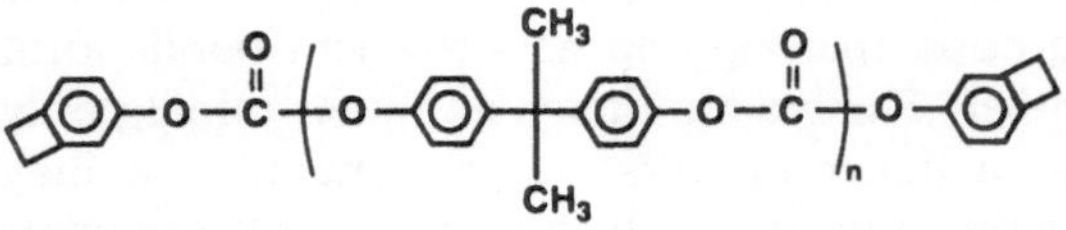

with segmental M_n as low as a thousand and as high as 12 000, and poly-dispersities of about 2.5 [585]. Due to their entropic flexibility, the polycarbonate segments are not stiff despite their substantial aromaticity, but the high T_g of the polycarbonate networks, about 158 °C [584], and the well-known ductility of polycarbonate resins makes them interesting although out of the scope of this review.

Stiffer segments end-capped with BCB are, for example [581]

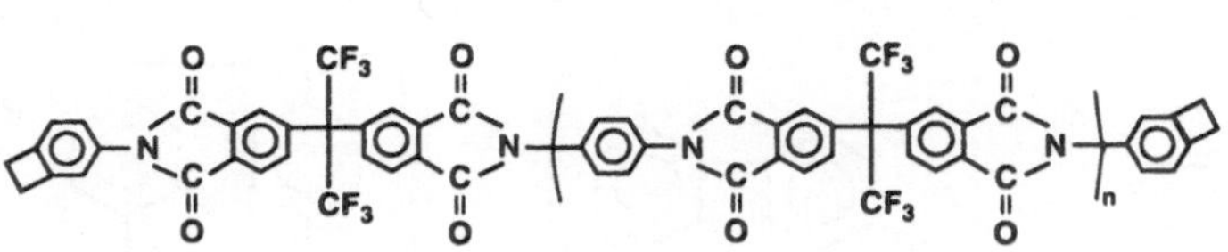

and

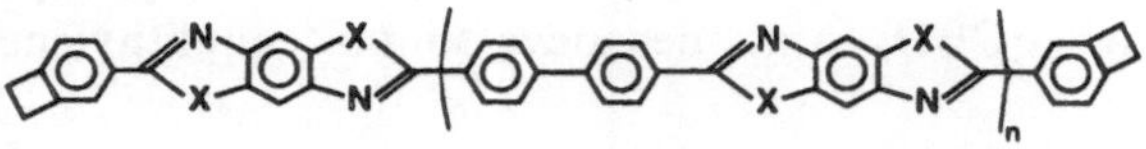

where X may be NH, C_6H_5–N, –O– or –S–.

From its inception, the work with biphenylene end-caps was directed toward the creation of polymeric matrices for composites possessing high thermal stability and good mechanical properties. Therefore systems rich in condensed aromatic rings were preferentially investigated. Among them, the polyquinoline oligomers were pursued by Stille and associates and gained a measure of prominence. In the category of biphenylene end-capped segments thus prepared, we find

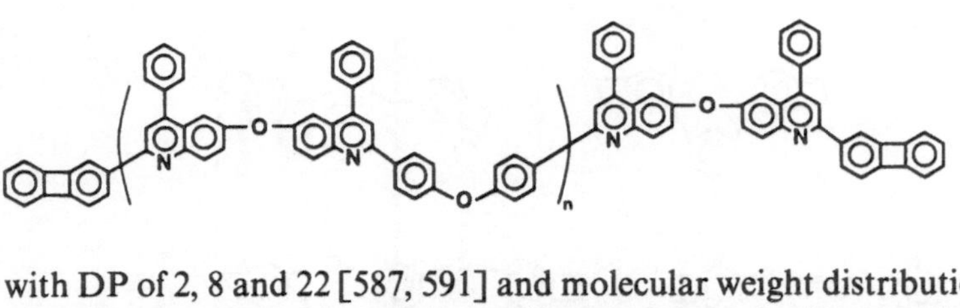

with DP of 2, 8 and 22 [587, 591] and molecular weight distribution in the range of $2.9 \leq M_w/M_n \leq 5.7$ [587]. To increase the crosslink density of the poly-quinone network progenies, biphenylene end-capped polyquinoline oligomers containing biphenylene pendants were prepared [590]. The length of the oligomer corresponds to an average $\overline{DP}$ of 22 and the number of pendant biphenylenes ranges from one to about three per molecule:

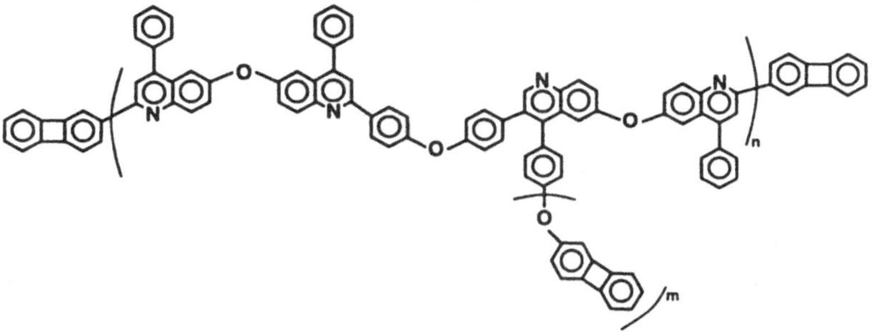

Somewhat more flexible polyimide [588, 589]

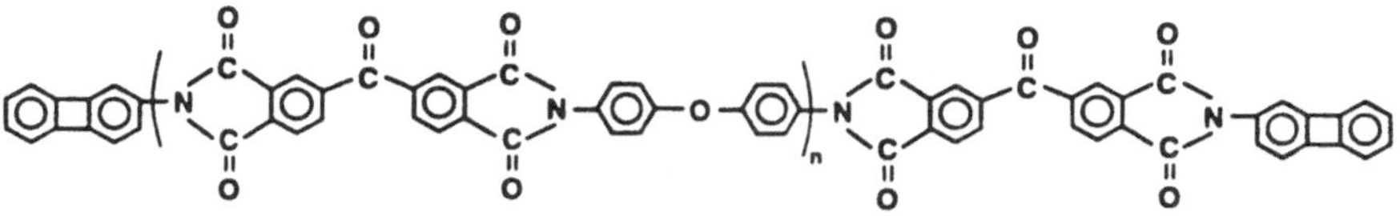

and poly(ether keto sulfone) [588] segments were prepared,

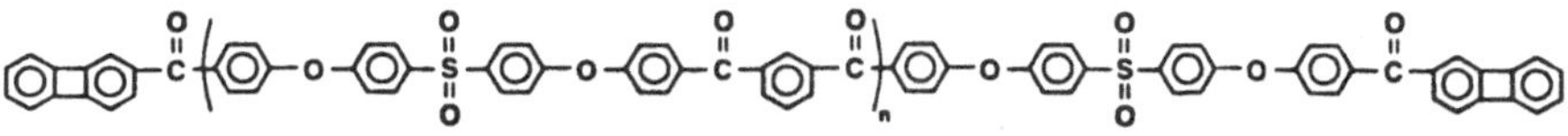

as well as the stiffer polyquinoxalines [588]:

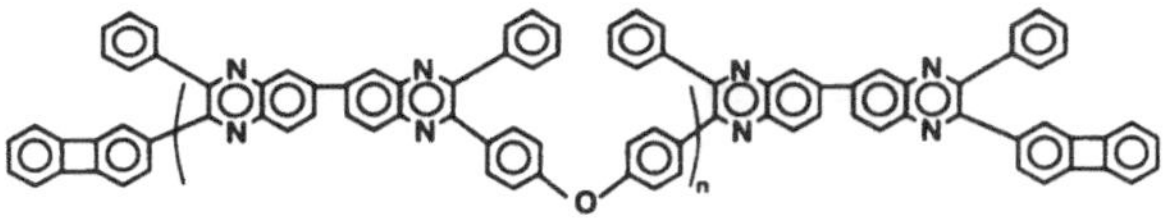

Polyquinoxaline network segments were also claimed by Arnold et al. [581] but with BCB end-caps instead of BP end-caps.

It has been demonstrated that the first step in the formation of junction points from BCB during the cure cycle is a ring opening of the benzocyclobutene [582, 583, 585] to produce the corresponding

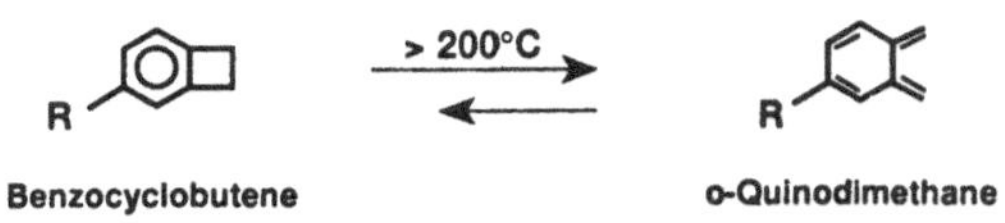

Benzocyclobutene   **o-Quinodimethane**

o–quinodimethane, where R is one end of the network segment. Then, the o–quinodimethanes add to one another in a Diels-Alder sequence of additions:

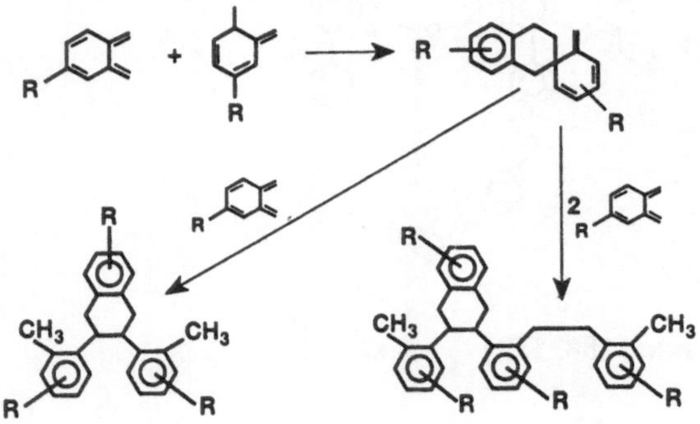

to produce aromatic-aliphatic junction points of varying degrees of complexity and junction functionality [583, 585]. A great variety of similar and intermediate junctions were also identified [583, 585]. When other sites of unsaturation are present in the system, besides BCB, then Diels-Alder type addition may take place leading to a new crosslinked junction. Thus, for example, BCB end-group may react with a maleimide pendant-group to form a junction:

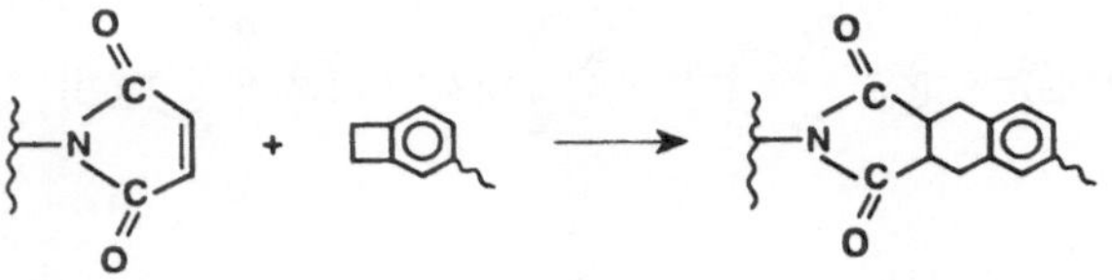

When linear oligomers end-capped with BCB and maleimide react together, they form a linear, extended-chain structure [582] as may well be expected.

When the biphenylene end-capped segments homopolymerize, the resulting product is mostly an extended linear chain connected together by tetraphenylene moieties:

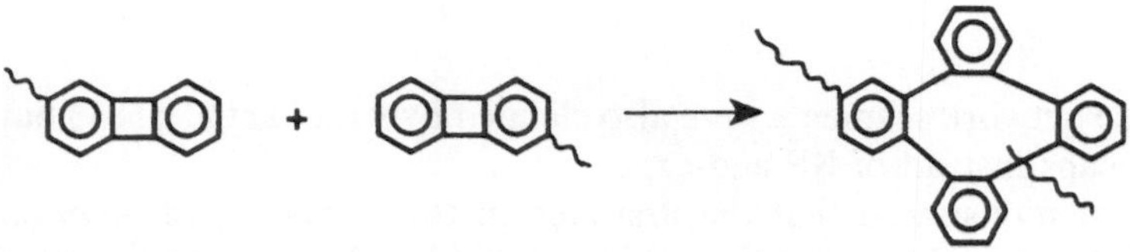

The good properties associated with the cured material are largely due to the inherent thermodynamic stiffness and the very high T_g of the aromatic backbone. Furthermore, the increased chain length allows for sufficient interchain entanglements, positioning the material in the entanglement concentration range conducive to high mechanical performance [170]. For a true network to be formed upon curing, segments end-capped with BP and containing BP pendant groups should be used. This is indeed the case [590]. The mechanical properties of the highly aromatic networks were rather close to those of the linear chain-extended polymer, both above and below T_g. The T_gs of the end-capped oligomers were all in the range of 230 °C and after curing and crosslinking they increased by no more than 30 K. The most disappointing characteristic of the BP-containing systems was their thermooxidative instability, which was traced to the poor stability of the tetraphenylene reaction product [589, 590]. The poor stability of the BP-capped polyquinolines led to the preparation of analogous oligomers of $\overline{DP}$ = 11 and 22, in which varying amounts of the ether groups along the chain were replaced by ethynyl groups [591]. The reaction between the biphenylene and ethynyl groups, requiring nickel catalyst and rather high temperature, generates a fully aromatic branchpoint:

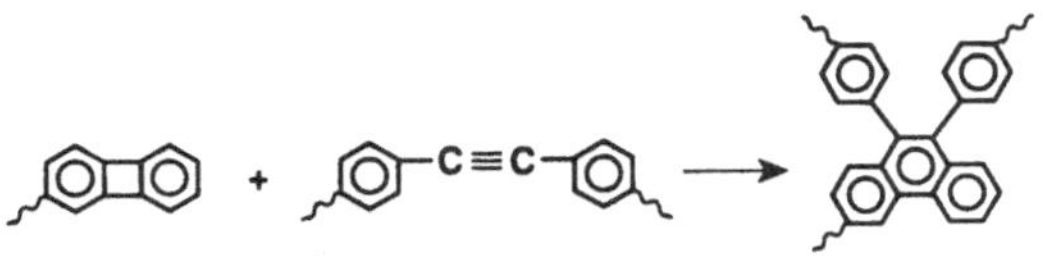

and, thus, a true highly aromatic permanent network.

A different approach to the creation of highly aromatic crosslinked networks, using strained-ring precursors to the junction points, was first taken by Meyers et al. [593], then followed by Marvel [594–596] and later by Stille [586, 592]. Here, the highly strained [2.2]paracyclophane (PCP) group opens up at elevated temperatures and the diradical combines with one or more PCP units to produce ethylenic crosslinks. When the PCP groups appear in the system as end-caps only [586, 592], they react together to yield an extended chain:

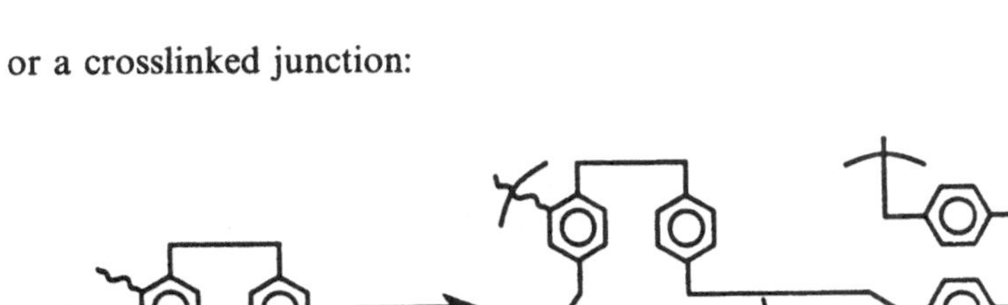

or a crosslinked junction:

When the PCP units are built into the chain [593–596], then the addition of a mere two of them is sufficient to create a crosslinked junction:

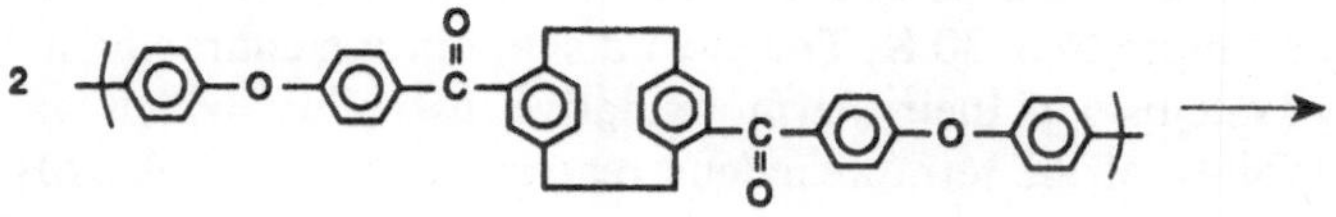

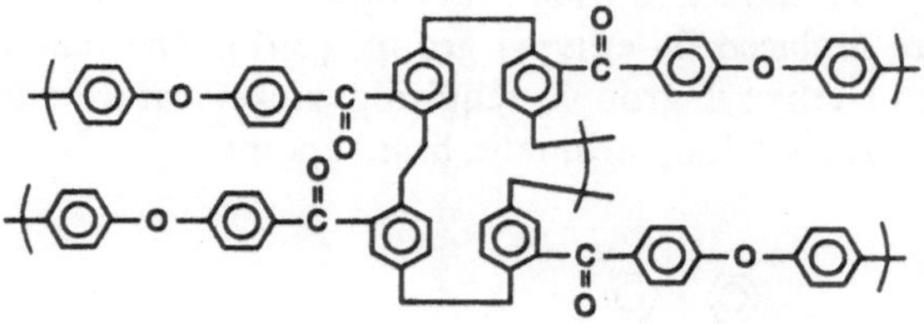

A more stable version of the paracyclophane, consisting of octafluorinated PCP, has also been described in the literature [592, 596] but appears not to be dramatically better than PCP itself.

Because the polymers containing BP and PCP reactive end-groups, and most of those carrying the BCB end-caps, were designed to serve as matrix materials in composites and laminates meant to withstand extreme thermal and oxidative environments, and to have as high as possible T_gs, they are generally very aromatic in nature. This adversely affects both their solubility characteristics and their molecular weight. As a family, the stiff aromatic polymers described here are low molecular weight oligomers which, upon crosslinking, produce rigid networks comprising relatively short segments between the junctions. Due to the stiffness of the segments, the networks do not swell much and their behavior is not expected to be similar to the behavior of common flexible-chain networks and gels.

3.1.6.3 Networks Crosslinked by Ethynyl End-Caps and Pendant Groups

The technologically-driven development of networks crosslinked by reactions of ethynyl groups with one another is similar in many respects to that of networks created by reactions of strained-ring species. In both instances, most of the effort is concentrated on the creation of matrices for composite materials desired by the aerospece industry. Thus, a combination of properties was specified, including high fluidity of the oligomers and their "dope" for ease of penetration, coverage and adhesion, a very high T_g coupled with a measure of ductility and crack resistance of the cured product combined with good thermooxidative stability, and reasonable pressure/temperature/time conditions for the cure process during which the solvents, if present, are first removed and the oligomers then crosslinked to form a permanent network. Development of microvoids due to the evaporation of crosslinking reaction by-products, or severe brittleness were not deemed acceptable. Therefore, it is not surprising to find that most of the oligomers we encountered in Sect. 3.1.6.2 and some of those mentioned in Sect. 3.1.6.1 were also studied with ethynyl end-caps or pendants [597, 598].

For the most demanding conditions poly(phenylquinoxaline) and polyimide oligomers were created, some end-capped only and others containing ethynyl pendant groups. Typical of the end-capped poly(phenylquinoxalines) are [597, 598]:

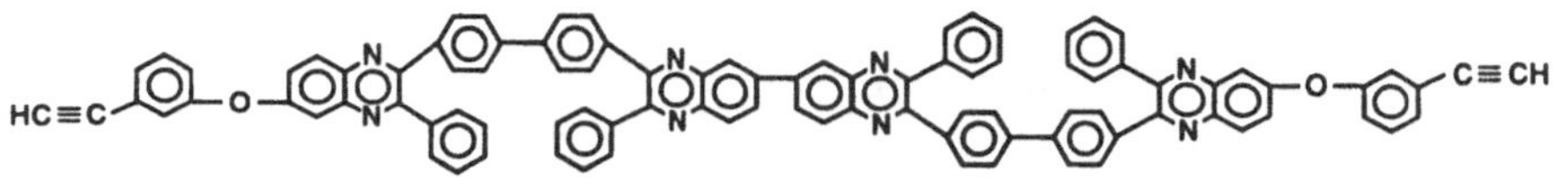

and [598, 599]:

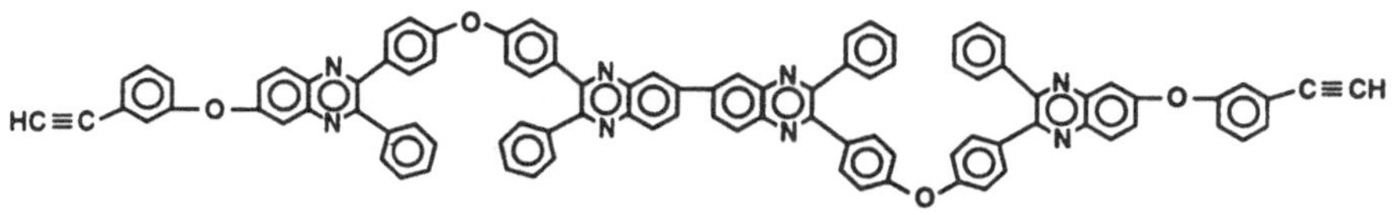

In attempts to improve the processibility of the poly(phenylquinoxaline) oligomers, some were altered by changing the bridging moieties between phenyl rings and introducing more flexible groups into the chain [598–600]:

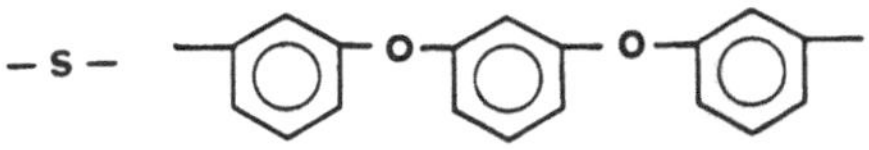

and

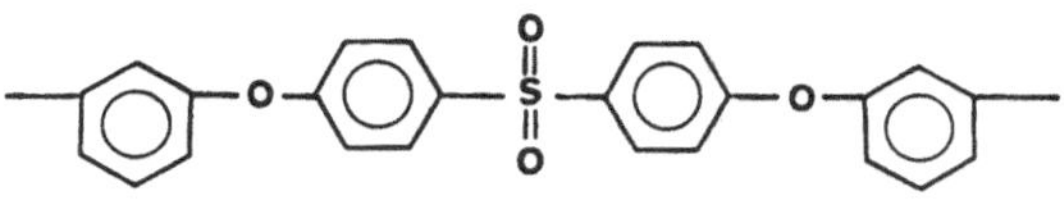

while in others the alkyne group was replaced by the more sterically hindered phenylethynyl and other aryloxyethynyl end groups [600, 601] typified by

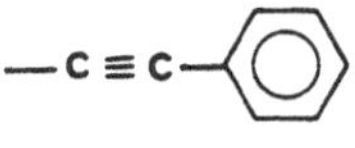

and

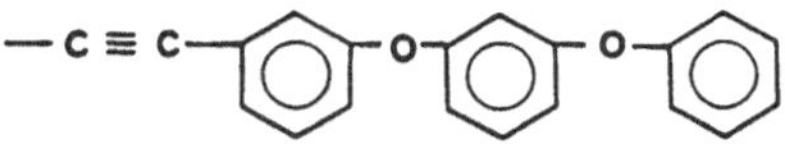

Poly(phenylquinoxaline) oligomers containing pendant ethynyl groups were prepared, in which the ethynyls are either hindered [601] or not [602], and their number per oligomer molecule may be altered to best fit the desired processing and performance requirements:

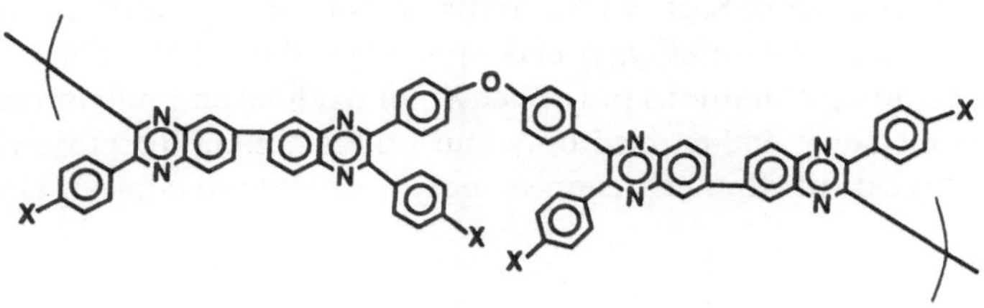

where X may be H, -o- , —c≡cH , —c≡c- ,

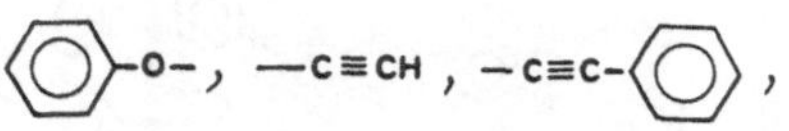

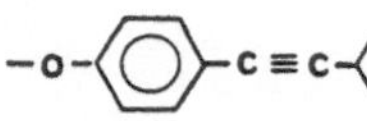

—o--c≡c- ,   or any combination of these.

Despite all the synthetic and processing efforts, composite materials containing poly(phenylquinoxaline) network matrices remain in very limited use. This is
due to several reasons. Among them, one can enumerate the costs of monomers
and of synthesis, lower than expected thermooxidative stability [598], reduced
demand from the "defense" industry and, most important, the suspect carcinogenicity of the pivotal monomer 3,3'-diaminobenzidine [597]. The verified or
suspect carcinogenicity of essentially all aromatic ortho-diamines makes it
highly unlikely for poly(phenylquinoline) polymers and reactive oligomers to be
in broad use in the future unless a new synthetic route to quinoxalines is found
in which aromatic ortho-diamines are not used.

An ethynyl-containing system that reached commercialization is the one
consisting of polyimide oligomers end-capped with ethynyl [597, 598] and
hindered ethynyl [600] reactive groups. Typical of the reactive oligomers
marketed by Gulf Specialty Chemicals under the tradename Thermid is Thermid-M whose idealized structure is

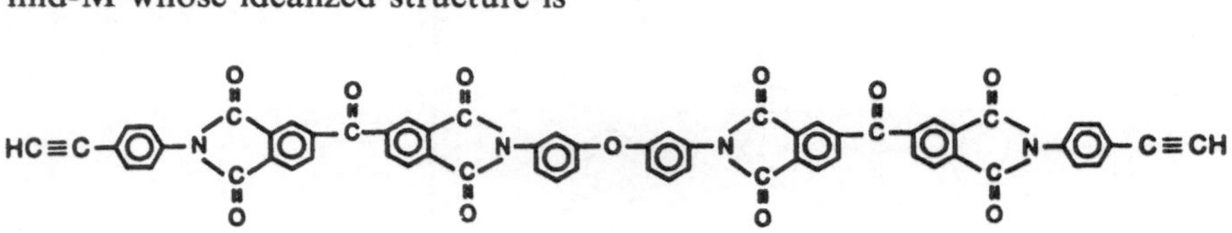

In its imidized from, this material poses major processing hurdles balanced by
encouraging properties of the cured network. The material is only slightly
soluble and has a short gel time at elevated temperatures [597]. This makes the
fabrication of large elements from it practically impossible. To improve its
processibility, or to replace it altogether, fluorinated imide monomeric species

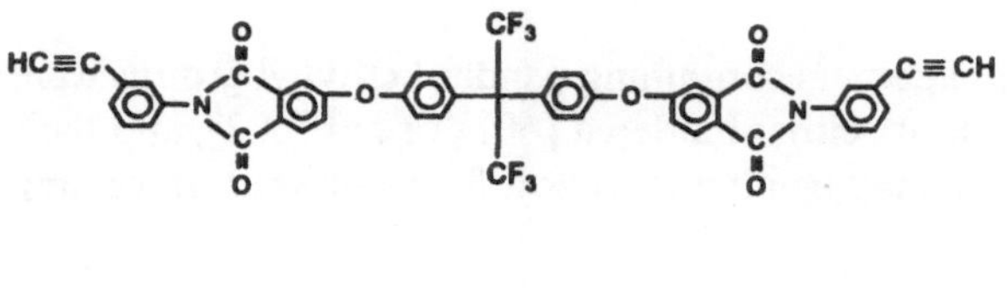

were prepared and then mixed together with Thermid-M to act as plasticizers [597], or crosslinked with themselves [600] to form networks:

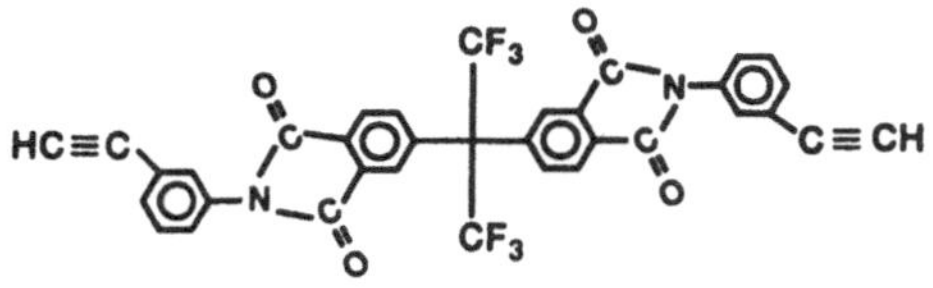

and

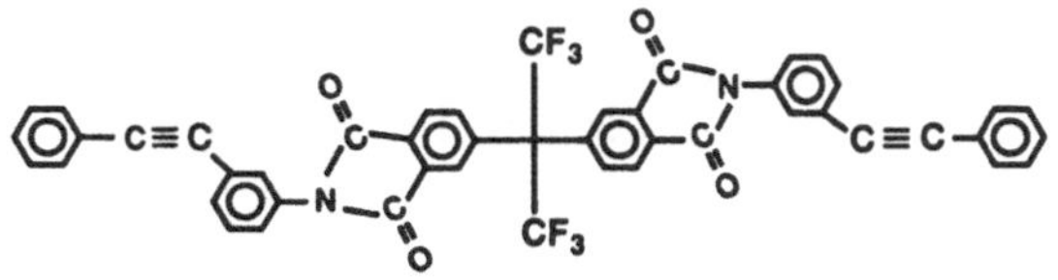

Another ethynyl-containing system is the one containing sulfone bridging groups [597, 598, 600, 603]. Ethynyl-terminated aromatic sulfone oligomers usually also contain ether and ketone bridging groups. The flexibility thus built into the oligomers reduces the T_g of the cured network by about 100 K relative to networks made from phenylquinoxaline segments [598]. Typical end-capped sulfone oligomers are [597]:

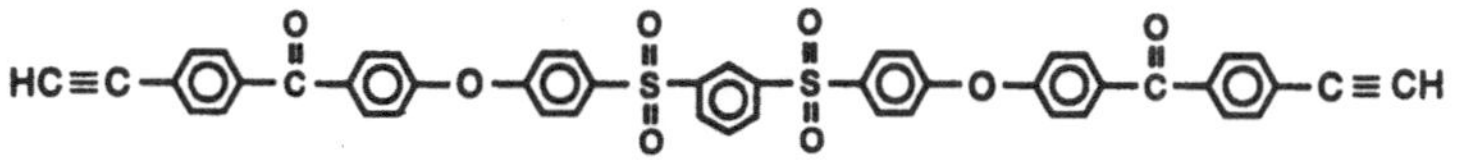

and [600]

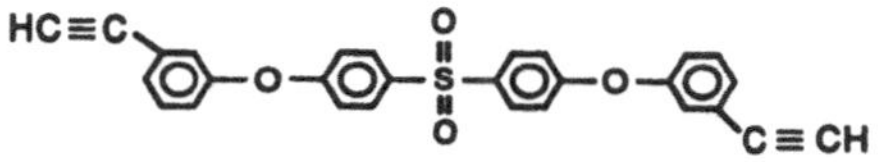

Modest molecular weight poly(ether ketone sulfone) polymers with pendant alkylene groups were also prepared and crosslinked into a network in a separate curing step [603]. These will be discussed below.

Aromatic monomers and oligomers containing ethynyl groups appear continuously in the patent literature [604]. In most instances they are variations on the theme of oligomeric polyimides containing hindered or unhindered ethynyls in the end-groups or along the chain. Ocassionally a new species appears, such as [604c, 604d]:

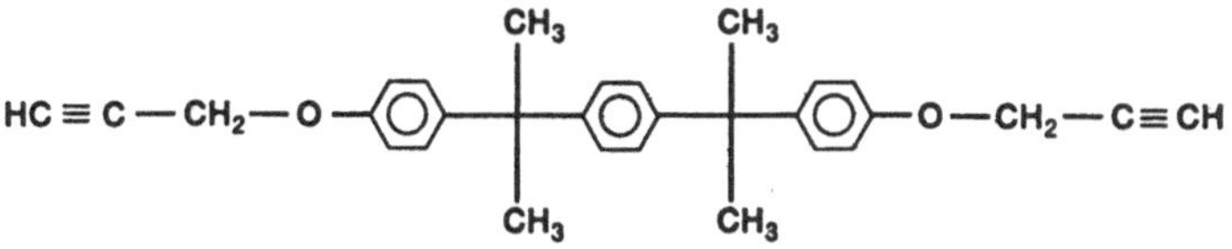

even though this one appears to be a take-off of the much older bispropargyl ether of bisphenol A and others [605].

Initially it was assumed that the heat-cure reaction leading to network formation involved a simple trimerization of three ethynyl groups [597, 603, 606]:

$$\underset{HC\equiv C-}{\overset{\diagup C\equiv CH}{\underset{}{}}} + \overset{\diagdown C \equiv CH}{} \longrightarrow \bigcirc$$

leading to a thermally stable trisubstituted phenyl branchpoint. It was repeatedly found, however, that the ethynyl-containing segments and the cured networks prepared from them are less stable against thermooxidative attack than the corresponding segments without the ethynyl groups [597, 598, 601–603]. Thermal and spectroscopic investigations of the curing process were combined with model compound studies and revealed that, although all the ethynylic character was lost during the cure process [598, 607], the majority of the crosslink junctions were not the phenylene product of ethynyl trimerization but were more complex in nature. Among the proposed crosslinking reactions one finds:

(a) $-C\equiv CH + HC\equiv C- \overset{\Delta}{\longrightarrow} -C\equiv C-C\equiv C-$ (Glaser coupling [608])

(b) $-C\equiv CH + HC\equiv C- \overset{\Delta}{\longrightarrow} -C\equiv C-CH=CH-$ (Strauss coupling [609])

$$-C\equiv CH + \bigcirc\!\!-\!\!\bigcirc- \overset{\Delta}{\longrightarrow} \bigcirc\!\!-\!\!\bigcirc-$$

and, upon exposure to higher temperatures, additional rearrangement and aromatization of the crosslinking groups. Reactions with oligomers and model compounds [599, 601, 602] indicated that cyclotrimerization accounts for only about 20% of the available ethynyl groups while most of them contributed to more complex oligomeric and polymeric products. We believe that the complex products originate from secondary reactions of the conjugated ene-yne bonds created by Strauss coupling [609] of two ethynyl groups. Our belief is grounded in the facts that when acetylenic groups interact together they, more often than not, generate the ene-yne system. This, for example, was demonstrated in the case when polymeric diacetylenes were crosslinked by intense irradiation [610, 611]:

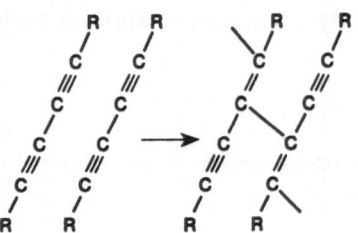

Infrared measurements of Lando and associates reported by Baughman and Chance [612] indicate that most, but not all, of the diacetylene functionalities in the crosslinked polymer were consumed during the crosslinking process. The crosslinked product contained mostly the ene-yne bonds [610, 611]. The thermogravimetric curves of the rate of the thermal degradation process of polymethacrylates containing conjugated acetylene groups in their side-chains, pass through a major peak at temperatures close to 200 °C [613]. At this temperature, conjugated ene-yne bonds appear in the system and replace the conjugated diacetylenes [613]:

```
  CH3        CH3        CH3                    CH3        CH3        CH3
   |          |          |                      |          |          |
 -C-CH2 ---- C-CH2 ---- C-CH2-                -C-CH2 ---- C-CH2 ---- C-CH2-
   |          |          |                      |          |          |
  C=O        C=O        C=O                    C=O        C=O        C=O
   |          |          |                      |          |          |
   O          O          O                      O          O          O
   |          |          |                      |          |          |
   R          R          R                      R          R          R
   |          |          |                      |          |          |
   C          C          C                      C ======= C          C ==
   |||        |||        |||          Δ         |          |          |
   C          C          C          ----->      C          C          C
   |          |          |                      |||        |||        |||
   C          C          C                      C          C          C
   |||        |||        |||                 == C          C ======= C
   C          C          C                      |          |          |
   |          |          |                      R          R          R
   R          R          R                      |          |          |
   |          |          |                      O          O          O
   O          O          O                      |          |          |
   |          |          |                      C=O        C=O        C=O
  C=O        C=O        C=O                      |          |          |
   |          |          |                    -C-CH2 ---- C-CH2 ---- C-CH2-
 -C-CH2 ---- C-CH2 ---- C-CH2-                   |          |          |
   |          |          |                      CH3        CH3        CH3
  CH3        CH3        CH3
```

The preference for the creation of an ene-yne conjugated system whenever conjugated diacetylene groups are exposed to thermal or shorter wavelength irradiation supports our belief that the thermally induced crosslinking of ethynyl groups passes mostly through the ene-yne conjugated group created by the Strauss coupling [609]. The presence of the sensitive ene-yne group and its non-aromatic yet probably conjugated progeny in the cured networks is most likely responsible for the unexpectedly low thermal and thermooxidative stability of these networks mentioned above.

3.2 Two-Step Aromatic Networks

In two-step aromatic networks the precursor chains are much longer than the average length between crosslink points:

$$L \gg l_0.$$

Unlike one-step networks, where all stiff segments projecting from each branch-point are completely or almost completely indistinguishable from each other, in the two-step networks the units connecting the precursor chains are more often than not somewhat different from the precursor chains. In several instances, such as in the case of two-step polyamide networks, the units connecting the chains are chemically identical with the precursor chains and all segments propagating from a given junction are indistinguishable from one another. In other cases the junction point consists of only a few atoms and bonds, linking otherwise uninterrupted chains. In this and any other case where the junction point and/or connecting units are chemically different from the precursor chain segments emanating from them, the junctions or connecting units will be called crosslinks. In this respect they are similar, say, to the crosslinks provided by sulfide bridges between polyisoprene chains in flexible elastomeric networks.

3.2.1 Two-Step Aromatic Polyamide Networks

There are in the literature three reported methods to prepare two-step aromatic polyamide networks. Two of them use the reduction of a pendant nitro group to a reactive amine as a means to create reactive sites along the precursor chain. A U.S. patent by Erhan [614] claims the preparation of "high molecular weight" linear polyamide by either interfacial or solution polymerization of a single AB-type monomer with itself:

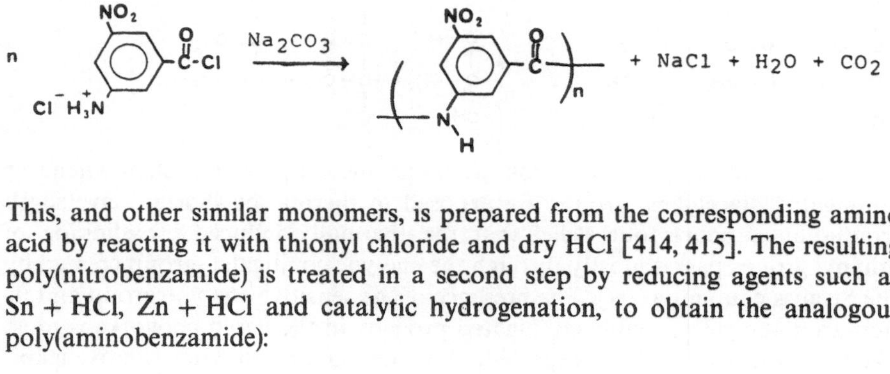

This, and other similar monomers, is prepared from the corresponding amino acid by reacting it with thionyl chloride and dry HCl [414, 415]. The resulting poly(nitrobenzamide) is treated in a second step by reducing agents such as Sn + HCl, Zn + HCl and catalytic hydrogenation, to obtain the analogous poly(aminobenzamide):

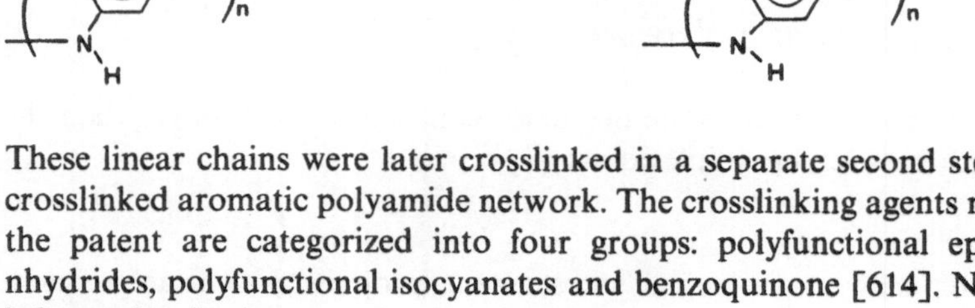

These linear chains were later crosslinked in a separate second step to form a crosslinked aromatic polyamide network. The crosslinking agents mentioned in the patent are categorized into four groups: polyfunctional epoxides, dianhydrides, polyfunctional isocyanates and benzoquinone [614]. No additional information is given.

It is interesting to note that Erhan [614] claims the polymerization of monomers such as 3-amino-5-nitro-diphenylcarboxylic acid

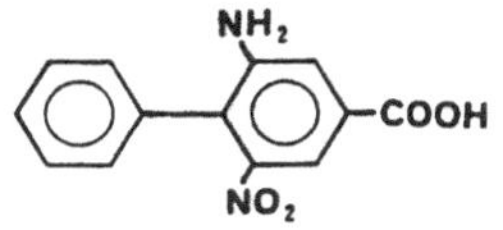

by polycondensation in NMP in the presence of TPP and pyridine. We have attempted (see Sect. 3.1.2) the polymerization of several nitroaminobenzoic acids by this method and all failed. Similar attempts by Professor Preston [615] also failed. We believe, however, that monomers such as [614]

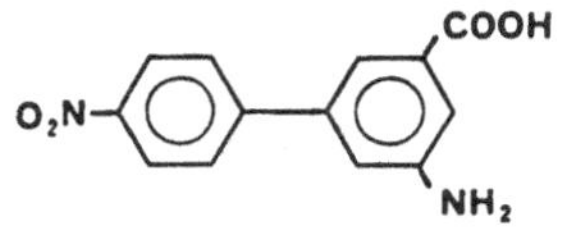

will homopolymerize after being activated by aryl phosphite. The difference in reactivity is a clear manifestation of the strong deactivating effect a nitro group exerts on an amine group when they are both attached to the same phenylene ring, and the reduced deactivating effect when the nitro group is on one ring and the amine group on the next. The fact that the aromatic rings in the biphenyl group are not coplanar [208, 616, 617] explains the decoupling of the nitro from the amine in this case.

The second reported preparation of two-step aromatic polyamide networks is by Aharoni [277]. According to this method, long and stiff linear aromatic polyamide chains are first prepared by the Yamazaki procedure [394]. Because the nitro-bearing group is the AA-type nitroterephthalic- or nitroisophthalic-acid residue, the nitro group may appear on every second ring or at larger distances. The procedure of Aharoni [277] is, therefore, less flexible than that of Erhan [614], but the monomers used by Aharoni are all commercially available, unlike those used by Erhan. After the polycondensation is completed and the stiff linear chains isolated, the nitro groups are reduced to amines by the use of a $SnCl_2 \cdot 2H_2O/HCl$ mixture or an alkaline solution of sodium hydrosulfite ($Na_2S_2O_4$) [618]. The resulting polymer, typified by the repeat unit

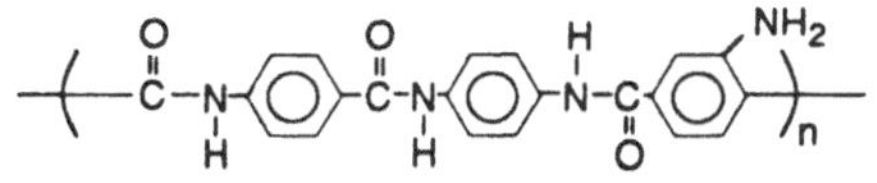

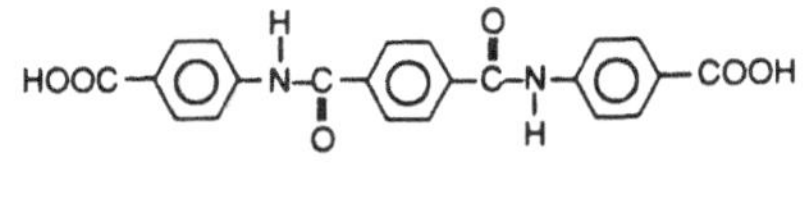

is then reacted in solution in a second Yamazaki-type reaction with monomeric species such as to form gelled networks in which the chemical composition and structure of the segments originating from the precursor linear chains are identical with those of the segments originating from the monomeric species added in the second step. When the connecting species between chains are rigid, we call them struts. Because all segments connected at a given crosslink point are identical, we call this point a branchpoint and not a crosslink point.

A different method by Aharoni to create two-step rigid polyamide networks employs pendant carboxy groups on the precursor linear chains, instead of nitro or amine. One way of preparing such chains requires the preparation of diacid chloride monomeric species from excess terephthaloyl dichloride and DABA:

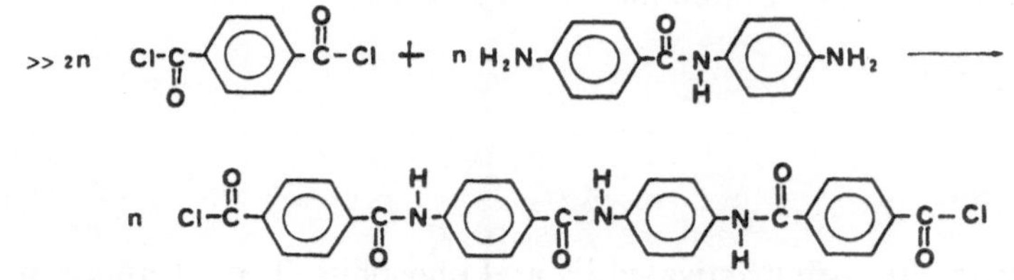

This species is then reacted under Schotten-Baumann-type conditions, in DMAc and the presence of pyridine, with 3,5-diaminobenzoic acid to yield a linear polyamide with 120° bends every fifth ring:

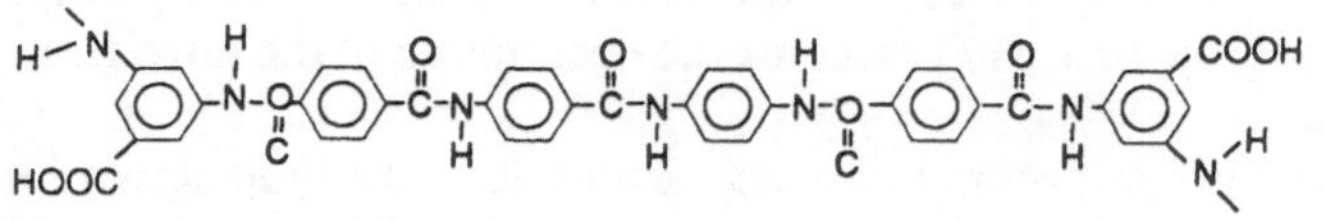

Chains with molecular weight of 10000 and higher were easily obtained. Another way to create aromatic polyamide chains with pendant carboxylic acid groups involves the preparation of the monomer mixture

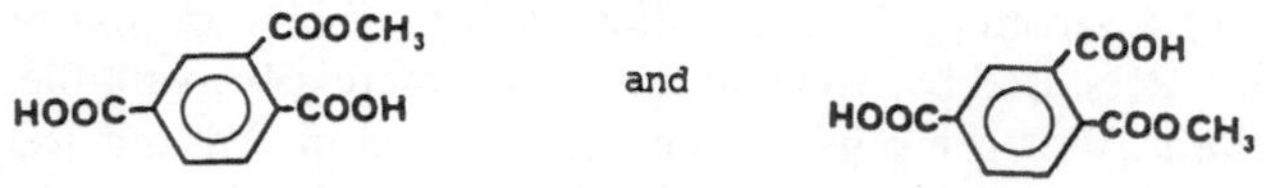

and

from the 1,2,4-benzenetricarboxylic acid anhydride

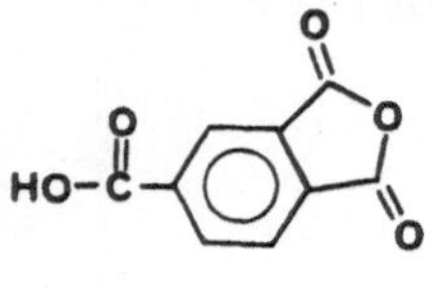

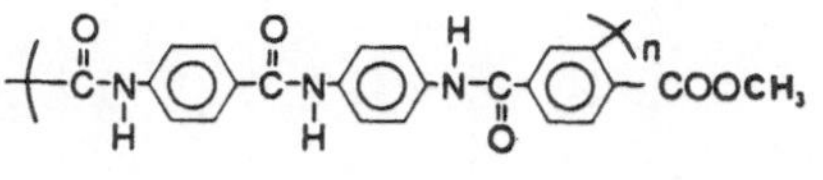

by treating it with hot methanol. The monomer mixture is then reacted by the Yamazaki procedure with equimolar amount of DABA or with 2:1 molar mixture of DABA and nitroterephthalic acid to obtain the respective polymers

and

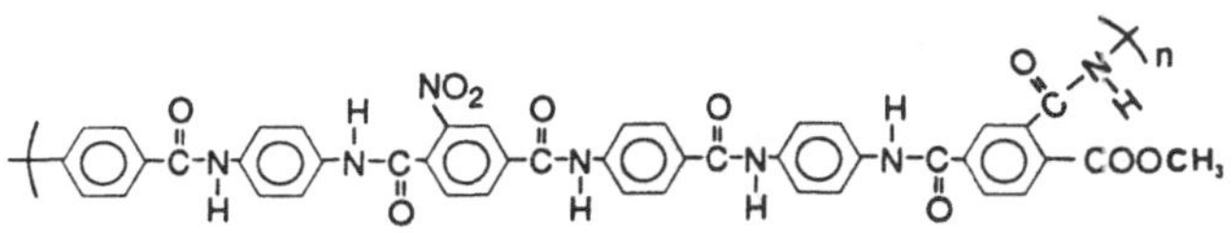

In these schemes only one of the positional isomers is shown, but analysis indicated that about 70% of the methyl ester is on the number 1 position and 30% on the number 2 position of the 1,2,4-benzenetricarboxylic acid monomer mixture. Therefore, in reality, the length of the straight backbone sections between bends is on average 30% longer than shown. The ester groups are then selectively hydrolyzed by treating the hot solutions of the polymers with LiI to produce the respective carboxy-decorated polyamides without any measurable reduction in chain length.

The linear polyamides with pendant carboxy groups prepared by both methods are reacted with rigid aromatic diamines, such as DABA, in solution under Yamazaki conditions to yield gels of two-step rigid aromatic polyamide networks such as [619]:

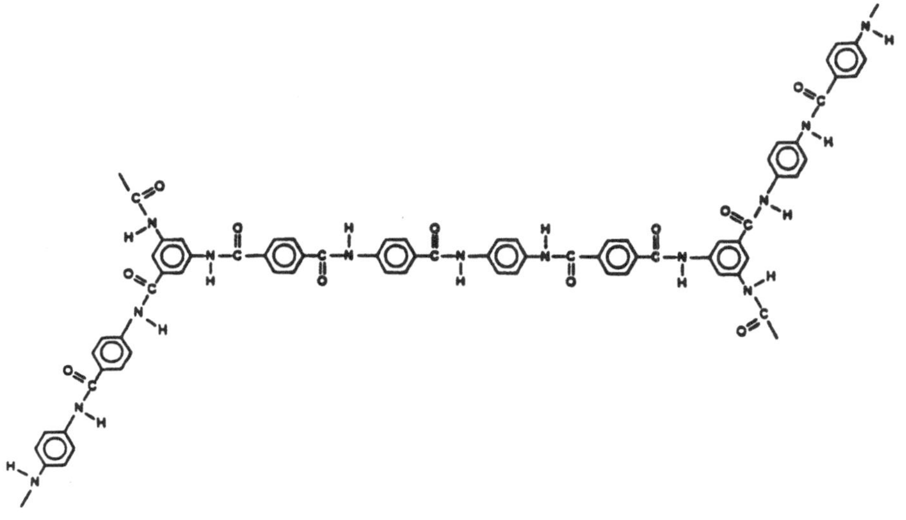

In all the polyamide networks prepared by Aharoni in two separate steps, the lengths of the precursor linear polyamide chains fell in the interval between several dozens to several hundreds nanometers. These lengths are far larger than the average diameter of the precursor fractal polyamides in the one-step polymerizations right before they reach the gel point [277, 619]. When equimolar amounts of accessible reactive sites on the precursor chains and on the ends of the shorter rigid connecting species are used in the second step of network formation, they react with very high efficiency, creating a large majority of rigid struts with only a small minority of unconsumed reactive sites coming from both sources. The very large number of struts and the length of the precursor chains combine together to create gels of networks with minimal levels of defects. These more "flawless" gelled networks exhibit very low levels of swelling upon immersion in good solvents, and much lower brittleness and higher ductility than gelled networks of comparable concentration and distance between branchpoints created by the one-step procedure. The higher brittleness of the one-step gelled rigid networks reflects, we believe, the substantially higher level of defects present in such systems.

The third method to create two-step aromatic polyamide networks is chemically similar to the one discussed above in Sect. 3.1.6.2. Here, a fraction of the terephthalic monomer in the reaction solution is replaced by 2,3-cyclobuta-terephthalic acid (619a). Upon polymerization a lyotropic aromatic polyamide is obtained which, when heated to 380 °C or higher, randomly crosslinks along the stiff chains to form a two-step network. Not much else is known at present about this system.

3.2.2 Two-Step Networks from Poly(Ether Ketone Ketone) (PEKK)

Two-step networks from PEKK were recently described by David [620]. Although not specifically mentioned, it is implied that the same crosslinking procedures are applicable also to poly(ether ether ketone) (PEEK). In the case of PEKK, the network is created after the polymer chains were previously prepared in a separate step.

As was mentioned in Sect. 3.1.4, linear PEKK may be prepared [476, 477, 485] by reactions such as

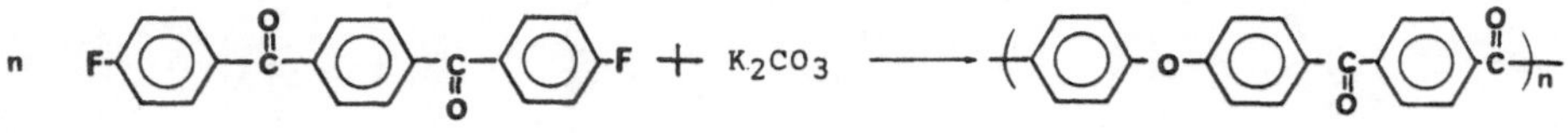

or

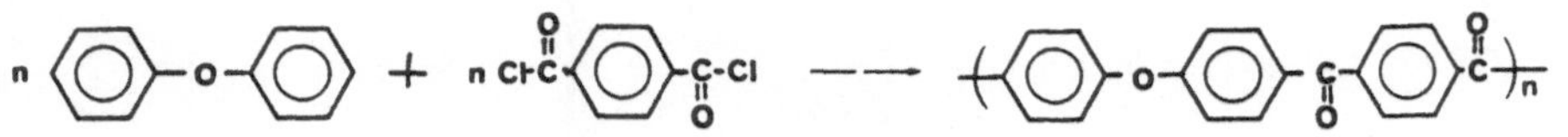

Two thermally-induced crosslinking reactions were identified [620]. One involves the reduction of carbonyl groups to carbinols

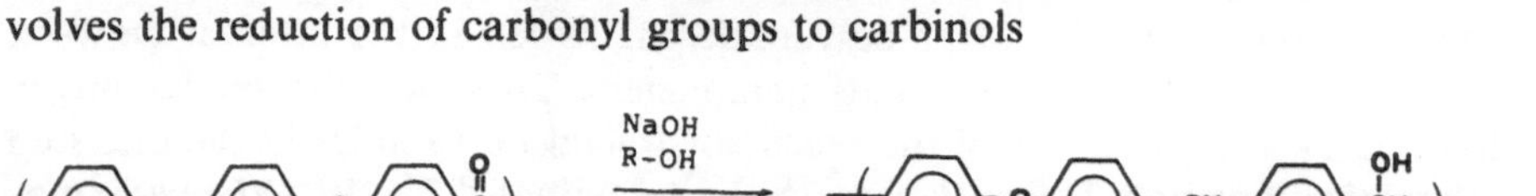

followed by several reactions taking place at temperatures higher than 200 °C and leading to various crosslinking junctions:

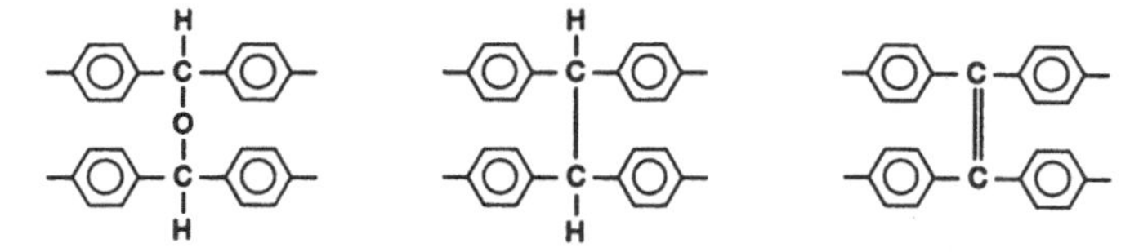

The swelling ratios of the crosslinked PEKKs are rather large, 300% and above, indicating that the crosslinking reaction is not as efficient as may be expected from the above scheme. Another crosslinking procedure for PEKK was to expose it to alkyl dihalides under autogeneous pressure in an autoclave at temperatures in the order of 250–270 °C. In this case the reactions proceeded to a much smaller degree than the crosslinking induced by the alkaline alcohol reaction. Studies with model compounds showed the reaction to involve only ether phenols and not ones connected to carbonyl groups. The model studies further showed the reactions to be alkylation mostly in the para position. Combining the above, we propose that the crosslinking of PEKK by this method requires the participation of chain ends and may be idealized as below:

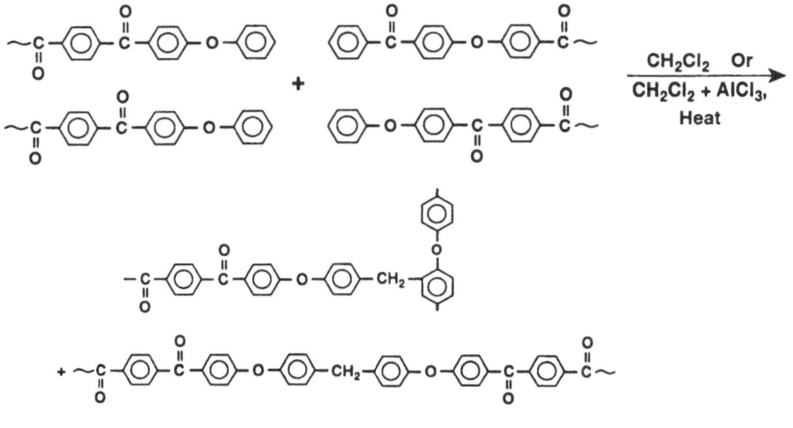

When the crosslinking occurred throughout the volume of the sample and not only on its surface, a polymer with tensile properties, dimensional stability and corrosion resistance properties significantly higher than those of the uncrosslinked analogue was obtained. No other information on the exact nature and structure of the crosslinked PEKK network is presently available.

3.2.3 Networks from Aromatic Linear Chains Created by Reacting Backbone Diacetylene or Pendant Acetylene Groups

The topochemical nature of the polymerization of diacetylene monomers into polydiacetylene is well known [612, 621–623]. Studies of these polymers revealed that they may undergo crosslinking ("crosspolymerization") when neighboring chains are well aligned in crystalline or semi-crystalline domains

[611, 612, 624]. The crosslinking may be initiated by heating or irradiation, and generates no by-products. It was subsequently shown that crosslinking between diacetylene residues can be similarly effected even when the diacetylenes are part of copolymers containing other groups such as thiophenes [625–628], urethanes, etc. [613, 629–633]. It was recently found that the same thermally initiated crosslinking reaction can take place not only between diacetylene residues in the chain backbone, but also when these groups are attached as pendants to flexible backbones such as polyacrylates or polymethacrylates [613].

Polyurethanes containing diacetylene groups are generally of elastomeric nature and often contain hard and soft segments that may separate into two coexisting microphases [629, 630]. They therefore fall outside the scope of this article. Systems containing thiophenes and similar aromatic groups are, however, highly aromatic and fall within the sphere of this article. Building on earlier work with ethynyl end-capped segments, Professor Stille [627, 628] and associates prepared a series of linear poly(diethynylthiophenes)

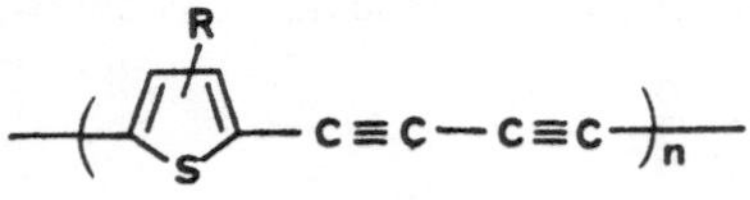

with molecular weights in the order of $M_n \cong 7000$ and $M_w \lesssim 20\,000$ [628]. The solubility of the polymers depended on the nature of the -R group: when -R = H the polymers were partly soluble and when -R was an alkyl group such as n—C_4H_9 or n—C_6H_{13} the polymers were fully soluble in solvents such as nitrobenzene. The polymers undergo extensive crosslinking by thermolysis at the moderate temperatures of $180 - 225\,°C$. Solid-state Carbon-13 NMR analysis [627] indicates that the crosslinked repeat unit may be represented by

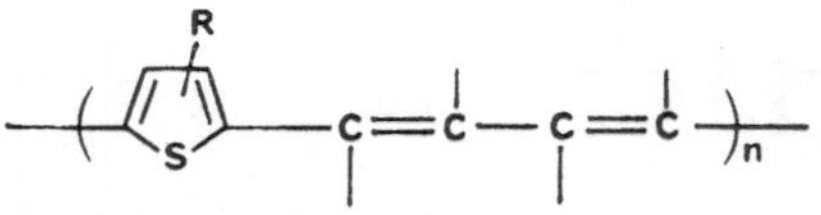

and not by the structures

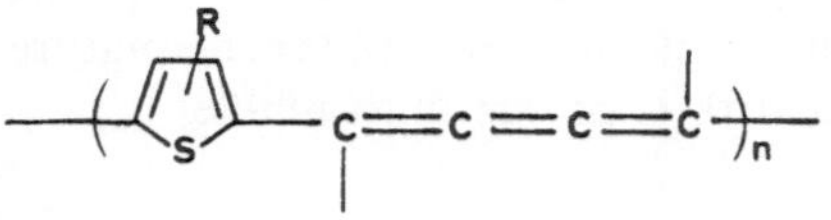

or

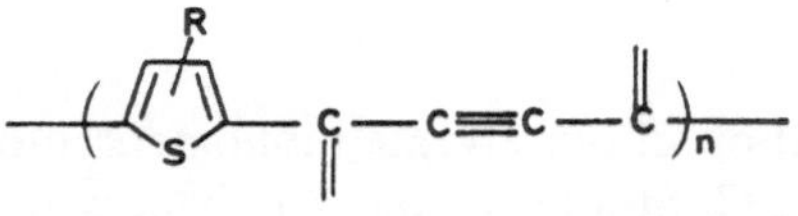

that were claimed in the literature [612, 634]. The polymers prepared by Neenan, Whitesides and co-workers [625, 626, 635] fall in the same range of molecular weights and indicate the same [626] crosslinking species as described by Rutherford and Stille [627].

Two-step polydiacetylene networks in which the aromatic repeat unit is other than thiophene were also prepared [628]. Aromatic (Ar) units such as

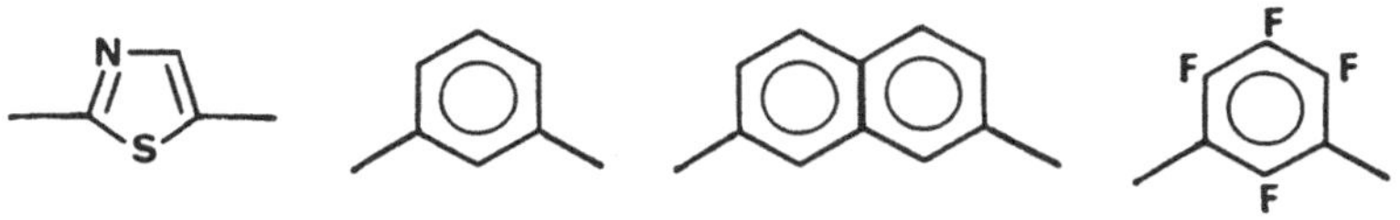

were successfully incorporated into the linear backbones and these were then converted into the respective conjugated networks by thermolytic crosslinking [625]:

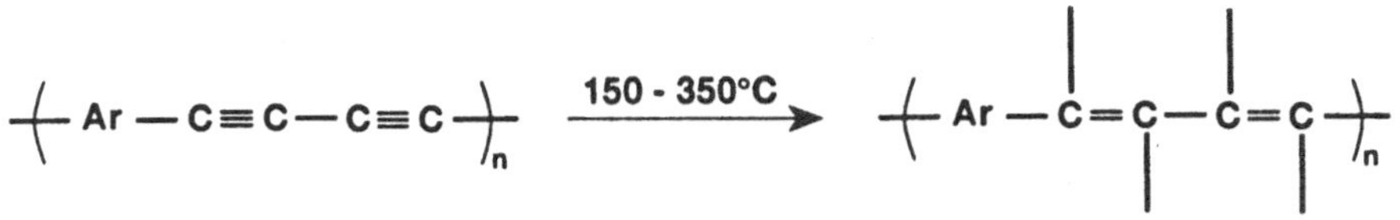

$$+\!\!\!-\text{Ar}-\text{C}\!\equiv\!\text{C}-\text{C}\!\equiv\!\text{C}-\!\!\!+_n \quad \xrightarrow{150\text{ - }350°C} \quad +\!\!\!-\text{Ar}-\overset{|}{\text{C}}\!=\!\text{C}-\overset{|}{\text{C}}\!=\!\text{C}-\!\!\!+_n$$

The two-step aromatic networks originating from linear polydiacetylene were barely characterized. We know the nature of the crosslinking units, the Young's and shear moduli, and density of several such polymers [625], some electrical conductivity measurements [628] and the thermal degradation characteristics of many [625–627] of them. In general the crosslinked networks are rather brittle at room temperature. Besides these few characteristics, not much else is known about this family of aromatic polymer networks. Two such points are worth a special mention. One is the ability of the linear polydiacetylenes to undergo reactions with primary amines and form linear polymers containing alternating thiophene and pyrrole units [627]:

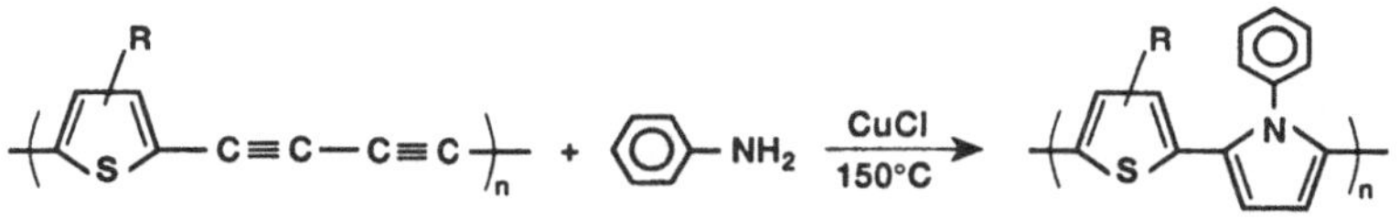

By the use of aromatic diamines, we believe such reactions can be employed in the creation of highly crosslinked fully aromatic and conjugated two-step networks:

(See Structure on page no 116)

A second interesting point is the fact that some of the thiophene-containing polydiacetylenes were shown to behave as lyotropic liquid crystals in concentrated solutions [627]. This observation may open the door to the creation of "liquid crystal" networks from such polymers by varying the level of interchain crosslinks. Such crosslinking can be effected either by heating the polymers neat or in solution, or by reacting them with reactive species such as primary diamines. On the other hand, exposure to heat of "liquid crystal" networks containing stiff segments with diacetylene groups in them may destroy their mesomorphicity.

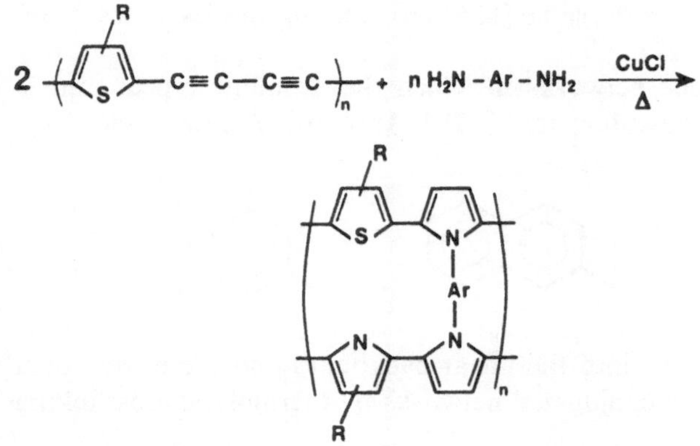

There exists in the literature at least one [603] reference describing the two-step preparation of highly aromatic networks from linear aromatic polymer chains decorated with occasional pendant acetylene group. The precursor linear polymer:

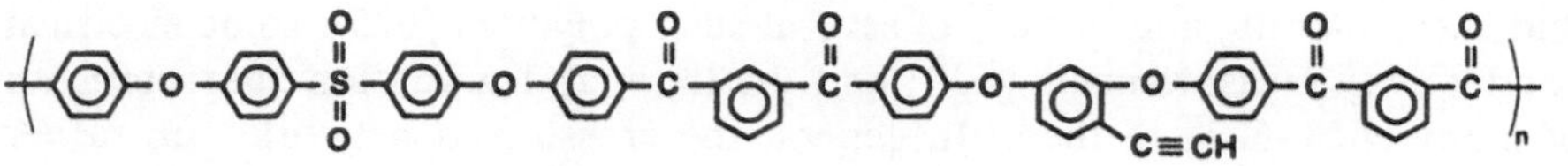

was prepared in multiple steps from the appropriate monomer mixture. The polymerization step used Friedel-Crafts reaction with $AlCl_3$ as catalyst. Then the polymeric product containing acetyl groups was converted into one containing β-chloro unsaturated aldehydes by treatment with $POCl_3$ in DMF, and into acetylene-carrying polymer by treating the last product with KOH in DMF–ethanol mixture [603]. The soluble polymers were subsequently cross-linked to varying degrees by exposing them to high temperature in the presence of several catalysts. Except for thermogravimetric studies, no attempt was made to further characterize the crosslinked networks.

3.2.4 Potential Two-Step Polyarylene Networks

Building on the previous work of Wegner and associates on soluble poly (p-phenylenes) [241, 242, 244, 636], Scherf and Müllen [637–639] described two two-step routes to soluble ladder polymers. One route involves the preparation of an aldehyde-bearing pre-polymer:

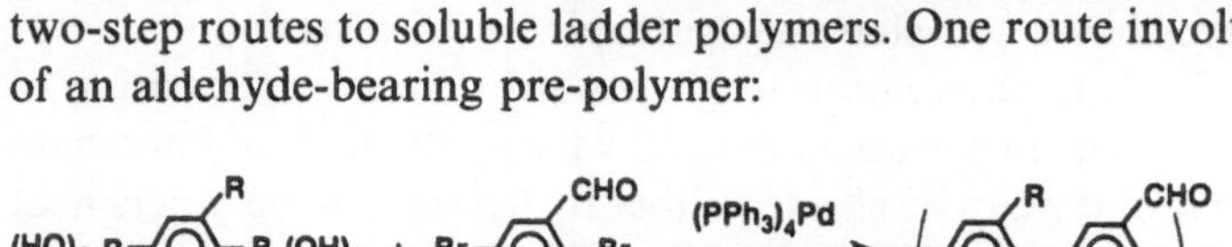

This soluble polymer is obtained with molecular weights in the range of $M_n \cong 5000$, corresponding to a chain length of about 25 rings [639]. When the reaction conditions are carefully chosen, no side reactions take place. Instead of continuing with the procedure of Scherf and Müllen for ladder polymers, one may use poly(p-arylene dialdehydes) to form two-step aromatic networks. A straightforward procedure may require the dissolution of the linear precursor polymer in solvents such as methylene chloride or toluene, followed by reaction with aromatic diamines to form interchain aromatic crosslinks containing two imine ($-CH = N-$) residues each. It appears that this system may lend itself very well to the preparation of rigid aromatic gels of essentially polyphenylene networks. The degree of crosslinking may easily be controlled by the replacement of the dialdehyde monomer in the linear precursor by other p-dibromoaryl monomers carrying inert solubilizing moieties such as alkyl groups.

3.2.5 Supernetworks

By the term supernetworks we mean networks constructed from large-molecule precursor units held together in the network in a manner such that the character of the individual precursor units is maintained to a significant degree. Among these large-molecule precursor units we enumerate polyamide FPs similar to the structures described in Sect. 3.1.2, polyester FPs similar to the structures described in Sect. 3.1.3, and hyperbranched FPs similar to the polyphenylenes described in Sect. 3.1.1 and to the intriguing structures of Sect. 3.1.4.5. We also believe that using the latest synthetic advances, C_{60} units commonly called "fullerenes" could be built into covalently bonded permanent networks and gels.

Until now, only one family of supernetworks had been prepared, namely a rigid polyamide network consisting of highly branched rigid aromatic fractal polyamides held together by stiff polyamide struts [350]. The synthetic strategy adopted is as follows: unlike the polyamide networks described in Sect. 3.1.2, the FPs prepared for the supernetworks were made exclusively from AB_n aromatic amino acid monomers. Here the A may be a carboxy acid and B an aromatic amine, or vice-versa, and n may be 1 or more such that the functionality of the branchpoint is determined by $f = n + 1$. To control the average distance between branchpoints, monomer mixtures varying in composition of AB_1 and AB_n (n > 1) may be used. In our case, only n = 2 was used thus far. By the use of only AB_n or a mixture of $AB + AB_n$, we insure that the exterior parts of the growing FPs will always be decorated with a large majority of B units, making it essentially impossible for these FPs to finally coalesce together and form covalently bonded permanent networks. This is unlike the stoichiometric mixtures of AA and BB monomers used to create the previously described polyamide networks, specifically designed to form FPs whose exteriors are decorated by about equal numbers of reactive A and B groups, maximizing the chances for inter-FP covalent bond formation and the creation of permanent "infinite" networks. To better control the molecular weight distribution of the

FPs prepared for the supernetworks, nuclei were used such as

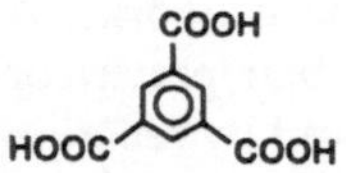

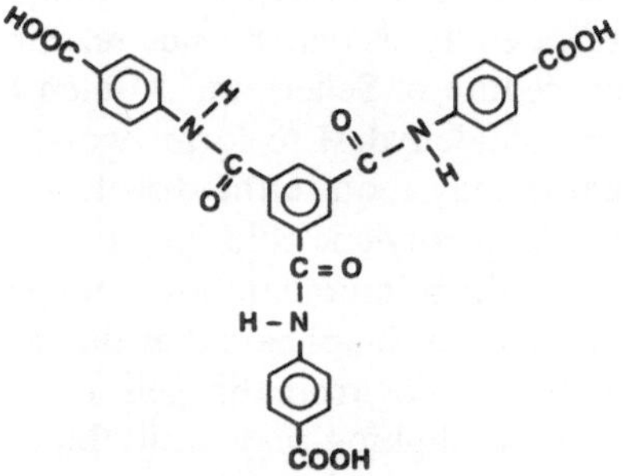

for FPs decorated on their exteriors by carboxy groups, and

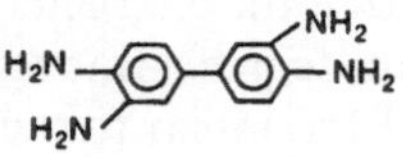

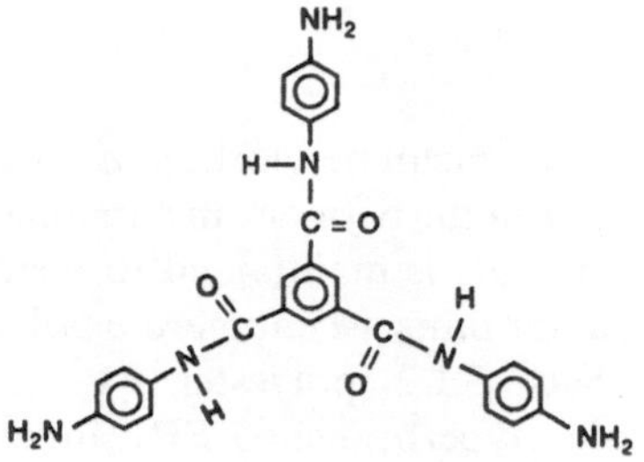

for FPs decorated by amine groups. For both kinds of FPs, the AB monomer was 4-aminobenzoic acid. For the carboxy-decorated FPs, the AB_n monomer was 5-aminoisophthalic acid and for the amine-decorated FPs the AB_n monomer was 3,5-diaminobenzoic acid. The polycondensation was the Yamazaki procedure, carried at about 115 °C. As expected, the polymeric products showed no tendency to form networks and gel. After workup they were analyzed and were found to have exclusively the desired exterior reactive group, very high M_w by LALS and very low intrinsic viscosity reflecting the highly branched nature of the FPs. The average number of reactive groups per macromolecule was estimated from the ratio of nuclei to monomers in the system, with ratios leading to 48 and 64 reactive groups per macromolecule most often used. A two-dimensional schematic description of a 48-amine-terminated FP is shown in Fig. 15.

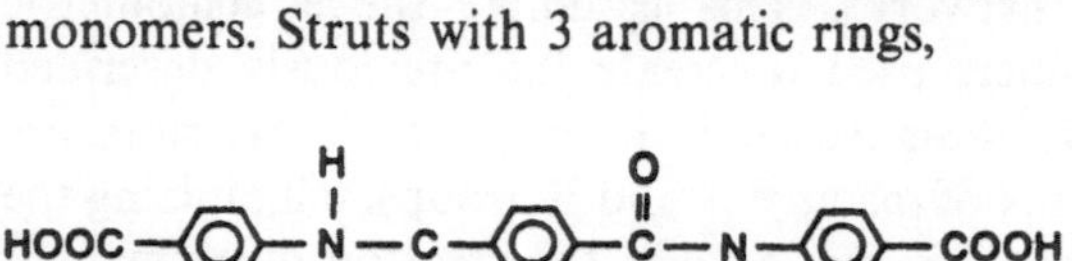

The rigid struts between the FPs in the supernetworks were prepared by either Yamazaki or Schotten–Baumann-type reactions from the appropriate monomers. Struts with 3 aromatic rings, were prepared by Schotten–Baumann-type reaction from 1:2 molar terephthaloyl chloride and 4-aminobenzoic acid. Here, all struts were of identical length. Struts with average length of 7 rings were prepared by the Yamazaki procedure from 3:2 molar mixture of nitroterephthalic acid and DABA:

$$
\text{HOOC}-\bigcirc-\overset{\text{O}}{\underset{}{\text{C}}}-\text{N}-\bigcirc-\overset{\text{O}}{\underset{}{\text{C}}}-\text{N}-\bigcirc-\text{N}-\overset{\text{NO}_2}{\underset{\text{O}}{\text{C}}}-\bigcirc-\overset{\text{O}}{\underset{}{\text{C}}}-\text{N}-\bigcirc-\text{N}-\overset{\text{NO}_2}{\underset{\text{O}}{\text{C}}}-\bigcirc-\text{COOH}
$$

Struts of both lengths were separately used with several amine-decorated FPs to create rigid aromatic polyamide networks, such as shown schematically in Fig. 16, Sect. 3.1.2. It is important to emphasize here that, because of the highly ramified exterior of the precursor FPs, it is practically impossible to insure that all the rigid struts in the supernetworks will be reacted at both ends. In fact, we expect that a not insignificant fraction of these species will end up reacted at only one end and will not contribute proportionately to the load-bearing and other mechanical properties of the supernetworks and their gels.

It is important to recognize here that permanent polyamide rigid networks can be prepared from mixtures of exclusively amine- and exclusively carboxy-terminated FPs. In fact, we have prepared one just like this from 48-amine- and 64-carboxy-decorated FPs present in equal weights in a $C_o = 10.0\%$ concentration in the reaction mixture. The reaction took longer than usual to reach the gel point and the gel product was rigid and brittle as expected, but extremely weak. The point is, however, that once the network formed, all components were identical and there was no possibility of differentiation as the one existing between the FPs and struts in the polyamide supernetworks. Therefore, unless there exists some unique differences between macromolecular-size components in a network, it should not be called a supernetwork.

The methodology we applied to create the polyamide supernetworks can also be used to make polyester and polyphenylene supernetworks. Supernetworks made from mixtures of, say, polyamide and polyester FPs can be made with remarkable ease.

Buckminsterfullerenes are spherical aromatic molecules, mostly C_{60}, which have recently caught the imagination of the scientific community [640–645]. The very recent reports on functionalization [646–658] and crosslinking [659] of fullerenes suggest, we believe, that supernetworks containing fullerenes are within reach. These may be achieved by the use of difunctional groups to functionalize fullerenes, instead of the monofunctional species used hitherto. If, in addition to being difunctional, these species will be stiff and rodlike and all of identical length, then an idealized supernetwork is conceivable with fixed distances between the fullerene precursors. A representation of such a supernetwork with face centered cubic (FCC) motif is shown in Fig. 18. Many other ordered and disordered supernetworks containing fullerenes are conceivable. Other spherical and tube-like macromolecular species, organic and inorganic [660–660c], may also be inserted into supernetworks in the future.

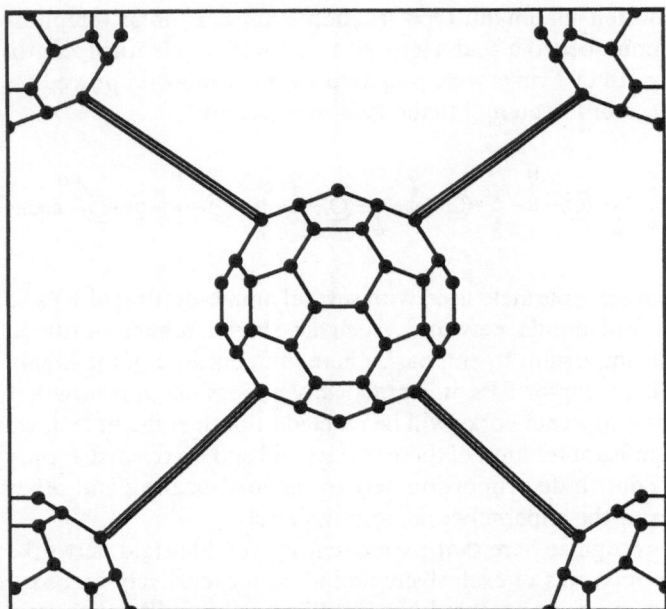

Fig. 18. Conceivable supernetwork created from fullerenes connected by equal-length rigid struts. A face centered cubic motif was selected for this picture

4 Features

4.1 Pre-Gel State

4.1.1. Fractal Nature of Growing One-Step Species and Comparison with Two-Step Species

Here, we deal mostly with stiff, aromatic chains or segments, leading to rigid networks. In the Introduction we defined one-step polycondensation as starting with a solution of the appropriate mixture of difunctional and higher functionality monomers, defining both the branchpoint functionality and the average length of the stiff segments between these branchpoints. The segment length may be defined in terms of its axial ratio, $\bar{x}$. Because the chain thickness of aromatic polyamides and polyesters is very close to the monomeric repeat distance, this definition is especially convenient because each additional aromatic residue along the chain increases $\bar{x}$ by about 1. Two-step network formation requires the preparation in a first step of long stiff chains decorated with a multiplicity of reactive sites, and in a second separate step reacting the long stiff chains with rigid difunctional bridging units that when reacted at both ends serve as rigid struts between the precursor long chains. Because the struts and the chains are chemically identical, it is impossible to identify which network segments belonged to the precursor long chains and which to the struts. Because of the initial length of the precursor chains, networks prepared from them are expected to behave differently from networks prepared from the monomers by a one-step procedure. We shall now concentrate on one-step systems, using aromatic polyamides with $f = 3$ as examples. We first discuss the nature of and problems with the growth process of the pre-gel species in solution. Since we have demonstrated by SAXS that these species behave as surface fractals [311], we feel justified in calling them fractal polymers (FPs).

A comparison between FPs and dendrimers [331–349] is most instructive. Dendrimers are prepared in a multi-step sequence or generations. In each generation a large molar excess of monomers are used which are blocked from reacting with one another or with same-generation monomers already attached to the growing dendrimer. This encourages all the de-blocked reactive sites of the penultimate generation of the dendrimer to react with the oncoming monomers and leads to rather densely packed polymeric species. In the case of the highly branched fractal polymers, the situation is vastly different. Here, the monomers in the reaction bath are not blocked and can react with one another and with monomers already attached to the growing species. The dimers and other small growing species react with one another, with monomers and with the larger species, hindered only by diffusion and geometric constraints. Unlike the case of dendrimers, the ability to react at random with species of any size in the reaction medium results in highly branched structures which are much more open and which retain many more unreacted reactive sites than dendrimers of the same composition and molecular weight. Figures 19 and 20 show, respectively, a dendrimer and a fractal polyamide of the same chemical composition and comparable molecular weight. Both structures were computer-generated by Dr W.B. Hammond [661] using the same, energy minimized, molecular modeling program. The high porosity and large radius of gyration of the FP are clearly evident when compared with the relatively compact appearance and small R_G of the dendrimer.

To appreciate the growth characteristics of highly branched rigid aromatic FPs obtained by one-step polycondensation, we have to understand the nature of the structural elements involved. When linear polymers are prepared by polycondensation from difunctional monomers AA and BB, aromatic diacid and aromatic diamine for instance, the number average degree of polymerization, $\overline{DP}$, is related to the fraction, p, of consumed monomer AA (or BB) by [39]

$$\overline{DP} = 1/(1 - p). \tag{20}$$

This means that when half the monomers in the reaction bath are consumed, the growing species are on average only dimers. When 75% of the monomers are consumed, $\overline{DP} = 4$ and when $\overline{DP} = 6$, 83.3% of the monomers are gone. In

Dendrimer
Fifth Generation

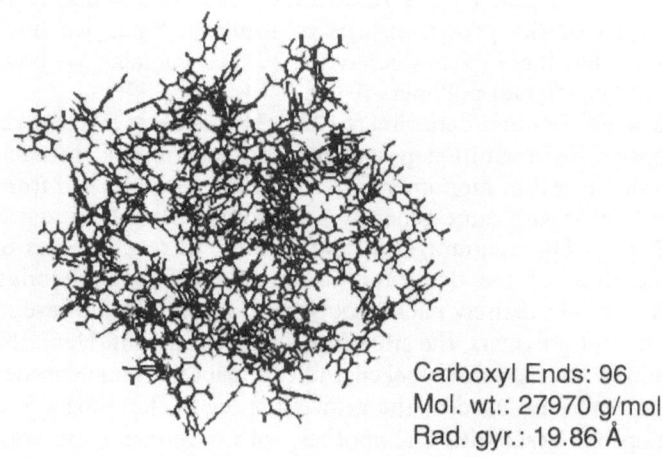

Carboxyl Ends: 96
Mol. wt.: 27970 g/mol
Rad. gyr.: 19.86 Å

Fig. 19. Appearance of 5-generations dendrimer obtained from the monomer 5-aminoisophthalic acid by computer simulation. In each generation monomers were allowed to react only with the previous generation of the growing dendrimer. The polymeric material occupies about 75% of the dendrimer's volume of gyration

ABₙ Fractal
Seed +3,5 - Diaminobenzoic Acid

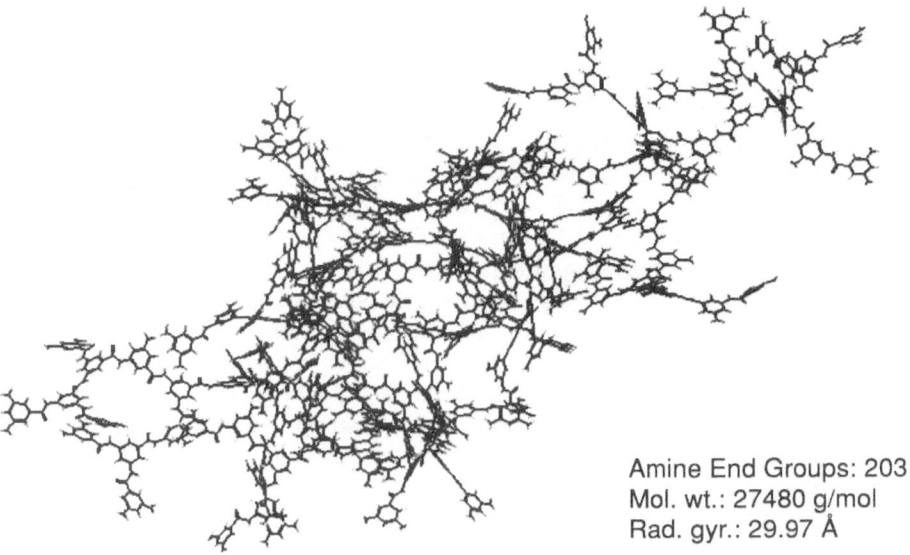

Amine End Groups: 203
Mol. wt.: 27480 g/mol
Rad. gyr.: 29.97 Å

Fig. 20. Appearance of a fractal polyamide obtained by computer simulation from the monomer 3,5-diaminobenzoic acid. Monomers were allowed to react with all reactive species in the reaction mixture. Polymeric material occupies about 24% of the FP volume of gyration. Molecular weight comparable with that of the dendrimer in Fig. 19

general, for linear polymers obtained by polycondensation, for $\overline{DP}$ = x,

$$p = (x - 1)/x \tag{21}$$

and the amount of unreacted monomer left in the reaction bath is

$$1 - p = 1/x. \tag{22}$$

In branched polymers the results are rather close [39]. For instance, for a linear polymer of $x = 6$, $p = 0.833$. For a polymer of $x = 6$ but containing one branchpoint with $f = 3$, $p = 0.80$.

The number of reactive ends, Z_m, of the unreacted difunctional monomers left in the reaction bath is

$$Z_m = 2(1 - p). \tag{23}$$

Combining Eqs. (21) and (23) we obtain

$$Z_m = 2[1 - (x - 1)/x] = 2 - 2(x - 1)/x. \tag{24}$$

The number of reactive ends of the growing linear species, Z_f, depends of course on the number of such species which, in turn, depends on their size distribution. When averaged over all growing species, it equals twice the fraction of consumed monomer divided by the average degree of polymerization, x:

$$Z_f = 2p/x. \tag{25}$$

A combination of Eqs. (21) and (25) yields

$$Z_f = 2(x - 1)/x^2. \tag{26}$$

Comparison of Eqs. (24) and (26) reveals that the number fraction of reactive ends belonging to the unreacted monomers is not dramatically different from the number fraction of reactive ends belonging to the growing linear species. For instance, when the average DP reached $x = 3$, $Z_m = 0.666$ and $Z_f = 0.444$. When $x = 5$, $Z_m = 0.4$ and $Z_f = 0.32$, for $x = 10$, $Z_m = 0.2$ and $Z_f = 0.18$, and so on. During the polycondensation, the number of reactive ends belonging to monomers is being depleted at a rate comparable with the depletion of the reactive ends of the growing species. In the case of branched systems undergoing polycondensation reactions, the mathematics is more complicated but the results are of similar magnitudes. To make the following more concise, we define network fragments as all non-monomeric small structural components that float in solution not yet attached to any of the large growing fractal polymers or to the network. (In subsequent discussions network fragments will also include small parts of the FPs or the whole network.) The above discussion leads us to an important conclusion: *in the case of pre-gel branched particles growing in solution by polycondensation, we have monomers and small network fragments competing with one another on reactable sites on the exterior and interior of the larger FP particles. The number of reactive ends belonging to monomers is of the same order of magnitude as the number belonging to small network fragments not connected to the larger FPs. Once an "infinite" network is formed, a similar competition on reactive sites continues throughout its volume between monomers and small network fragments.* The structure of the growing FPs, and final network, is very porous, posing no or almost no hindrance to the approach of monomers and small network fragments to reactive sites in the interior of the FPs, especially in the pre-gel state. This can be easily seen from Figs. 21 and 22, depicting two stages in the growth of the same fractal polymer. Its monomer composition is 1:1 p-aminobenzoic acid and 3,5-diaminobenzoic acid, resulting on average in branchpoints every second aromatic ring. The respective M_n of both FPs are 6830 and 26008 and their radii of gyration are 23.11 Å and 32.09 Å, respectively. Computer calculations revealed that the van der Waals volume actually occupied by the smaller FP is only 0.13 of the volume corresponding to its R_G, and in the case of the larger FP the volume fraction occupied by the polymer is only 0.18 of the volume of gyration. The relatively small volume fraction taken by the polymer mass is a clear indication of the porosity of the growing FPs and final network. The porosity can be increased or decreased at will by respective increases or decreases in the ratio of difunctional to trifunctional monomers in the polycondensation reaction mixture.

AB$_n$ Fractals

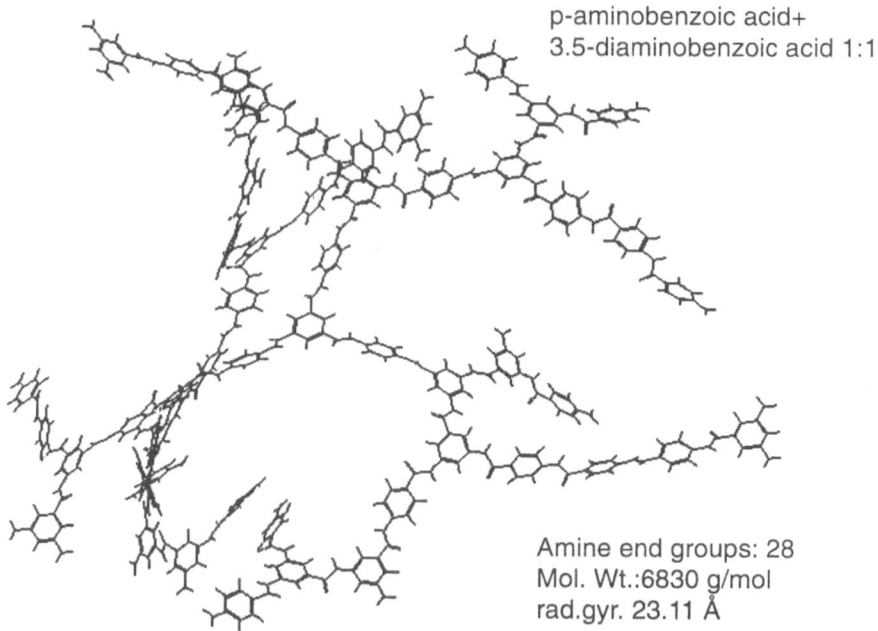

p-aminobenzoic acid+
3.5-diaminobenzoic acid 1:1

Amine end groups: 28
Mol. Wt.:6830 g/mol
rad.gyr. 23.11 Å

Fig. 21. Small AB$_n$ fractal from 1:1 *p*-aminobenzoic acid and 3,5-diaminobenzoic acid. 13% occupied volume

Therefore, when we devise a theory to explain the growth and nature of highly branched rigid aromatic pre-gel and post-gel species, we must take into consideration the following two important facts: (a) in the reaction bath the number of reactive ends belonging to monomers is relatively close to the number belonging to larger network fragments, such as FPs; and (b) the inflexibility of all constituents in the system limits the accessibility of reaction sites on the growing fractal polymers and network, and the efficiency of the condensation reactions.

The situation with flexible FPs and networks is expected to be more complex than with their rigid analogues. The segmental stiffness and branchpoint or junction rigidity in the growing rigid FPs either completely eliminates or minimizes intramolecular cyclization. This is not the case with the flexible species where entropy greatly increases the probability of intramolecular cyclization. Furthermore, unlike the rigid FPs, flexible species are rather sensitive to solution temperature and solvent quality effects, resulting in substantial swelling or collapse of the growing particles with modest changes in temperature or solvent quality. During the collapse or swelling, the concentration of accessible reactive sites in the interior and exterior of the growing flexible FPs is expected

AB$_n$ Fractals

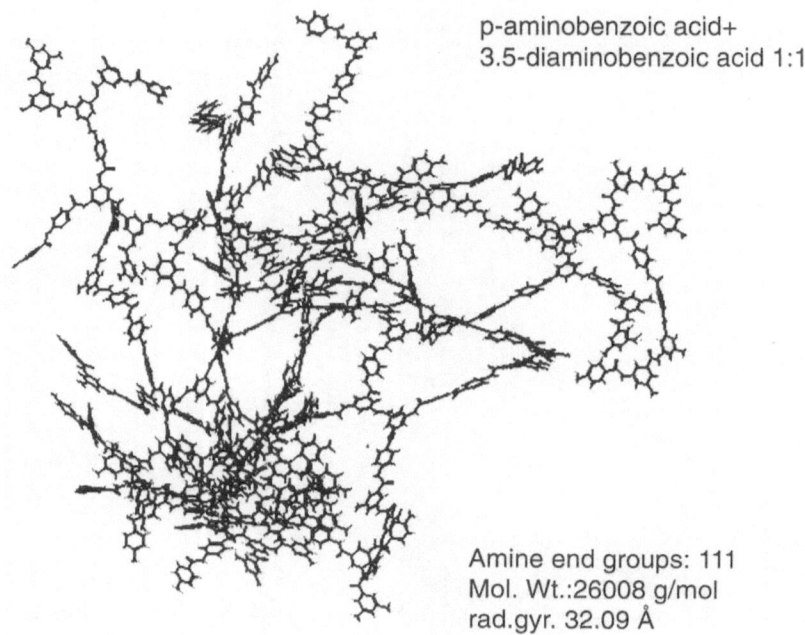

p-aminobenzoic acid+
3.5-diaminobenzoic acid 1:1

Amine end groups: 111
Mol. Wt.:26008 g/mol
rad.gyr. 32.09 Å

Fig. 22. Larger AB$_n$ fractal of same composition as in Fig. 21. 18% occupied volume

to change. This is expected to affect the level and rate of FP growth and, through it, the gelation process and its kinetics. In contradistinction, our experience with rigid polyamide FPs and networks instructs us that the rigid species are rather insensitive to modest changes in solvent quality and temperature, producing at different reaction temperatures and in modestly varying solvent quality rigid networks with essentially the same modulus. As they fall out of the scope of this work, entropically flexible networks will not be discussed further.

Levels of segmental stiffness in networks may be estimated from a measure of the persistence length, a, of the analogous isolated linear chain. It is well known that many para-substituted aromatic polyesters and polyamides exhibit thermotropic and/or lyotropic liquid crystallinity. In fact this appears to be their most useful property. A review of literature data [166, 168, 170, 175] clearly demonstrates the very large sizes of a or A associated with polymer liquid crystallinity, as compared with the small corresponding lengths found in non-mesogenic flexible polymers. Concentrating on aromatic polyamides, we find that, for monodisperse or highly fractionated samples, the persistence lengths are of the order of 150 to 300 Å [213]. With repeat units being about 6.45 Å in length and 6 Å in thickness, this translates to an axial ratio x̄ of about 25 to 45. The effects of the onset of chain curvature on dilute solution behavior of aromatic polyamides were felt at $\bar{x}$ values as low as $10 < \bar{x} < 13$ [56, 662]. The persistence lengths of common flexible chains are an order of magnitude smaller than those of the aromatic polyamides and polyesters. They usually fall in the range of 10 to 30Å [166, 168, 170, 175]. Chain curvature is observed already over the length of two repeat units which may be as short as 5 Å. While the high flexibility of common polymers derives from their negligible enthalpic rigidity and from large reduction in their entropic rigidity upon dissolution in good solvent or temperature increase, the stiff polymers owe this characteristic to their high enthalpic rigidity. In molecular simulation studies [663], it was found that segments of aromatic polyamides are not rodlike but exhibit a limited measure of flexibility. The segmental deformation was found to be more complex than initially expected because deformation of part of the structure became possible above a certain thermal threshold while other features of the structure remained fixed at all temperatures used in our study. The commonest mode of deformation was one in which torsional angles between the aromatic rings and the amine groups departed in the same direction and at the same time from the minima of their respective energy wells. For each repeat unit the movement is similar to thermal oscillation, but when several are operating in the same direction at the same time, their departure from the energy minima results in a combined and complex twist-bend motion. In addition to this twist-bend deformation, two additional modes of segment deformation were observed with decreasing frequency. The more frequent one was a 60 degree aromatic ring flip from $\pm 30°$ relative to the plane of the amide groups, to $\mp 30°$. This creates a bend in the segment and relaxes much of the stresses along it, allowing many of the other torsional angles to drop back to their minima. A less frequent mode of deformation was observed on occasion involving the creation of bends of about 20° in the segment, caused by amide group flipping by 180°. This changes the placement of two adjacent amide groups across an aromatic ring from anti- to syn- or from syn- to anti-. Such a bending mechanism was proposed previously for aromatic polyamide segments in rigid networks and gels [48, 412, 664]. While the first mode of deformation may reflect the thermodynamic flexibility of aromatic polyamide chains, the last two modes of deformation are the reflections of the thermodynamic rigidity of the same stiff chains. All three modes result not in a pure bending motion but in various levels of twist-bend motions. Their relative weights appear to depend on the heights of the energy barriers to torsional rotations, the higher the barriers the lower the weight of the concerted deviation from energy minima. It is important to recognize here that the above three modes of deformation may not be the only ones present in aromatic polyamide chains, and may be supplanted altogether in other kinds of stiff chains.

An important outcome of the above was the realization [664] that, because there is almost no energetic difference between syn- and anti-placements [412, 665] in stiff aromatic polyamides, there is practically no energetic advantage to either placement. Therefore, during the one-step polycondensation, monomers add to stiff segments and these add to network fragments, FPs and finally the infinite network, in either the anti- or the syn-configurations, resulting in the network segments being straight or containing one or more 20° kinks. In the growing and final network we have a system, then, where stiff segments may be straight or contain bends due to anti-syn placements. Here we broaden the above to state that the segments may show limited deformation capability due to concerted kinetic oscillations of torsional angles and/or ring flips and amide-group interconversions. Despite this, their stiffness far outweighs their modest flexibility. It is important to recognize that stiff polyamide segments connected to the network at only one end were found to deform more easily by means of ring flips and amide-group interconversions than their analogues of the same length firmly attached to the network at both ends.

The discussion above leads to an important conclusion, that during the creation by one-step polycondensation of rigid networks comprising stiff segments connected by rigid branchpoints, the stiffness of the segments and branchpoints affect the network growth process and results. This is dramatically different from the comparable features of flexible networks. When flexible networks grow, the inherent kinetic flexibility of the monomers, network fragments, segments and crosslink junctions all allow rapid thermal oscillations, changes in position and trajectory, and for segments and junctions to move out of the way and allow monomers and network fragments to penetrate deep into fractal polymers or post-gel networks. Segments, network fragments and FPs twist and bend and stretch, and junction points change distances as a conse-quence of both thermal oscillations and applied stress. This results in the process of flexible network formation by polycondensation being indifferent to the flexibility characteristics of the constituent monomers and network fragments, and being controlled by the nature of the chemical reactions and the statistics of monomer composition and functionality [39]. The essential independence of network formation from the flexible chain degree of flexibility led to the "principle of equal reactivity" according to which the reactivity of a monomeric reactive group and a reactive group at the end of a growing chain are the same [39]. As will be shown below, this does not hold in cases where rigid networks are prepared by one-step polycondensation.

The effects of concentration and trajectory on the polymerization of rigid rodlike species were recently investigated experimentally [666] as well as theoretically [667]. The problem of trajectory is also mentioned in an excellent recent paper by Warner [668] dealing theoretically with the growth of rigid fractal polymers. In this paper, the terms one-step and two-step growth models are used but, unlike our definition, they mean the growth of FPs from monomer mix and from small network fragments, respectively. These are the two extreme cases for what we describe as the one-step growth by polycondensation, where monomers and network fragments compete with each other. Even though they deal with only parts of a more complex reality, the models are revealing and lead to important conclusions. Arguing that steric frustration occludes the growth of rigid segments, Warner concluded that the volume fraction of segments in the

FP does not grow with distance from its nucleus, "but attains a value where rod attachment is balanced by frustration allowing steady (constant density) outward growth" [668]. This is consistent with our experimental [311] and molecular modeling [661] results, where we found that aromatic polyamide FPs behave as surface fractals and their internal segmental density was coarsely uniform and far lower than in the case of dendrimers.

According to Warner [668], the species being attached to the growing cluster, or FP, are either monomeric or of segmental length and shape. More complex species, such as network fragments, are not considered. As was described above, a typical segment length in our polyamide networks comprised only about six monomers. This means that small branched species appear in the reaction bath rather early and at that time the number of reactive groups belonging to the growing fragments is not very different from those belonging to the depleting monomers. Of course, the longer the average stiff segment between branchpoints, the lower the probability of finding branched species in the pre-gel and post-gel systems. The growth of rigid fractal polymers by a one-step polycondensation in solution is perceived by us as follows: nucleation of growth occurs at random as long as there are monomers available, or may be initiated by seeding the reaction bath first with multifunctional species followed by the addition of the rest of the monomers. The length of an average segment is dictated by the ratio of difunctional to higher functionality monomers, with the latter serving as branchpoints. In the initial step of the condensation, the number of monomers rapidly drops and when the average oligomeric species is, say, only six monomers long, 80% of the monomers are already consumed. At that early stage some of the growing species are branched while others are still linear. Already at this stage some of the growing species meet in solution and coalesce to form larger units, network fragments. The bonds between the coalescing entities are covalent and may occur directly between two such units or with the consumption of one or more monomers which are invested in the creation of a new segment. In a typical polyamide network with $x \cong 6$ and rigid trifunctional symmetrical branchpoints, a network mesh size averaging about 50 Å is gradually being created. We have demonstrated by solution properties studies [311] that FPs having such characteristics behave as if they are almost fully draining, that is, they are almost transparent to the passage of solvent-size molecules through them. Because the size of monomer molecule is about twice that of a solvent molecule, the FPs are expected to allow the flow of monomeric species through them with almost the same ease. This means that monomers can easily penetrate to the interior of any network fragment or growing FP and condense with available reaction sites anywhere in the reaction bath with almost the same ease. For networks with very short segments and very small mesh size, the ease of translation of monomeric species may increasingly be hampered. Note that we have no problem of the diffusion limited aggregation type because during condensation polymerizations under usual conditions the number of available monomers and unattached small oligomers swamps the reactive sites on the growing fractal polymers.

The situation is different for network fragments. Because of their own and that of the growing FPs rigidity, the larger the network fragments the smaller the probability that a reactive site on them can penetrate deep into the growing FP, reach to and react with a reactive site in the FP. Furthermore, the chances for such a reaction vary with the location of the reactive group in the network fragment. If a reactive group is at the tip of a long segment of the network fragment, it has a better chance of reaching deep into the FP and reacting there than if the reactive group is at an underutilized branchpoint between two long segments. The difference is schematically shown in Fig. 23. Thus, in the case of one-step rigid network creation by polycondensation, the principle of equal reactivity fails. In this case, the monomers are more reactable than reactive groups on network fragments, and from among these, those on segment tips and those belonging to smaller fragments have access to more reactive sites than those on large fragments or on fractal exteriors, especially when they are at branchpoints and not on segment tips.

In addition to the problem of obstructed access, we also have to consider the unique problems associated with trajectories of approaching rigid species. When the approaching species is a monomer or even a dimer, it can easily change trajectories, even when present in solution or a gelled network. After some Brownian motion, it will find itself in the correct approach for the chemical condensation reaction to take place. In the case of stiff segments and network fragments, the situation is more complicated. To approach a reactive site deep in a growing FP or network, the stiff segment or network fragment must align itself such that it may move toward this site along a very specific trajectory. Because of the rigid network mesh structure, the long stiff segment or network fragment may be prevented from significant lateral movement and may not be able to use

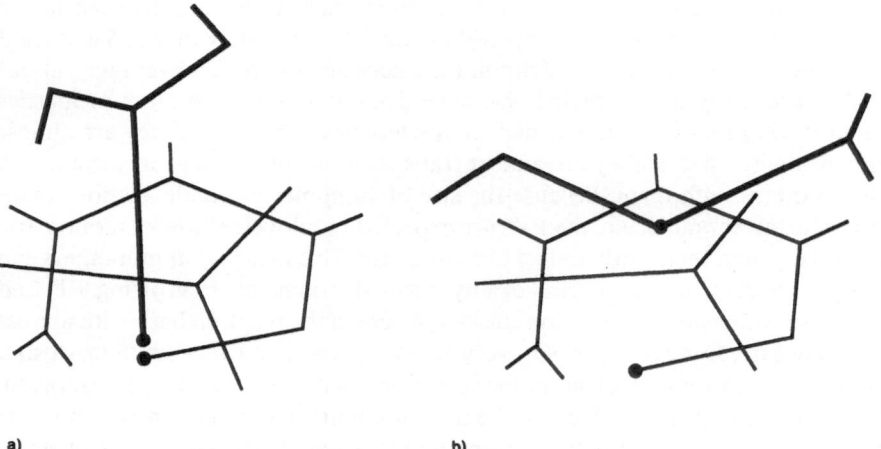

a)   b)

Fig. 23a. Accessibility of network reactive site to reactive group on segment tip. **b** Inaccessibility of the same network reactive site to reactive group in underutilized branchpoint

its Brownian motion to reach the correct trajectory. Furthermore, following the arguments of Agarwal and Khakhar [667], the ability of the long segments and especially network fragments to undergo rotational diffusion is substantially reduced due to their own size and to the interference by stiff components of the rigid FP or network. We are thus confronted with a situation in which the ability of long stiff segments and especially network fragments to undergo both rotational and translational diffusion, and reach in the correct trajectory reactive sites not right on the exterior of the growing FPs or network, is much smaller than that of stiff monomers or that of comparable flexible entities in flexible networks.

In the light of the above, it is most likely that stiff aromatic polyamide, and probably polyester, fractal polymers and their network progeny grow by a relatively rapid aggregation of network fragments of various sizes, some segments and some difunctional and higher functionality monomers. At that point the average segment length in the participating network fragment was already reached, since it is dictated by the ratio of difunctional to higher functionality monomers. The mesh size in the growing FPs and subsequent network is determined from the beginning by the length of the average stiff segment. Concomitantly with the growth of full size FPs, network fragments and some individual segments continue to grow and a small number of new growth nuclei may appear among the rapidly depleting monomers. Because of the reduced ability of the larger network fragments to penetrate significantly into the growing FPs, most of the growth of the FPs, especially when they approach gelation, is directed outwards. A small amount of inwardly directed growth is possible, with the building material being monomers, small segments and fragments, limited in size to be smaller than the network mesh such that they may diffuse into the growing FPs.

On occasion two FPs or an FP and a network fragment may grow together by starting from two close nuclei and getting intertwined rather early in the growth process. This can easily happen if one FP is displaced by, say, a quarter or a half the average mesh size. With time, the growth of one FP may be sufficiently frustrated by segments from the other FP, that it will stop growing. The other FP will continue to grow and eventually coalesce with others and form an infinite network. The FP that stopped growing will remain nestled in the completed network and, because of the rigidity of the system, will not be removable even by sol-gel fractionation techniques.

In summary, during one-step polycondensation of polyamides, starting with a solution of the appropriate monomer mixture and ending with a rigid gelled network comprising stiff aromatic segments connected by rigid branchpoints, the number of reactive sites belonging to monomers rapidly decreases in the initial steps of the reaction. From then on, a competition on reactive sites exists between the depleting monomers and the evolving larger structural elements such as segments, network fragments and, finally, fractal polymers. This competition affects the nature of network growth during one-step polymerizations. The fact that both the growing network and the participating structural elements are all inflexible significantly affects the location of attachment to the network of these elements according to their size. While monomers and small species may link to the FPs and network anywhere throughout their volume, to both interior and exterior sites, larger structural elements such as large network fragments and FPs, cannot penetrate into an existing highly branched entity and may attach themselves only to its exterior. This means that when rigid networks are constructed by the one-step method of polycondensation, the "principle of equal reactivity" [39] becomes invalid, dramatically different to flexible networks made from flexible monomers. A simple schematic description of the above discussion is given in Fig. 24.

It is well-known that, when the solution concentration, c, of long stiff chains or segments beyond a critical length $\bar{x}^*$ is gradually increased, the solution passes from a random isotropic state, through a biphasic interval, to a single anisotropic phase in which the stiff chains or segments are aligned in more or less parallel arrays. We have found [61, 62] that when the chains are constructed

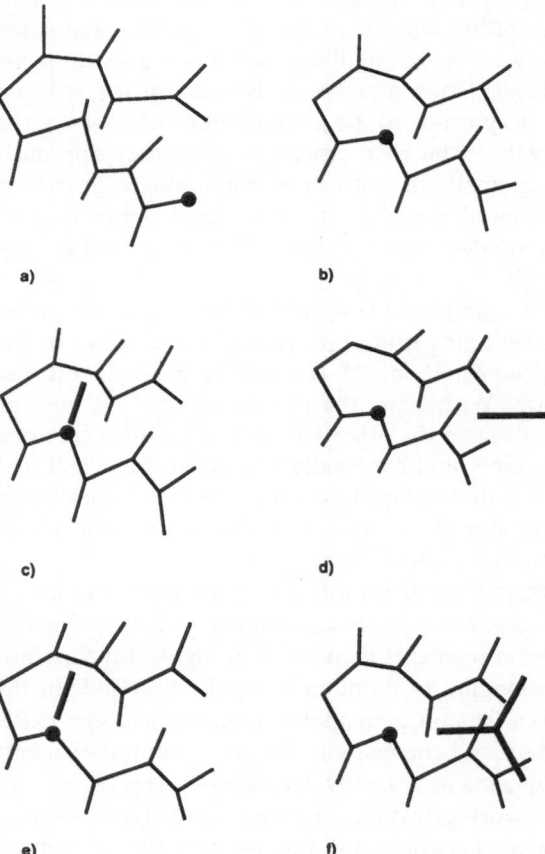

a)

b)

c)

d)

e)

f)

Fig. 24a. More accessible reactive site. **b** Less accessible reactive site. **c** Short segment reaches reactive site. **d** Long segment cannot approach accessible reactive site at the correct angle for condensation to take place. **e** Unsubstituted reactive segment may reach accessible reactive site. **f** Substituted segment may tangle in network and not reach site

such that they are forced into a zigzag shape, the transition to the liquid crystal state does not occur even when the segments between the zigzags are longer than $\bar{x}^*$. With increased concentration the system retains its isotropicity until finally it gels or vitrifies to produce an amorphous phase in which the stiff zigzag polymer chains are randomly tangled. A very poor alignment capability was noted by us in networks and gels of stiff zigzag polyamides [409]. In IR and solid-state NMR studies of dry aromatic polyamide networks it was found [412] that not only were the systems amorphous, but their ability to form intersegmental H-bonds was greatly affected by the average length of the stiff segments between branchpoints. Thus, segments with $DP = x = 6$ contained about 25% "free" amide groups, a fraction that increased to 50% when x dropped to 5. In analogous linear polyamides, 100% of the amide groups are intersegmentally H-bonded and no "free" amide is detectable. All this and previous direct observations [393] instruct us that rigid branching of stiff molecules obstructs the parallel alignment of segments and that this obstruction increases with the concentration of branchpoints. Flexible junctions between long stiff segments do allow, however, for parallel alignment to take place [402]. The poor packing ability of rigid branched segments or network fragments at high solution concentration or in the absence of solvent leads us to believe that attempts to create by polycondensation one-step networks comprising stiff segments of modest x and rigid branchpoints are bound either to yield at best extremely defective networks or to fail altogether. One should recall that under such concentration conditions, the ability of the growing network fragments to undergo translational and rotational diffusion is greatly reduced so that reactive groups on any species substantially larger than a monomer may not be able to approach one another in the correct distances and trajectories for the coupling reaction to take place. From the synthetic point of view, we also have to remember that the monomers, segments and network fragments may not be soluble at high concentrations, or miscible with each other in the bulk, and may start to phase separate out of the reaction mixture prior to the formation of a uniform and complete network. This phase separation may be manifested by either crystallization of monomers or segments, or microsyneresis of the larger network fragments or fractal polymers.

Rigid aromatic networks with very short segments suffer some additional problems. It was demonstrated by Aharoni et al. [311] that, while the critical concentration, C_0^*, for "infinite" network formation of polyamides with $l_0 = 65$ Å and $f = 3$ was only about 2%, the value of C_0^* rapidly shoots up with decreasing x and reaches $C_0^* \geqslant 30\%$ for networks made from only trifunctional monomers where $l_0 = 6.5$ Å. To create networks with higher integrity and not merely to reach the gel-point, higher concentration, C_0, is needed but is unreachable because of solubility limitations. To circumvent such experimental problems two approaches are used, often together. One is simply to run the polycondensation reactions at very high temperatures. This lowers the solvent viscosity when and if it plays a rôle in the reaction, maximizes the kinetic flexibility of the monomeric and polymeric species, increases the Brownian motions in the system, improves the solubility or miscibility of the ingredients and reaction products and, most importantly, increases the overall polycondensation reaction rate. The second approach is to introduce a measure of thermodynamic flexibility in the constituent monomers and polymeric products. This is commonly done by the insertion of methylene, ether or similar short bridging unit, allowing the attached aromatic rings at either side a measure of torsional rotation freedom. In addition to the torsional freedom of the aromatic units, the bridging groups also facilitate a substantial decoupling of the motions of one aromatic ring from the neighboring ones. The increased thermodynamic flexibility of these monomers and polymeric products tends to reduce the viscosity of the reaction mixture and improve the probability for the condensation reaction to take place. It also tends to reduce the T_g of the products, a T_g that is being elevated by the increased crosslink density in the very same products. The conclusion is that the creation at very high concentrations, or in the bulk of low-defect aromatic rigid networks having short stiff segments, is very hard to achieve by polycondensation using segments devoid of at least some thermodynamic flexibility.

We have previously demonstrated [311, 312, 669] that in one-step polycondensations the pre-gel growing polymeric species are fractal in nature. We have shown that the size of these particles right before the gel point was of the order of 300 Å in diameter and their molecular weight exceeded 700 000 [311]. Even smaller FPs were capable of forming networks by undergoing polycondensation reactions with each other in the absence of any monomers, oligomers or small network fragments [312, 669]. For networks to form from such fractal polymers, especially networks and gels with substantial mechanical moduli, each FP must be connected to each of its FP neighbors by at least one and preferably multiple covalent bonds. In the case of two-step network formation from preformed long stiff chains, the situation is strikingly different [277, 619]. In this case, stiff chains of lengths in the interval of 1000 Å to over 1500 Å are present in solution. In the case of aromatic polyamides, they are worm-like in shape. A network is formed when each chain is connected to at least two others, but one crosslink only is sufficient to connect each pair. The crosslinks can be effected at any location along the chain, and because there are so few needed and the chains are long, the average distance between two crosslinks along each chain may be measured in hundreds of Angstroms, at least at the gel point. We have found that in two-step polycondensations leading to rigid networks under ideal conditions and correct stoichiometry, the crosslinking step is remarkably fast. It may be that during this short duration the branching and crosslinking species have fractal characteristics, but at present no method is available for their detection. After an "infinite" network was created, filling the whole reaction medium, additional long chains can be attached anywhere by rigid struts. In fact, based on the observation that the struts can form almost stoichiometrically and their number depends on the ratio of reactive sites along the stiff chains to difunctional rigid crosslinking molecules [277], we believe that the number of such additional chains may be relatively high. As can be seen from the schematic description in Fig. 25, the struts may sometimes act as crosslinking agents by being attached to two separate chains and sometimes remain attached to only one chain as barbs on a barbwire. We must caution, however, that in light of the demonstrated limited flexibility [277, 619] of wormlike chains, at least aromatic polyamide ones, it is most likely that even after the network has formed, struts continue to evolve connecting stiff chains in the network. The maturing network may thus develop local in-homogeneities in interchain strut concentration, but we do not expect to find fractal characteristics in mature rigid networks and gels comprising stiff pre-cursor chains and chemically identical crosslinking rigid struts.

4.1.2 C_0^*, the Critical Concentration for Gelation

Stiff aromatic polymer chains are generally described as worm-like and charac-terized by large persistence lengths. Aromatic polyamide chains in which all rings are para-substituted have persistence lengths of the order of 200 to 300 Å [56, 213]. Such lengths correspond to less than 50 repeat units. Representing each of them by a virtual bond traversing the aromatic ring from one amide backbone atom to another, and by another, the actual central amide bond, the persistence lengths involve a total of less than 100 bonds. In aromatic polyamide networks, where the distance between branchpoints is measured in tens of Angstroms instead of hundreds, the number of bonds between branchpoints may be as low as two and as high as twenty or so. Other aromatic polymers, even flexible ones containing ether or methylene bridgepoints, are also typified by a very small number of virtual and actual bonds per unit length of linear chain or between branchpoints in a network. Because of the very small number of bonds in stiff, and even in semi-flexible, aromatic network segments, such segments cannot be treated by the statistical methods usually reserved for Gaussian chains or segments containing a large number of bonds between crosslink junctions [400].

In addition to the small number of repeat units usually found in aromatic polymer networks such as the polyamides, the relatively high energy barriers to torsional rotations around the actual and virtual bonds render the segments rather stiff. This does not mean that all segments are fully extended, but that even when bent the bend angle and the number of bends per segment are relatively small [48, 400, 664]. This means that upon swelling in a good solvent the stiff segments cannot extend much and upon shrinking their end-to-end distance cannot decrease much.

The stiffness and relative shortness of most aromatic polymer segments combine with the rigid branchpoints in the resulting networks to prevent them from significantly swelling or collapsing upon equilibration in good or poor solvent, respectively [48, 664]. When the branchpoints are not fully rigid, these restrictions are somewhat relaxed with slightly larger changes in volume upon swelling or collapsing. In the pre-gel state this leads to a relative constancy in

Fig. 25. Schematic appearance of rigid two-step network. A molar excess of connecting species was used, resulting in rigid struts connected to stiff chains at both ends and rigid "barbs" attached to stiff chains at only one end

size of the precursor FPs. For each particular aromatic polymer, the average radius of the FPs at constant molecular weight is dictated by the ratio of difunctional to higher functionality monomers in the final FP: the higher the fraction of difunctional residues, the larger its radius. Naturally the ratio of di- to higher-functionality monomers is also reflected in the average segment length of the FPs and their network progeny.

When the axial ratio of the segments forming rigid aromatic FPs is very small, of the order of 1 or 2, a significant crowding effect becomes evident in the increasing number of unreacted functionalities, slowed FP growth rate and large increases in C_0^* with decreasing l_0. This is clearly evident in the solid-line branch of the curve in Fig. 26, describing the dependence of C_0^* on l_0 [311]. Plots of C_0^* against $1/l_0$ are more revealing. Such a plot for aromatic polyamides is shown in Fig. 27. Here we find that the beginning of the precipitous climb in the dependence of C_0^* on $1/l_0$ starts at $l_0 = 13$ Å which, in this case, corresponds to $x = 2$. As the axial ratio reaches unity, the dependence of C_0^* on l_0 becomes much more pronounced, strongly reflecting the effects of crowding and growth frustration. When the segmental axial ratio is sufficiently large for the crowding effects not to be noticed any more, a linear dependence of C_0^* on $1/l_0$ appears to exist. In the case of aromatic polyamides in Fig. 27, this dependence obeys the relationship

$$C_0^* = 0.0125 + 0.28 \text{ Å}/l_0$$

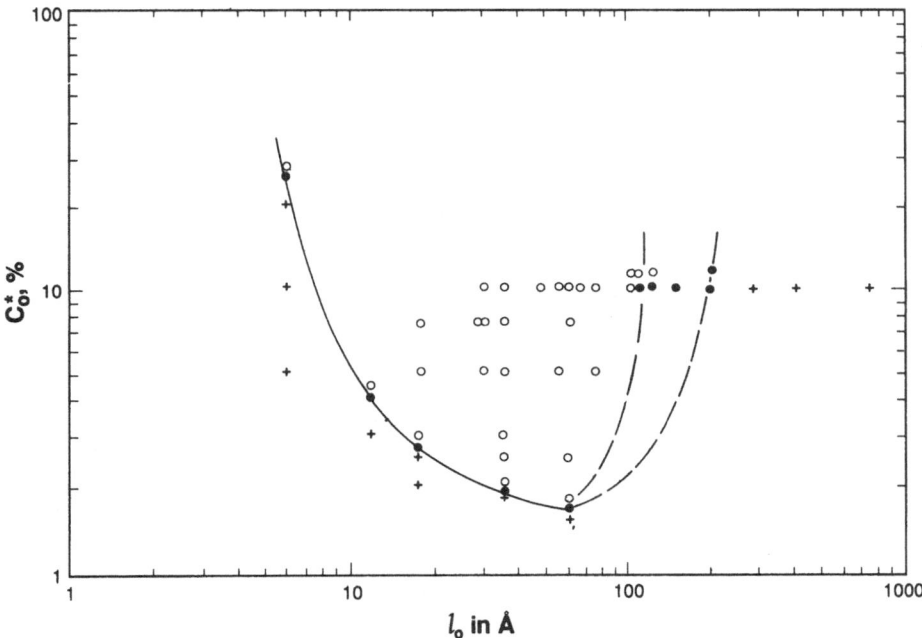

Fig. 26. Dependence of C_0^* on average segment length, l_0

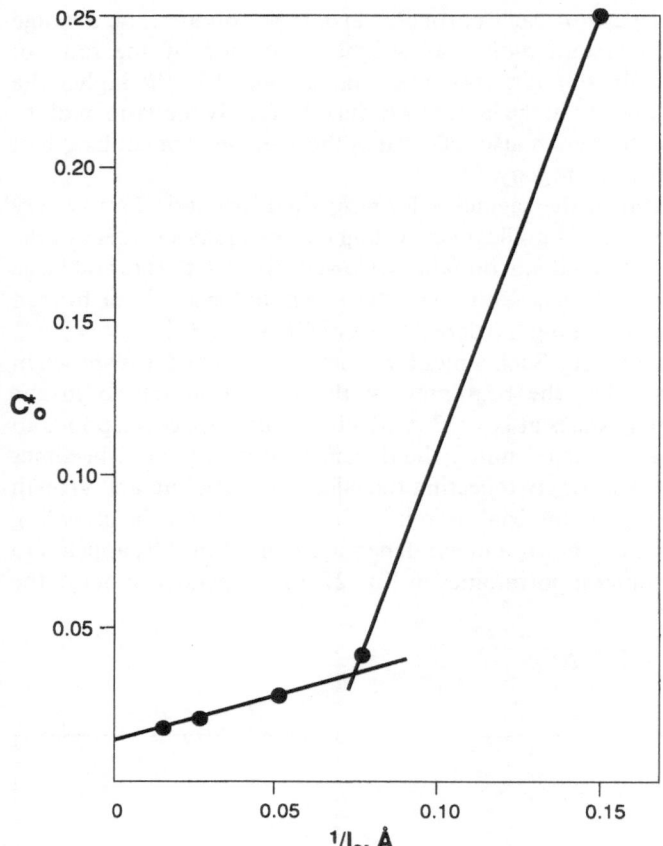

Fig. 27. Inverse dependence of C_0^* on l_0 for the aromatic polyamide networks in the lefthand branch of the curve in Fig. 26

with l_0 being measured in Angstrom units. The value $C_0^* = 0.0125$ implies that at infinite segment length the critical concentration for network creation is 1.25%. We have found, however, that when the average value of l_0 exceeds 100 Å, gravity wins and the swollen gels lose their ability to keep their shape. In such cases, depicted by the dashed lines in Fig. 26, gelled specimens prepared in the shape of tall cylinders could not keep their cylindrical shape once the supporting vessel was removed, and flattened down to appear as thick pancakes. It is important to emphasize here that, despite the dramatic change in shape of the gelled network, its total volume did not change and no solvent was secreted out. We have found that, in the case of rigid aromatic network gels, C_0^* values of about 1.75% were the smallest ones where gelled "infinite" polyamide networks may exist and maintain their shape without support.

An interesting dependence of C_0^* on the molecular weight of precursor FPs was found in the case of rigid aromatic polyamides. A series of FPs was prepared, purified and characterized, all with identical chemical composition, $f = 3$ and $l_0 = 38.5$ Å. They were designated as series 45X and their characteristics are given in Tables I and III in [311]. Then, quantities of each of the FPs were polymerized into networks in the absence of any low molecular weight species. The polymerizations were separately conducted in solutions of decreasing concentrations. The last concentration, below which no "infinite" network could be prepared, was defined as C_0^* for the particular size FPs. The dependencies of C_0^* on $M_w^{\frac{1}{2}}$ and on R_H of the precursor FPs are shown in Fig. 28 top and bottom, respectively. In both instances,

$$C_0^* \propto (M^{\frac{1}{2}})^{0.37}$$

and

$$C_0^* \propto R_H^{0.37}.$$

It is interesting to note that, although we could not prepare networks from FPs at $C_0 < C_0^*$, networks prepared at $C_0 \geq C_0^*$ could be swollen in a good solvent

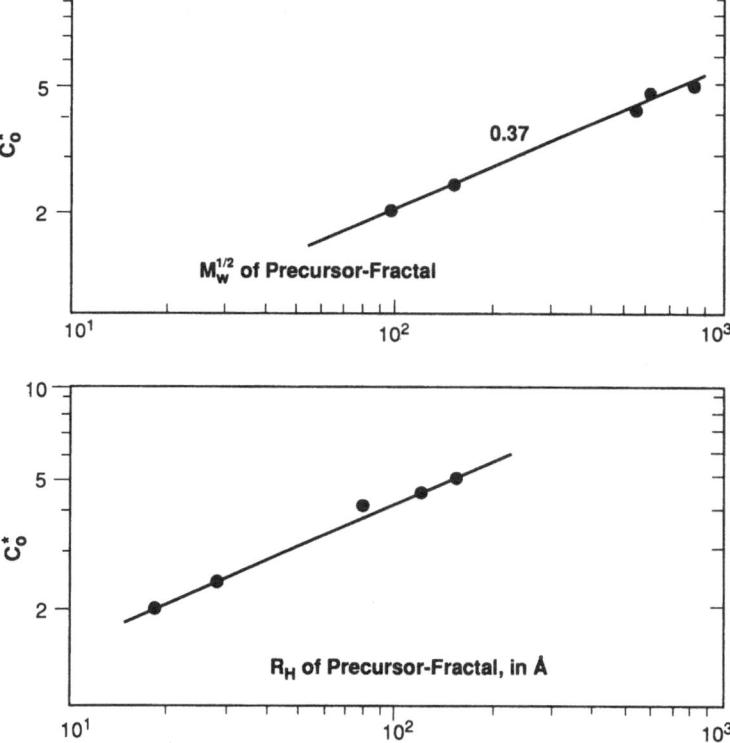

Fig. 28. *Top*: power dependence of C_0^* on $M_w^{\frac{1}{2}}$ of precursor FPs. *Bottom*: power dependence of C_0^* on R_H of precursor FPs

and reach equilibrium at $c < C_0^*$. In no case, however, could we swell rigid networks prepared as gels at $C_0 \geqslant C_0^*$ to reach concentrations lower than $c \simeq 1.75\%$. The relationship between the network concentration and the M_w of the precursor polyamide FPs is shown in Fig. 29.

The mathematical problem of when a network becomes possible is treated at present in terms of percolation theory. A reader interested in the subject should consult a textbook such as that of Stauffer and Aharony [670] or Grimmett [671]. Because of their enthalpic nature, growing rigid FPs are expected to follow the theory closely. The situation is less clear where flexible, entropic FPs are concerned. Because of the flexibility and length of segments between junctions, changes in temperature or solvent quality may cause significant changes in the volumes pervaded by the growing FPs. Such changes may affect C_0^* and render its exact location unclear. Viscosity measurements aimed at following the approach with time to the gel point may also be affected by the flexible nature of the growing FPs and their network progeny in a fashion different from the effects of rigid FPs. The viscosity vs reaction time curves in

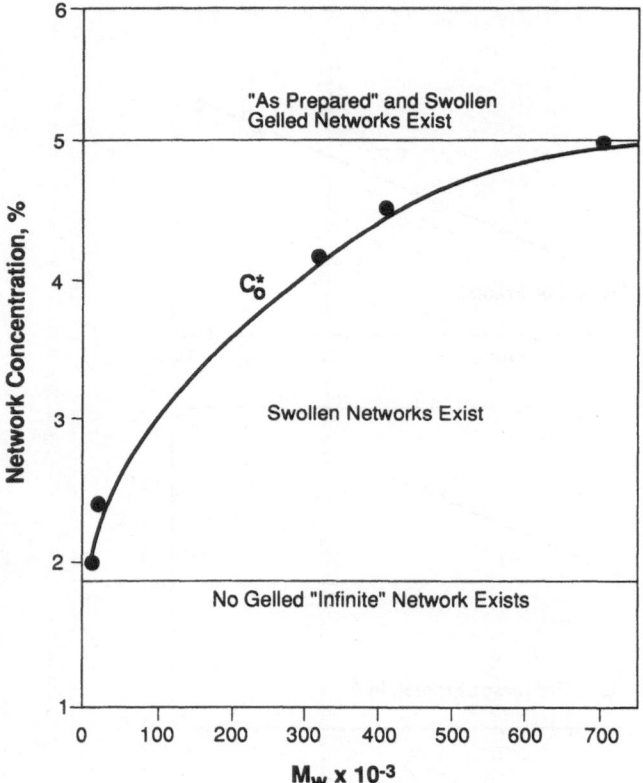

Fig. 29. Relationship between network concentration and M_w of precursor FPs

Fig. 30 may serve as an example. Here, polyamide networks were prepared using trifunctional branchpoints, all at $C_0 = 7.5\%$. The reactivity of the condensing species was evaluated and found to be the same for all networks. The viscosity of the systems was measured kinematically during polymerization using an oscillating sphere method operating at a natural frequency of ca. 4000 hertz. The measurements were stopped at the upper limit of the viscometer, at 140 000 centistokes. The data were converted to centipoises before plotting the curves. In all cases an incubation period of around 10 min was observed before the viscosity started increasing rapidly. The very precipitous curves terminating before 20 min reaction time all belong to rigid networks. The less steep curve extending to 100 min belongs to a flexible network. The gel point of all systems was traversed when the viscosity of the reaction mixture reached about 200 centipoises. The abrupt onset of viscosity increase and steepness of the curves of the

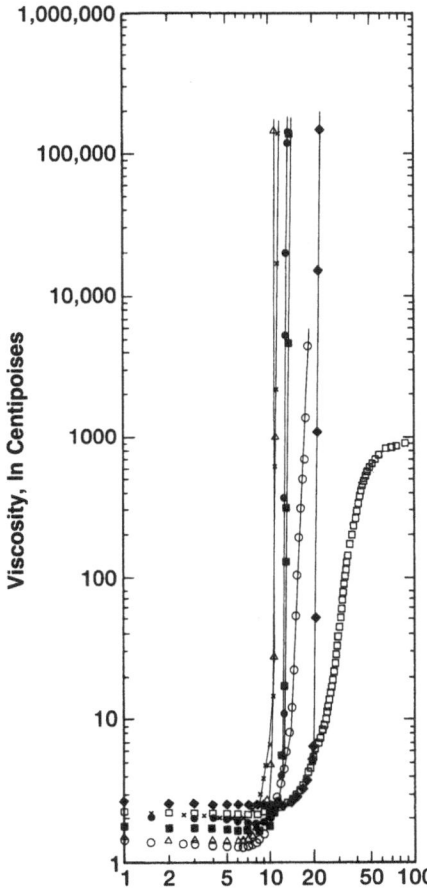

Fig. 30. Conformity of viscosity curves around the gel point with the percolation model. *Steep curves* are for rigid network gels and the *less steep curve* is for flexible network. All systems are aromatic polyamide networks formed in a one-step polycondensation procedure

rigid systems are in close agreement with expectations from percolation theory. The gentler slope of the viscosity curve of the flexible system reflects the fact that the viscosities of flexible polymers in both the pre-gel and post-gel states are dramatically smaller than the corresponding viscosities of stiff polymers and rigid network gels. In fact, static moduli about two orders of magnitude larger than those for flexible network gels were measured by us [311, 400] on rigid network gels with the same branchpoint functionality and average segment length. Changes in reaction temperature and monomer concentration, C_0, appear to affect the length of the incubation period before the viscosity of rigid systems takes off, but do not appear to affect measurably the steepness of the precipitous slope of viscosity increases upon approaching the gel point. In Fig. 30, the righthand-most curve from the very precipitous ones belongs to a system polymerized at a temperature 12 K lower than the rest. The incubation period increased from ca. 10 to ca. 20 min. The steep slope was not affected. In Fig. 17, a rigid network was prepared at $C_0 = 3.0\%$. The incubation period now extended to about 40 min, but the slope of the precipitous part of the curve was not affected measurably. In the case of flexible FPs, we expect the converse to hold true. Because in these systems, excluded volume interactions are concentration dependent [292], the approach to the gel point is expected to be more concentration- and temperature-dependent than in the case of rigid FPs, and the change in viscosity with reaction time less precipitous.

There are present in the literature many treatments of the gel point and the effects of pre-gel particle swelling on it. Among them one may enumerate the seminal papers of Flory [8, 9, 672], deGennes' book [673] and review articles such as those by Rempp et al. [22], Candau et al. [28] and Stauffer et al. [30]. They all deal, however, with flexible networks only, which will not be discussed below.

4.2 Post-Gel State

4.2.1 Fractal Nature of Gelled Rigid Networks

The use of the fractal model to describe polymer networks is now becoming more and more accepted. When scrutinized more closely, one finds that, for covalently bonded organic polymers, the very large majority of the publications deal with and describe the properties of the growing pre-gel flexible entities and the final flexible networks at or very close to the gel point [308–310, 670, 674–681]. Papers dealing with the fractal nature of mature gelled organic polymer networks, far beyond the gel-point, are scarce. Some such papers deal with rather rigid networks, such as resorcinol–formaldehyde ones [516, 682, 683]. In these instances the fractality was found to be present over size ranges far larger than molecular size, with the primary structural units in the network being dense colloidal particles of diameters in the order of 50 Å or thereabouts. The surfaces of these primary units were found to be rather smooth. Naturally the density and smoothness of the primary particles indicate that each of them contains a huge number of molecular repeat units, most likely in a collapsed version of a more porous entity that probably existed in solution before the solvent was removed. There are only a few papers in which the fractal nature of polymeric networks was morphologically demonstrated on a size scale corresponding to individual highly branched macromolecules [311, 312].

Studies on inorganic polymer structures, such as silica or alumina aerogels and electrodeposited metals, reveal them to be fractal on a coarse scale and spherical or columnar dense colloidal particles on the primary particle size scale. The fractal networks appear as branchlike aggregates of the dense colloidal primary particles [296, 521, 684–687]. Here, too, the relatively large size of the primary particles indicate that each contains a huge number of monomeric species and may contain many highly-branched pre-gel polymeric entities.

Other systems, such as reversible polymeric gels and various aggregates, were also described in terms of the fractal model but here, again, the primary particles forming the fractals are of relatively large size [688–695].

Unlike the above, we are interested in fractal networks in which the fractality is evident on the same size scale of individual highly-branched rigid macromolecules. We picture gelled rigid one-step networks as created from the impingement and aggregation of rigid precursor highly-branched polymers and the concurrent creation of covalent bonds between them. It was demonstrated by others using flexible polymers and by us using rigid polyamides [311, 312] that up to the gel point the precursor polymers conform with the fractal model. The growing aggregates of covalently bonded and occasionally entwined FPs continue to coalesce with one another until a point is reached where they form a single giant molecule traversing the whole reaction medium. This is the gel point. If the supply of monomers and small oligomers or network fragments is not depleted at the gel point, additional segments and branchpoints will be added to the network. A small fraction of these will be added in the interiors of the precursor FPs, but the majority will be added in the exteriors of the FPs, in the "interfacial" zones between FPs filling the interstitial voids caused by random packing of the coarsely spherical FPs prior to and at the gel point. Some FPs are also undoubtedly either enmeshed or occluded in the network without being covalently bonded to it. Because of the rigidity of the system, these FPs cannot be removed from the "infinite" network in a fashion similar to sol-gel extraction of flexible networks.

We have demonstrated by SAXS studies [311] that, in the case of rigid aromatic polyamides, the fractal character of the pre-gel highly branched polymeric entities is retained in the post-gel network even after it was allowed to continue to grow and mature for a long time after the gel point. The fractal nature was evident in both the swollen gel and in the fully dried network. As was the case with the pre-gel FPs, the fractality of the final network was found to be surface and not mass fractality, meaning that the exteriors of the FPs are highly ramified even when incorporated into the final network, and the mass density in the interior of the FPs is more or less uniform, albeit low, and does not decrease

gradually and smoothly with distance from the fractal center. Remnant percola-
tive disorder in highly-cured somewhat stiff networks was very recently described by Adolf et al. [695a].

Recalling that in polycondensation reactions most low-M species are already consumed at the gel point and that from this point on the reaction rate is
diffusion controlled, we recognize that the interstitial and interfacial regions of
the one-step network are less dense in polymer material than the intrafractal
volumes and that a higher concentration of network defects are present in the
interstitial and interfacial areas. When the network is constructed from only
rigid FPs without the aid of monomeric "mortar", the effects of interstitial flaws
are greatly exacerbated, as was clearly demonstrated by the inverse power
dependence of the modulus of gelled rigid networks on the radius of their
precursor FPs [312].

At the gel point, and even in the aggregates of FPs right before the gel point,
most of the oligomeric species not yet consumed and a lot of the monomers will
be pushed into the interstitial volumes by the corrugated growth fronts of the
growing FPs. This is similar to the movement of unincorporated material ahead
of the growth fronts of many polymer crystals. Furthermore, in the interstices
the least amount of steric and geometric constraints are present against the
approach and correct alignment of the reactable small species and their
attachment to accessible segmental tips or branchpoints of the FPs proper.
Within the FPs, and even in the interfacial regions, the hindrances are significantly larger, especially for the largest oligomers and network fragments, making
it less and less likely for such entities to react with the FPs.

The reactivity of monomers in the FPs and the interstitial regions at around
the gel point and thereafter is mostly controlled by diffusion. Because the highest
concentration of reactive sites belonging to FPs is present at the segmental tips
and unreacted branchpoints at the FP exteriors, the monomers will, mostly but
not exclusively, be attached in these regions. After the gel point, it is reasonable
to assume that the attachment will increasingly take place in the interstitial
regions. Therefore, even when we start with more or less spherical FPs, in the
final network they will evolve to become closer in shape to polyhedra, just as
when deformable spheres are compressed together and gradually attain the
shape of multifaceted polyhedra. This shape transformation is especially true, we
believe, in the case of rigid networks that were allowed to mature and continue
their growth long after the gel point. Because of the broad size distribution of the
precursor FPs right before the gel point, the polyhedra associated with the
precursor FPs will also have a broad size distribution. The volume of the gelled
one-step rigid network at the gel point and beyond is, hence, divided into a large
number of polyhedra, each evolving from a precursor FP. The size distribution
of the polyhedra corresponds to that of the precursor FPs right at the gel point
and immediately thereafter.

The appearance of the fracture surfaces of solvent-exchanged and then fully
dried rigid aromatic polyamide networks in Fig. 31 strongly supports the above
description. In the figure, four photomicrographs are shown, obtained in a

scanning electron microscope at a magnification of 60 000 times. The panels in Fig. 31a–c belong to a rigid network coded 45XD, and in Fig. 31d belongs to the corresponding pre-gel fractal polymer coded 45XB. Both systems were prepared by a single step polycondensation and the reaction was stopped shortly after the gel point was reached. Both were previously described by Aharoni et al. [311]. WAXD scans of dry and solvent-swollen samples of both systems produced scattering exponents consistent with surface, and not mass, fractality [311]. The fracture surface of 45XD in Fig. 31a–c looks as if made from soft deformable fuzzy balls packed together. In many places the balls are compressed together to such an extent that the boundary between adjacent balls appears as a straight line. This is fully consistent with the polyhedra picture we painted above. The shape and size similarity of the primary particles in the fracture surface of 45XD and the pre-gel FPs in 45XB implies that both substances consist of the same primary units. The sizes of the visible primary particles in the network and the precursor FPs fall in the interval from ca. 200 Å to around 800 Å. These sizes are comparable with the sizes of pre-gel FPs just before gelation, as measured by light scattering of their solutions [311]. The observed primary particles in the fracture surfaces are, then, expected to be individual FPs in the developing network. The boundaries between the fractal polymers in the network are very clear. The nature of the fracture surface of 45XD implies that the fracture had propagated by following the weaker interfractal boundaries and did not pass through the stronger FPs themselves. Both observations support the assumption that, in the rigid network being studied, where the polymerization was not allowed to continue for long after the gel point, the material density between FPs is substantially lower than inside the FPs. By the same token, it is expected that when the polycondensation will be allowed to continue long after the gel point, then the interfractal boundaries will gradually be filled with additional network material and lose their definition. Finally, it is important to mention here that when the networks or their precursor FPs are flexible in nature, they tend to collapse during sample preparation with the apparent loss of definition of the individual fractal polymers [311].

Geometrically, the evolving network is a system in which the coarsely spherical FPs are replaced by a 3-dimensional space-filling ensemble of polyhedra with a substantial size distribution. In this system, the coarsely curved precursor fractal boundaries are replaced by flat surfaces bounded by straight lines connecting three-dimensional vertices. In such systems, deviations from straight lines and flat surfaces may be considered as rapidly relaxing degrees of freedom that fade away in favor of slowly moving vertices. The two-dimensional analogues of the above three-dimensional gelled network model are commonly known as Voronoi [696, 697] or Dirichlet [698] tilings and the process of creating such non-uniform mosaics is often called tessellation or construction. Each participating unit may be named a Voronoi cell. In two dimensions the average number of edges of each polygon is six. The Voronoi tessellation was used in the description and analysis of crystalline metal grains [699–701], froths and bubble rafts [702–707], random packing [708, 709] and optimization [710]

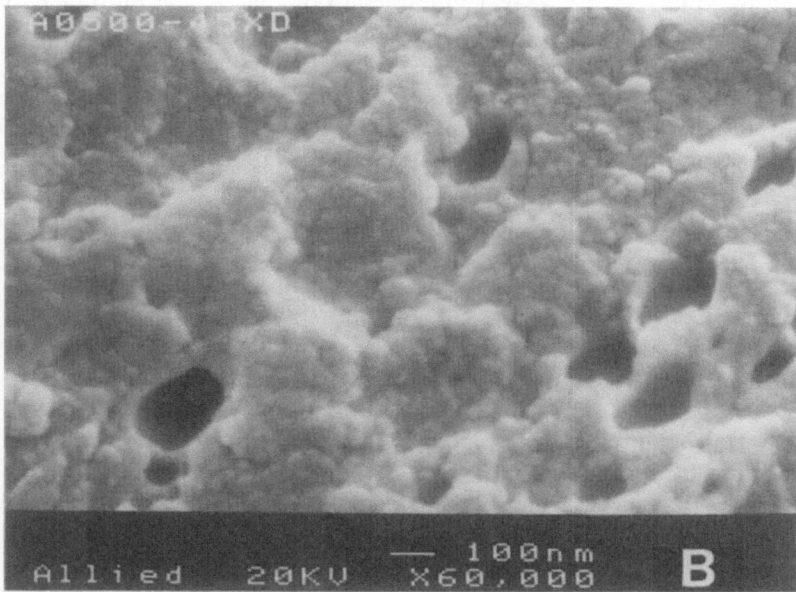

Fig. 31a–d. Fracture surface of rigid aromatic polyamide networks. All networks were prepared in one step, passing through the FP stage prior to gelation. The networks were fractured in the gel state and subsequently dried. Original magnification: 60 000x

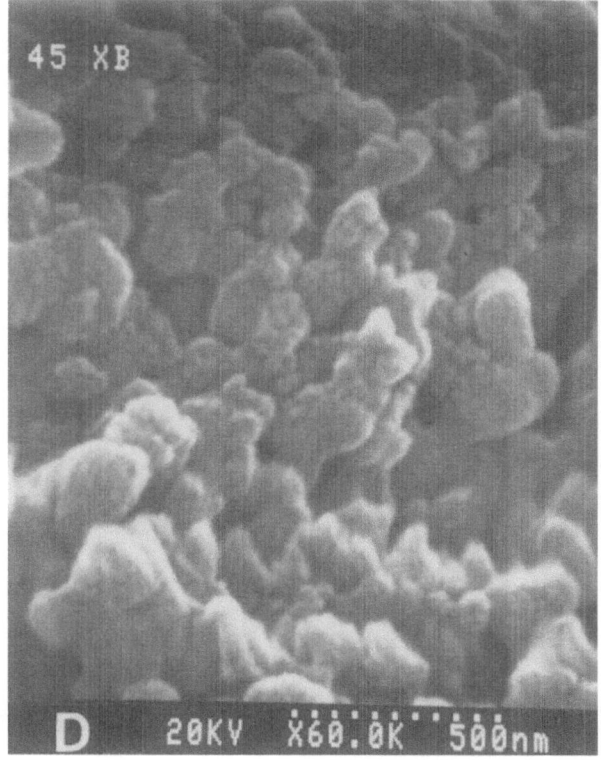

Fig. 31c, d

problems, among others. Most importantly, the Voronoi construction can be generalized to spaces other than the plane [706, 707] and are called metric Voronoi construction. The dual network connecting all nuclei of the Voronoi cells is called the Delaunay simplicial complex [707]. It should be mentioned, however, that some of the geometrical rules that simplify considerations of two-dimensional constructs are lost in three-dimensions, making the problem considerably more complex [706, 711, 712].

Using the Voronoi tessellation concept, we have constructed two-dimensional models of some pre-gel FPs and their aggregates, and of the mature gelled network. They are shown in Fig. 32a, b. In Fig. 32 the darkly shaded cells are those containing additional FPs, enmeshed or occluded in the network, but not part of it. The unshaded cells stand for voids created by local depletion of

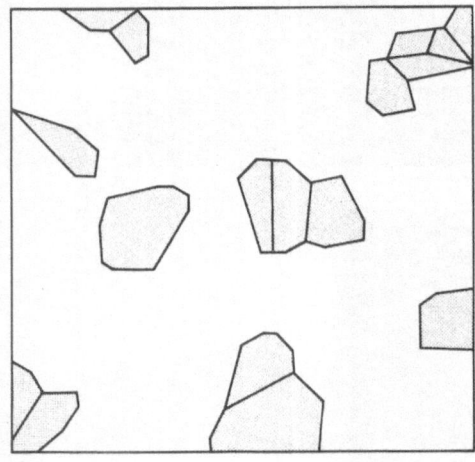

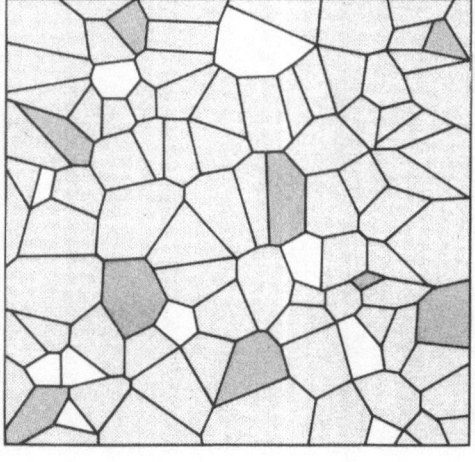

Fig. 32a, b. Two-dimensional Voronoi construction models: **a** pre-gel FPs and aggregates; **b** the final network

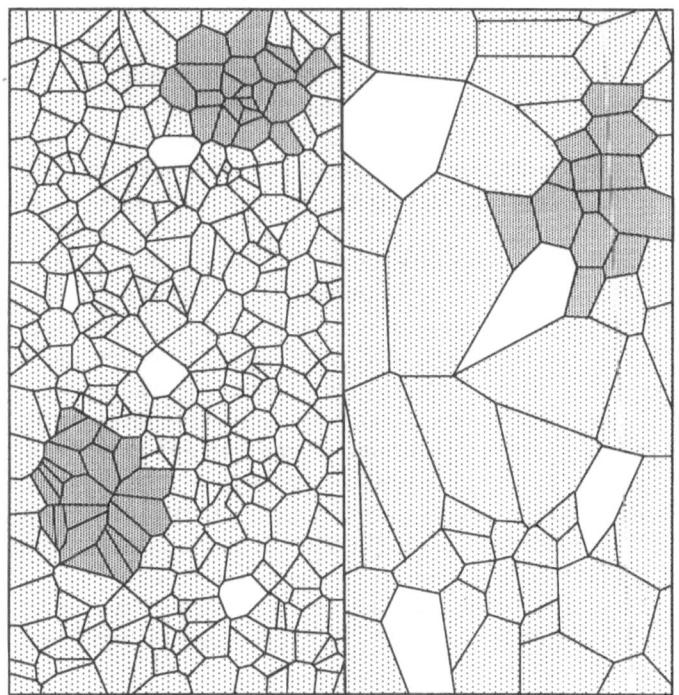

Fig. 33. Two-dimensional Voronoi constructions of two mature networks. *Lefthand side*: small precursor FPs. *Righthand side*: large precursors FPs

monomeric and oligomeric network-building material. In Fig. 33 we attempt to describe two mature networks, the left-hand one made from relatively small precursor FPs and the right-hand one formed from large FPs. Here the heavy dotted cells are those containing extra FPs in addition to those forming the infinite network, and the undotted cells stand for network voids. As will be shown below, we believe that a relationship exists between the average size of the FPs, the network voids, and the physical properties of the gelled network.

A 3-dimensional Voronoi construction was very recently used to model the diffusion of small molecules in polymeric glasses [713]. It will not be surprising if this modeling method will soon be widely accepted.

We believe that the fractal character of the precursor polymers and final gelled networks can well be accommodated within the framework of the Voronoi constructs, especially when the polymers and their networks are rigid.

4.2.2 Manifestations of Rigid Network Defects

The effects of polymer network defects are most frequently manifested in lower than expected moduli and/or higher than expected swelling upon immersion in good solvent. These were repeatedly described and explained in the literature on the macro- [13–18, 27, 39, 40, 46, 714] and molecular-scales [51, 319, 715]. Without exception, the explanations are in terms of networks composed of long flexible chains lightly crosslinked in a manner allowing the network segments to be treated by means of Gaussian statistics, and present at temperatures high above T_g or in sufficient dilution to allow substantial motional freedom of all network components. In such networks the defects are generally described as dangling chain-ends or as intramolecular cyclic structures, neither of which contribute to the mechanical properties of the networks. This stands in contra-distinction with rigid aromatic networks, especially those created in a one-step process and having precursor FPs. Because of segmental stiffness and branch-point rigidity, no or very few intrafractal cyclic units can form. When and if cyclic units form, they generally involve two growing FPs being joined together. Because of the same rigidity considerations, a segment connected to the network at only one end may be able to resist deformation and support some load. Typical of such loads are ones caused by other segments or branchpoints moving in such a way as to rest on the segment in question and apply to it pressure in order to bend it or deflect it out of the way. So in cases where rigid networks are involved, intramolecular cyclic structures may not be invoked as an explanation for network defects, and, unlike flexible systems, "dangling" segments may potentially contribute positively to the modulus of rigid net-works. We shall return to this point in Sect. 4.2.3.

Depending on the nature of network segments and junction points, and on the size and concentration of network defects, the flaws may or may not be observable in the form of network heterogeneities. Until very recently, elasto-meric networks were considered to be uniform on a molecular scale level, having no detectable heterogeneities associated with network defects such as dangling chain ends. Furthermore, no non-uniformities in crosslink density were ob-served in common amorphous elastomers such as rubbers or polyurethanes. A reason for the failure to detect non-uniformities is that such amorphous systems have very poor contrast when studied by scattering techniques such as WAXD or SAXS. Only recently, SANS afforded us an insight into the non-uniformity of amorphous elastomeric materials [28, 716] by virtue of the contrast existing between deuterated and protonated polymer chains. The inhomogeneities were shown to be enhanced when the elastomers were swollen to become gels or when they were stretched [717, 718]. A few light scattering studies [28, 719] also showed macromolecular-size inhomogeneities in elastomers. In all these, the observed inhomogeneities were explained as arising from non-uniform distribu-tion of crosslink junctions, being more concentrated in certain small islands and more dilute in the surrounding material [717, 720, 721].

Similar inhomogeneities, also attributable to local variations in crosslink density and the consequential network flaws, were observed in non-elastomeric networks in the glassy state. Some of these observations date from as early as the late 1950s [722, 723]. Although several families of polymers exhibited these inhomogeneities, they appear to have one common denominator: they are all rather intensely crosslinked with relatively short segments between junction points. Most of the systems showing non-uniformity were epoxies [723, 724], crosslinked polyesters [723, 725–728] and phenolic resins [722, 723]. In all instances the observed inhomogeneities were explained as arising from non-uniform crosslink density.

In the gelled rigid aromatic polyamide networks prepared by us, we could often see the effects of network defects. They were observed by their effects on the static modulus, G, of the gel, and on the swelling ratio C_0/c. When stiff segments are built into a flawless rigid network, some are bent and some are straight [664, 665]. It is safe to assume that the ratio of bent to straight segments, as well as the number of bends per segment, depend on the stiffness of the segment or of its analogous linear chain. That is, the more thermo-dynamically stiff the chain is the higher the energy barriers to bending and twist-bending, and the smaller will be the average number of bends per segment and the smaller the fraction of bent segments in the network. We can, therefore, correlate the swelling level of a flawless rigid network gel with the persistence length, a, of the corresponding linear stiff chain:

$$C_0/c \propto 1/a. \tag{27}$$

The average number of bends per segment is expected to be directly dependent on its average length, l_0. Therefore, the swelling ratio should be affected by l_0:

$$C_0/c \propto l_0. \tag{28}$$

Conversely, the swelling ratio is inversely dependent on the functionality of the rigid branchpoints, the higher the functionality the lower the swelling:

$$C_0/c \propto 1/f. \tag{29}$$

Assembling together the above proportionalities, we find that, for a gel of flawless rigid network:

$$C_0/c \propto l_0/a \cdot f. \tag{30}$$

We have found that in the case of aromatic polyamide networks approaching flawlessness, the swelling ratio from the "as-prepared" C_0 to the equilibrated c, amounted to only a few percent. This was observed in both one-step [311, 312, 400] and two-step [277] gelled networks. In most cases, however, the gelled networks were defective to varying degrees and, as a rule, the swelling ratios increased with the level of network defects. The simplest way to gain a qualitative estimate of network defects is to first synthesize a series of gelled networks of identical l_0 and f but at ever decreasing C_0, and then measure their swelling ratio C_0/c as a function of C_0. A clear inverse dependency of C_0/c on C_0 was observed in gelled rigid polyamide networks with f = 3 and "straight" stiff segments of $l_0 = 38.5$ Å [311] and in gelled rigid polyamide networks with f = 3 and zigzag stiff segments averaging 57 Å in length [409].

The most dramatic swelling effects of network defects were observed in gels of networks created from rigid fractal polyamides in the absence of monomers

[312] and in gels of networks created in two steps from long reactable stiff chains [277]. In the case of gelled networks made from FPs, the swelling ratio was found to be a direct power law of the measured radii of the precursor FPs [312]. In the two-step networks, the degree of swelling was found to be directly proportional to the magnitude of deviation from stoichiometry of the reactable crosslinking groups [277]. In both instances the swelling ratios are directly dependent on the degree of network flaws introduced by the changes in FPs diameters and stoichiometry in the respective gelled networks.

4.2.3 Mechanical Properties of Rigid Networks and Their Gels

Currently, various mechanical and swelling properties of flawless and faulty networks and their gels are being evaluated within the frameworks of two limiting models. According to one, the networks are made from closely cross-linked highly flexible chain segments. Each segment consists of many rotatable links or bonds and their number per segment is sufficiently large that the statistical distribution of the end-to-end distance of the segments reduces to the Gaussian form:

$$W(\underline{r}) = [(3/2)\pi\langle r^2 \rangle_0]^{3/2}\exp[-3r^2/2\langle r^2 \rangle_0] \tag{31}$$

where $\underline{r}$ is the chain end-to-end vector, $\langle r^2 \rangle_0$ is the mean-square magnitude of $\underline{r}$ averaged over all configurations and $r \equiv |\underline{r}|$ is the displacement length of the chain. The elastic free energy, A_{el}, of the chain characterized by $\underline{r}$ is

$$A_{el} = c(T) - kT\ln W(\underline{r}) \tag{32}$$

where $c(T)$ is a function of T and k is the Boltzmann constant. Substituting Eq. (31) into Eq. (32) we obtain the elastic free energy of a flexible segment with a fixed $\underline{r}$ in the network:

$$A_{el} = A^*(T) + (3kT/2\langle r^2 \rangle_0)r^2 \tag{33}$$

where $A^*(T)$ is a function of T alone. The magnitude of the average retractive force $\bar{f}$ exerted by the chain at fixed r is obtained by differentiating Eq. (33):

$$\bar{f} = 3kT\langle r^2 \rangle_0 - 1_r. \tag{34}$$

The elastic properties of a network of Gaussian segments follows from Eqs. (31–34) with the shear modulus G being directly dependent on the number of elastically effective network segments ν:

$$G \propto \nu RT. \tag{35}$$

The number concentration ν is obtained by dividing the density of the polymer, ρ, by the molecular weight of an average segment between branchpoints, M_c. A parameter $(1 - 2/f)$ was introduced into the theory of rubber elasticity [729, 730] reflecting the interconnectivity of the crosslink junctions, such that [731]

$$G = (\rho/M_c)(1 - 2/f)RT. \tag{36}$$

The Poisson ratio of elastomeric networks is usually very close to 0.5. When it is taken to be 0.5, a relationship between G and the tensile (Young's) modulus, E, is defined by

$$G = E/3 \tag{37}$$

and

$$E = (3\rho/M_c)(1 - 2/f)RT. \tag{38}$$

In the case of elastomeric networks where the segments obey Gaussian statistics, the moduli are directly dependent on the absolute temperature, T. From the knowledge of G or E, one may estimate M_c or v.

At the other extreme, we find a model in which the segments are all rodlike and the crosslinks are rigid and permanent [732]. Defining E_r as the rod material modulus and d the cross-sectional dimension, the rod behaves in the limit of small deformations as a spring with a spring constant S:

$$S \propto E_r d^4/L^3. \tag{39}$$

Approximating the segmental repeat unit by the chain diameter d, a situation very close to the reality of aromatic polyamide segments, we obtain for the tensile modulus of this network:

$$E \propto B/d^3N^2 = E_r d^3/d^3N^2 = E_r/N^2 \tag{40}$$

in which N is the number of repeat units of length and diameter d, and $Bd = E_r d^4$ [732]. For straight rigid segments, the elastic modulus scales as the volume concentration, ϕ, squared [732]:

$$E \propto \phi^2 \tag{41}$$

and is independent of temperature. This independence is a major differentiating factor between the fully rigid and fully flexible network models. It will be shown below that gelled networks having stiff segments and rigid branchpoints behave closer to expectations for the fully rigid than the fully flexible network model. However, because Eqs. (39) and (40) apply to rods of uniform diameter and spring constant, we believe that, with the possible exception of segments such as poly(p-phenylene), this model is inapplicable to stiff segments, especially those whose bending involves more complicated motions such as the twist-bending mode of deformation observed in aromatic polyamides and polyesters. For historical purposes it is worthwhile mentioning at this junction that treatments of polymer networks [17, 22, 38–40, 46, 319, 733–737] completely neglect both fully rigid and stiff networks and their gels.

In between the above extremes reside the networks of interest in this review. It appears that, in general, the glass transition temperature of intensively crosslinked networks is inversely proportional to the average molecular weight between branchpoints [731]:

$$T_g - T_{g\infty} = K/M_c. \tag{42}$$

Here, $T_{g\infty}$ is the glass transition of the uncrosslinked polymer and K is an empirical factor introduced by Fox [738]. For each polymer, $T_{g\infty}$ appears to be related to the difference in thermal expansivity above and below $T_{g\infty}$ [739]:

$$T_{g\infty}(\alpha_L - \alpha_G) = T_{g\infty}\Delta\alpha = 0.113. \tag{43}$$

This dependency of $T_{g\infty}$ on $\Delta\alpha$ is in reasonably good agreement with experimental results, but fails to explain the behavior of the polymer in terms of its own structural features. An attempt to relate the thermal transition of aromatic polymers to the characteristic ratio of many linear polymers was not altogether satisfactory, but revealed that the dependence of the transition on chain stiffness is similar in most flexible polymers and is dramatically different from the same dependence in stiff polymers [162]. An equilibrium theory, discussing the glass transition temperature of flexible chain polymers in terms of chain conformational and free volume hole energies, was put forth by Gibbs and Dimarzio years ago [740, 741], but some different analysis will have to be applied to rigid networks of stiff-chain polymers. It remains an important unsolved problem.

The description and understanding of the dependence of mechanical properties on the structure of highly crosslinked and, especially, stiff aromatic networks are also lacking. In most cases, the highly crosslinked networks originated from fully or largely flexible chain polymers and their moduli at temperatures above T_g were evaluated in conformity with Eqs. (35), (36) or (38) above. The dependence of G on T holds even for highly crosslinked flexible-chain networks, while the values of M_c appear to drop to far below their expected values once G reaches magnitudes greater than about 10^7 Pa [714]. A lot of effort was invested in attempts to explain the M_c discrepancy [731, 742–746] resulting in a qualitative appreciation of the fact that in highly crosslinked flexible-chain networks "the elastic energy storage appears to be based largely on an entropy effect even though the statistical theory based on Gaussian distributions must be modified for such short segments" [741]. In rigid networks comprising stiff segments, the energy storage is based not on entropic effects but on energetic considerations.

An approach that may lend itself to describing the dependence of G on the structure of networks made from relatively stiff segments originates with the dependency of G on network characteristics devised by Smith [747] for systems whose behavior is based on entropy effects:

$$G = \Phi\nu RT\Gamma(1/\bar{\lambda}_m). \tag{44}$$

In this equation, Φ is the ratio of the end-to-end distance of segments in the undeformed network to the end-to-end distance of an analogous free chain of equal length. ν is moles of chains per unit volume, RT is the product of the gas constant and absolute temperature and $\Gamma(1/\bar{\lambda}_m)$ is a function of $1/\bar{\lambda}_m$, the mean fractional extension of segments in the undeformed network. For stiff segments present in the network substantially in the extended state with only a few bends per bent segment, Φ is expected to be close to unity [743, 747] and will be neglected below:

$$1/\bar{\lambda}_m \propto \Phi C_n/n \tag{45}$$

where n is the number of actual and virtual bonds per segment and C_n is the characteristic ratio which, for short segments, is dependent on n. To simplify the relationship we replace C_n by C_∞ and use the equality [51, 166, 168, 748]

$$A = 2a \cong C_\infty \cdot l = (\langle h^2 \rangle_0 / L)_\infty \tag{46}$$

where l is the length of an average bond in the segment and a is the persistence length of the analogous linear chain. Removing the RT dependence which is invalid for stiff segments, we are left with the proportionality

$$G \propto v(a/nl) \tag{47}$$

meaning that the shear modulus is dependent on the concentration of network segments times the ratio of the persistence length to the contour length of the segments. In other words, G depends on the number of chains times a factor describing their curvature, the straighter the segment the closer the factor to 1.0. When the length of the stiff network segment

$$l_0 = nl \tag{48}$$

grows, then G becomes smaller even at constant v and, conversely, when v changes with constant nl, G changes with it. Equation (47) is consistent with our observations on gels of rigid polyamide networks containing rigid branchpoints with stiff linear [311, 400] or stiff zigzag [409] segments. The most important congruency is the independence of G from T. Other relationships consistent with Eq. (47) are the inverse dependence of G on l_0 and the direct dependence on v through the dependence on C_0. A comparison of the entries in Table V in [311] and Table VIII in [400] reveals that the modulus G of networks of functionality f = 3 and identical average segment length, $l_0 = 58$ Å, is much higher for rigid networks than for more flexible ones, even when the concentration of the rigid network is lower than the more flexible ones. All these consistencies with Eq. (47) invite a better theoretical grounding of this relationship.

4.2.4 Observations on Effects of Gravity and Flow-Stress During the Formation of Rigid Network Gels

The structure of rigid networks prepared in solution at $C_0 \gg C_0^*$ is very tenuous at the gel point and right after it. When the polymerization is allowed to continue, the networks add on more and more material and become increasingly robust. If the networks are prepared at $C_0 = C_0^* + \Delta c$, where Δc is a small increment in concentration, they remain rather tenuous after the gel point simply because the system is depleted of monomers to be added to the maturing structures. In this respect our one-step gelled rigid networks are similar to other systems, such as aerogels [521] and colloidal gels [693]. Over the years we have observed that, right after the gel point, one-step rigid networks prepared at C_0^* + Δc, especially, showed effects of external forces such as gravity and solution shear rate.

Our interest in the effects of gravity on network formation during poly-condensation in solution was sparked off by our previous observation linking gravity effects with the rate of phase separation in biphasic solutions of lyotropic polymer liquid crystals. When working with such solutions of polyisocyanates [192, 749, 750], we have observed that in the same solvent the rate of phase separation to a clear isotropic phase and opaque anisotropic phase depended on the difference in specific gravity between the polymer and solvent. The larger the difference, the faster the phase separation. It therefore occurred to us that in unstirred isothermal polycondensations, where thermal convection does not take place, we would expect that, if the densities of the pre-gel polymeric entities and the solvent are sufficiently different and a density gradient is established in the solution, the gelation will start in the regions of highest polymer concentration and propagate to fill gradually the whole reaction volume. In the case of aromatic polyamides, the bulk density of the amorphous polymers is around $1.4 \, g/cm^3$ [409] and that of the solvent DMAc/5% LiCl is very close to $1.00 \, g/cm^3$ [61]. If no LiCl is present in the reaction mixture, the difference in specific gravities will grow even further. It is expected, hence, that because of the density gradient established in solution, the gelation of the aromatic polyamide networks will start at the bottom of the unstirred isothermal reaction volume. If a sol phase remains at the end of the reaction, it should be concentrated at the top part of the reaction volume. These expectations were repeatedly observed in instances where we conducted polycondensations of aromatic polyamide net-works under isothermal conditions and without stirring, for the purpose of preparing gel articles of special shapes [212] and for testing the present hypothesis. In these cases, even when $C_0 \gg C_0^*$, not only the "infinite" network gel started forming at the bottom of the reaction vessel, but when the mature gel was later immersed in good solvent, the top part of the gel swelled more than the bottom, indicating that the bottom was more heavily crosslinked than the top. Since the same average functionality was used throughout, the differences in swelling are attributable to differences in polymer concentration: the higher the polymer concentration, the more intense the crosslinking and the lower is the swelling. The time to reach the gel point also varied with position in the unstirred isothermal reaction volume. In such cases, the gelation started at the bottom of the vessel and propagated upwards, reflecting the gradient in monomer and pre-gel polymer concentration. It is important to realize that, unlike the case for stirred solutions, as one reduces C_0 and approaches C_0^* from above, a very thin sol layer starts evolving at the top of the reaction mixture and the parts of the gelled network in contact with the sol phase are the most defective and mechanically weakest in the system. When swelled in good solvent they often develop a crumbly transluscent character. Such appearances were repeatedly observed [311] in studies aimed at determining C_0^* as a function of l_0 in rigid polyamide networks. The effects of gravity on evolving tenuous net-works, similar to our polyamides around the gel point at $C_0 = C_0^* + \Delta c$, were recently investigated theoretically [690, 694]. We believe, however, that this aspect of network formation merits further study.

When the reaction mixture is stirred, a completely different behavior is observed. For stirred reactions we generally used a magnetic stirring egg at the bottom of a round-bottom flask. Here, when $C_0 \gg C_0^*$, rigid gels started forming from the top of the reaction volume, first as a thick "O"-ring and then covering the vortex and stopping it. In a few seconds or minutes, depending on Δc, the temperature and the shear rate caused by the rotating stirring-egg, the gel reached the magnetic egg from above and later stopped its motion. As a rule, the weakest part of the final rigid gel, even after maturing for hours beyond the gel point, was the region close to the stirring-egg where the solution in the pre-gel stage experienced the highest shear. When C_0 was reduced and approached C_0^* from above, the volume fraction of rigid gel decreased and, on occasion, the region closest to the stirring-egg failed to gel. In such cases, the parts of the gel closer to the stirring egg that formed last were the most defective and weakest in each preparation. Upon swelling in a good solvent, they swelled most and showed the same transluscent appearance and consistency as the defective gels obtained at the top part of the unstirred reaction mixtures described above.

When C_0 approached C_0^* and Δc became small, the whole gelation process slowed noticeably and the effects of stirring rate became rather dramatic: the faster the stirring rate the slower was the gelation, the larger the volume near the stirrer that did not gel, and the larger the volume fraction of the gel being defective and mechanically weak. As expected, the closer the formed gel to the stirring-egg, the more defective it was. In cases where $C_0 \gtrsim C_0^*$, the fluid stresses due to stirring were qualitatively affected even by the shape of the reaction vessel. In taller vessels, where the top layer of the reaction mixture was farther removed from the stirring-egg, the gelation, especially on top, started substantially faster than in shallower vessels where a larger fraction of the reaction volume experiences higher shear stresses. Naturally, such effects were most dramatically noticeable in the vortex region: at $C_0 \gtrsim C_0^*$, the higher the flow rate and fluid stresses in this region, the slower the gelation there and the more defective and of weaker consistency the gelled network.

We are not aware of studies relating gravity and fluid stresses to the morphology and properties of gelled networks. Some computer simulation [751–753] and experimental [754–756] studies on flocculating colloidal particles were performed in the recent past and may be useful as a point of departure for future investigations. In these, parameters such as network segment flexibility and length, and overall network rigidity and concentration should be studied in addition to the parameters mentioned above.

4.2.5 High-Temperature Stability of Rigid Aromatic Networks

High-temperature stability is a very sought-after property of polymer networks intended for use in many high-technology applications. The stability is often defined according to the anticipated application. It may be the time to decomposition at a fixed temperature in inert or aggressive atmosphere, or the temperature of the onset of decomposition upon heating at a constant rate. In both instances, the sample weight loss is measured as function of time or temperature using thermogravimetric analysis (TGA) instruments.

Even though the interest in the thermal properties of polymers goes back over thirty years, no theory was hitherto developed to correlate thermal and structural properties. Currently, a series of structure-property correlations stands for theory. We believe the best group of correlations was assembled by Arnold [757] some time ago.

The strength of interatomic bonds in a polymer chain imposes an upper limit on the vibrational energy that the chain may store without bond scission. Since heat increases the vibrational energy, the heat stability of a polymer chain is directly related to the bond-dissociation energy of the chain's backbone bonds [757]. A further improvement in thermal stability may be achieved through resonance stabilization of structural units such as aromatic and heteroaromatic groups. Additional contributions to thermal stability come from secondary valence forces such as inter- or intra-chain hydrogen bonds and strong dipole–dipole interactions [757]. Additional requirements for heat stability are high melting or softening point and the removal of "weak links" in polymer chains and networks. The high melting or softening points are achieved by the creation of stiff, enthalpic chains consisting of as high an aromatic fraction as possible, and crosslinking them into three-dimensional rigid networks that do not melt and can withstand high temperatures. "Weak links" may have the form of reactable end-groups, reaction by products and chain-moieties, such as –C–H, easily attacked at elevated temperatures. Therefore, the most thermally stable systems are rigid aromatic networks in which the distances between aromatic branchpoints are very short and the aromatic segments are preferentially substituted in the para-positions. When bridges or swivels are present, care should be taken to minimize the number of hydrogenated aliphatic groups. These should either be phenylated instead of being hydrogenated, or completely replaced by groups such as –O–, –S–, amide or ketone.

Now, in order to rupture an unstressed bond a certain energy must be supplied. If, however, the bond is under tension due to a constant force pulling on either end, the bond rupture activation energy will be decreased by an amount equivalent to the work performed by the mechanical force over the stretching distance from the equilibrium position [758]. For a polymer network, the internal forces acting on a particular bond are the result of the deformation of some finite number of inter- and intra-segment bonds in the vicinity of the bond undergoing rupture [758]. Therefore, changes in the geometry of reactive sites, brought about by the application of external forces, may be reflected in increased rate of intra-network bond failure, in addition to the energy stored in the bonds [759]. Hence, a stressed rigid network will undergo high-temperature thermal degradation more rapidly and at lower temperature than its unstressed analogue; the higher the applied stress and network deformation, the lower the temperature for the onset of degradation and, under isothermal conditions, the more rapid the degradation takes place.

Finally, there were established several correlations between certain chain transition temperatures and the extent of network formation. Accordingly, transition temperatures such as T_g [760–762] or heat distortion temperature [763] show linear dependence on the level of network perfection, the less defective the network the higher the respective transition temperature within a range characteristic of the particular network composition. Based on our current level of understanding of the relationship between transition temperatures and concentration of network flaws [764], a significant amount of additional experimental, computer modeling and theoretical work remains to be done before a theory will evolve with predictive powers describing these relationships.

4.3 Some Properties of Liquid-Crystal Polymer Networks

The status of studies aimed at determining and understanding various properties of rigid main-chain LCP networks lags far behind similar investigations involving flexible main-chain and side-chain LCP networks. While in the case of rigid or stiff networks the studies are still at the learning stages for making such LCP networks, initial characterization and inducing permanent or reversible anisotropy, in the case of flexible LCP networks such studies occurred earlier and concurrently with the synthetic efforts. They were generally limited to network characterization and mapping on the temperatures of the transitions between the crystalline, various mesomorphic and isotropic phases. The liquid crystal behavior of these networks is caused by the ability of the mesogenic groups to align themselves in parallel or antiparallel arrays. Many examples of such reports were cited in Sects. 1 and 3 above and will not be repeated here.

Presently, the advanced studies of flexible LCP networks appear to concentrate on a group of properties owing their existence to a unique feature of the crystal unit cell of the polymer in the network: the property is present when and only when the crystal unit cell is non-centrosymmetric. Among these properties one finds ferroelectricity, piezoelectricity, pyroelectricity and nonlinear optical (NLO) responses to electrical or optical impulses. In all systems with these properties, there exists a transition from the non-centrosymmetric unit cell where the property exists, to a more symmetrical unit cell where the property ceases to exist. The transitions are reversible and are usually, but not exclusively, thermally induced. In many of the systems exhibiting any of the above properties, opposing polar domains appear spontaneously, balancing the electrical charges in the material. The direction of these domains may be reversed without changing the absolute magnitude of their polarization. When an external electric, magnetic or mechanical field is applied to the sample, then all domains and averaged dipoles may be pointing in one direction. When the field is removed, the domains may randomize again. This allows the "switching on" and "switching off" of the sample. The interested reader is directed to textbooks [765–772].

    The non-centrosymmetry of the crystal unit cell in LCP networks arises from
the ability of the mesogenic units and their dipoles to align themselves in more
or less parallel arrays with a unique sense. That is, the dipoles in the parallel
arrays all point in one direction and are not arranged in the anti-parallel
alignment so common in more symmetrical unit cells. The alignment, spontan-
eous or induced by an applied field, is greatly facilitated by the flexibility of
flexible main-chains and their spacers, and of spacers and tethers for pendant
mesogenic groups. In instances where the crosslinked network does not have
sufficient mobility to allow the alignment of the mesogenic groups after the
network is formed, an external field is applied to the precursor-polymers or pre-
polymers in the reaction vessel before the system is crosslinked. After alignment
is achieved, the crosslinking is effected while the external field is on. In this
fashion the highest possible induced anisotropy of the network may be obtained.
When the applied field is electric or magnetic, this process is often called poling.

    Elastomeric LCP networks with the potential to show piezoelectric behavior
were described by Zentel and co-workers [109, 110, 125, 773]. More recently,
Meier and Finkelmann [774, 775] created piezoelectric cholesteric elastomers
and described some of their properties. Polymers [776] and networks [126, 777]
with nonlinear optical responses were recently synthesized. Some of these
polymers are liquid crystalline [776]. We see no reason why mesogenic monomers and NLO responsive monomers may not soon be combined to form
permanently crosslinked NLO LCP networks.

    In several instances, flexible and semi-flexible LCP networks were created by
polymerizing mesogenic monomers in a liquid crystal state [140, 142]. In other
cases the liquid crystalline phase was oriented before the polymerization [142,
148]. It was demonstrated that liquid crystallinity and orientation are maintained in the final network but the kind of liquid crystallinity may change during
the polymerization [147]. Anisotropic networks were also obtained from worm-
like and stiffer long chains by crosslinking them either in a quiescent liquid
crystal state or when oriented under the influence of external fields [111,
272–276, 278, 280–281, 283, 419, 423, 567, 778, 779]. The base mesogenic worm-
like polymers were hydroxypropyl cellulose [272–276, 778], end-capped polyesters [111, 419, 423] and polyamides [567], poly(γ-benzyl-L-glutamate) [280,
281, 779], poly(diethylsiloxane) [283] and polyisocyanates [278]. Besides demonstrating the presence and kind of liquid crystallinity in the networks and the
effects of orientation, not much else was reported about the properties of the
above systems. Effects of swelling on the appearance and disappearance of
birefringence in gelled LCP networks were reported recently [112, 402].

    There have been several mechanical studies performed on gelled networks of
stiff LCPs, some where the liquid crystallinity of the precursors was retained in
the gels [278, 780] and others where stiff mesogenic segments were connected by
rigid branchpoints, leading to loss of mesomorphicity [277, 311, 312, 393, 400,
409]. Thus far, the results are limited to reported stress-strain curves [278, 393]
and to values of static shear modulus obtained by linear compression ex-

periments [277, 311, 312, 400, 409, 780]. In all cases the moduli of the LCP networks are of the same order of magnitude as those of non-liquid crystal networks created from similar stiff polyamide segments but connected by rigid branchpoints. All these networks have moduli about two orders of magnitude larger than flexible networks of comparable segment length and concentration [311, 312, 400, 402, 409, 780]. Not much more is presently known on the mechanical properties of these systems.

Recently there began a few theoretical attempts directed toward understanding the mechanisms involved in, and the prediction of correct magnitudes of, the responses of rigid networks to applied mechanical stress. Thus far these first steps are limited to attempted descriptions of the bending process of rodlike or stiff segments [48, 412, 663, 664, 732, 781]. At the other end of the spectrum, in the realm of flexible networks, work was initiated by Vilgis to describe theoretically defective and inhomogeneous networks as well as ones whose topology depends on the sequence of synthetic steps employed in their creation [782–784]. We believe that the concepts and analytical approach used to describe the flawed flexible networks may be translatable to isotropic and anisotropic rigid networks and their gels.

4.4 Modes of Segmental Deformation

4.4.1 Description of Segments

4.4.1.1 Mathematical Description of the Molecules

There are many cases in polymer science where a simple mathematical description of the polymer suffices to cover its physical properties and those of networks formed from it. The general theoretical situation will be reviewed in Sect. 5, but it is appropriate at this point to see if there is a reasonably concise way to describe the polymer. Clearly if a polymer varies all along its length and has properties dependent on that variation there is no easy way; e.g., it is useless to describe DNA or a protein in a simple way if their biological interactions are being studied. But the molecules described in the previous and discussed in this and the next Sections do have repeated properties and can be reasonably described.

The simplest attribute is clearly a place in space, so that if the polymer behaves like a line, a locus, in space the label R_n of a monomer goes over into an $\underline{R}(s)$ where $\underline{R}$ is the position of the s-th monomer. If a molecule is inextensible along its length, the rules of differential geometry will apply and Pythagoras' theorem gives $d\underline{R}^2 = ds^2$, i.e., $dx^2 + dy^2 + dz^2 = ds^2$. (Note that for fully flexible polymers it is convenient to take ds *not* the distance between monomers, but a distance already large enough for Gaussian statistics to apply.)

A different kind of mathematics then applies via the Wiener integral where the probability over large distances of finding a curve R(s) is

$$\exp\left(-\frac{3}{2l} \int_0^L \underline{R}'^2(s)ds \right).$$
(49)

For stiff or "stiffish" molecules this approach is not appropriate (more of this in Sect. 5).

In differential geometry it is shown that there is a set of three vectors, $\underline{t}$ tangent, $\underline{n}$ normal, $\underline{b}$ binormal which emerge from differentiating $\underline{R}(s)$ according to the Serret–Frenet formulae:

$$\underline{t} = d\underline{R}/ds \qquad\qquad t^2 = 1 \text{ (Pythagoras)}$$

$$\underline{n} = d^2\underline{R}/ds^2 (d^2\underline{R}/ds^2)^{-1}$$

or

$$\underline{R}'' = \kappa n$$

where κ is the curvature

$$\kappa^2 = \underline{R}''^2.$$

For example, a circle

$$\underline{r} = (a\cos\gamma s,\, a\sin\gamma s,\, 0) \text{ has}$$

$$\underline{t} = a\gamma(-\sin\gamma s,\, \cos\gamma s,\, 0) \text{ so that} \qquad a\gamma = 1$$

and

$$\underline{n} = (\cos\gamma s,\, \sin\gamma s,\, 0), \qquad \kappa = a^{-1}.$$

Finally,

$$d\underline{n}/ds = -(\kappa\underline{t} + \underline{b}\tau)$$

$$d\underline{b}/ds = \tau n$$

where τ is the torsion

$$\kappa^2\tau = \text{Det}\begin{vmatrix} x''' & y''' & z''' \\ x'' & y'' & z'' \\ x' & y' & z' \end{vmatrix}.$$

For example a helix

$$\underline{r} = (a\cos\gamma s,\, a\sin\gamma s,\, cS)$$

has

$$\underline{t} = (-a\gamma\sin\gamma s,\, a\gamma\cos\gamma s,\, c)$$

so that

$$1 = c^2 + a^2\gamma^2$$

and

$$\underline{n} = (\cos\gamma s, \sin\gamma s, 0)$$

and

$$\underline{b} = (-c\sin\gamma s, c\cos\gamma s, -a\gamma)$$

$$\kappa = \gamma = (1 - c^2)^{1/2}/a$$

$$\tau = c.$$

A curve is completely described by κ and τ and this description $\kappa(s)$, $\tau(s)$ is independent of any coordinate system, i.e., is intrinsic. The simple law of resistance to bending takes the form of an energy $H = \int \varepsilon R''^2 ds$ and therefore a weight $\exp(-\int \varepsilon R''^2 ds/kT)$.

Note that if a curve likes a shape different from the straight line $R'' = 0$, say $R_0(s)$, i.e., κ_0, τ_0, the energy will become

$$(\kappa - \kappa_0)^2 \text{ or } (\kappa^2 - \kappa_0^2)$$

and

$$(\tau - \tau_0)^2 \text{ or } (\tau^2 - \tau_0^2).$$

Such a description does not allow for the surmounting of an energy barrier. If the temperature allows for this to happen easily one is back to a flexible polymer. Intermediate situations are not simple. The interaction of different polymers will contain the simple

$$\iint U(R_1(s_1) - R_2(s_2)) ds_1 ds_2 \qquad (50)$$

but the relative orientation can also come in, and the simplest form is that already explored in liquid crystal theory, $\sin^2\theta$, with θ being the angle of intersection between two rodlike particles. This can be rearranged into

$$\iint ds_1 ds_2 (\underline{R}_1^1(s_1) \times \underline{R}_2^1(s_2))^2 W(R_1(s_1) - R_2(s_2)). \qquad (51)$$

Since there is clear evidence of nematicity in main-chain liquid crystal polymer networks from their quiescent birefringence, this term must be studied.

So far in this section we have treated the molecule as a straight line which can be bent into a smooth curve. It can also be twisted but at this point something extra is being added to differential geometry, for a straight rod can be twisted and remain straight. An additional vector $\underline{\phi}$ is therefore needed such that $d\phi/ds$ measures the twist. This is schematically shown in Fig. 34. The theory of the elasticity of twisting is given in textbooks of elasticity. A brief but physically clear version is given in Landau and Lifshitz [785], and a more mathematical treatment by Sokolnikoff [786]. The simplest form for the energy of twisting is proportional to $(d\phi/ds)^2$. There can be a relationship between ϕ' and κ, τ as is seen in the examples in the text. To understand mathematical models of this, one has to generalize the concept of a curve, for that mathematical concept on its own would make $\underline{\phi}$ quite independent of $\underline{t}$, $\underline{n}$ and $\underline{b}$. The simplest extension is to

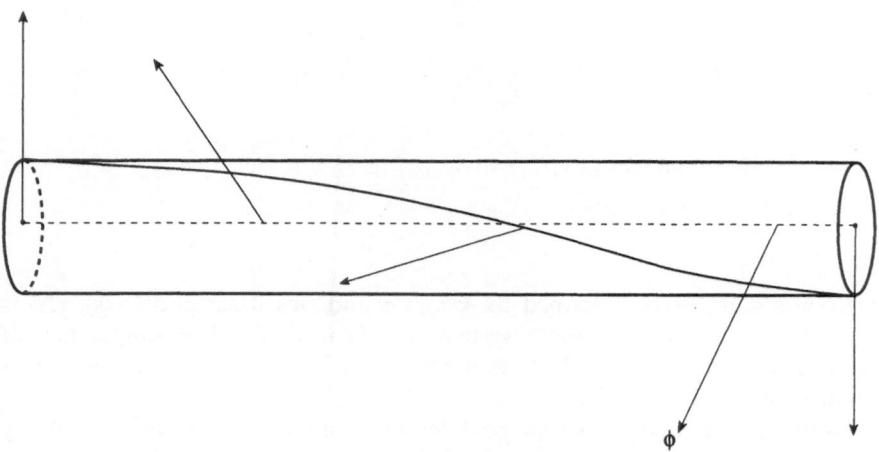

Fig. 34. Schematic of a stiff macromolecule as a twistable rod

think of the molecule as a strip. The energy stored in a twisted strip bent into an arc will not be the same as in an untwisted strip. It is not an issue of orientation but simply of the extension of the elastic material, i.e., the bonds in the actual molecule. Therefore, the energy will have the form $(\phi'^2 \kappa^2)$ and this form is a higher order term. Some examples are given by Sokolnikoff [786]. The matter first arose in the twisting and bending of aircraft wings by Griffiths and Taylor [787]. In general the formulae are very complex but one rather simple result relates the plane of deformation of a non-circularly cross-sectioned rod to the plane of the deformation forces. As usual with beam theory the key quantities are the moments of inertia, I_1 in the face of the rod along the z axis, and I_2 in the x and y directions. The angle θ between the planes of bending is $\tan^{-1}[(I_2/I_1)\tan\alpha]$ where α is the angle between the plane of the force and the xz plane. Although there is a considerable literature of elastic problems of this kind, for our purposes it is best to describe the deformation and twisting empirically.

4.4.1.2 General Empirical Description of Network Segments

It is far easier to visualize or trace the response to applied force of network segments, fragments and FPs when they are present in a swollen gel than when present in a glassy network in the absence of diluent. For ease of treatment, it is desirable for the rigid aromatic networks to be flawless where every functionality is fully reacted and all segments are connected to the network at both ends. In many or perhaps all cases, such networks are unattainable. In defective rigid networks and their gels, aromatic or otherwise, large-scale deformation or swelling are associated with non-affine smaller scale changes in shape and size of network defects, whole fractal polymers and network fragments. Network defects, which may be as small as a missing single covalent bond or as large as an absent whole fractal polymer, may appear or change their volume or change their shape upon gross deformation or swelling of a network gel. These changes can be brought about by translations of whole FPs or network fragments in the gel without significant deformations of network segments or branchpoints. Such translations would not occur had the network gel been flawless. The fact that in the presence of defects the translations do occur and are associated with only a small number of segment deformations reduces the resistance of flawed rigid networks to gross deformation or swelling, as compared with their flawless analogues. This means that the defects in rigid aromatic networks reduce the modulus of such networks and their resistance to swelling. Network deformation or swelling modes necessitating significant segment or branchpoint deformations will make their appearance only after all or most of the contributions due to defects are exhausted. With increased deformation defective rigid networks thus undergo strain hardening. At the same time, chain ruptures start appearing with increasing frequency, caused by the network rigidity and limited ability to deform affinely with the gross deformation of the network. When a sufficient number of old and new defects are present in the system, at least one of them may grow under the applied stress and become larger than the critical size for catastrophic failure. When this happens, the network will fail in a rather brittle manner.

When gels of flawless aromatic one-step or two-step networks are deformed, a significant displacement takes place in the position of rigid segments and branchpoints relative to their initial positions and to one another. In a flawless network the effects of gross deformation are felt immediately on a segmental size scale, without the passage through nanometer size scale effects caused by the presence of network defects. While positional translations of segments and branchpoints during gross deformation are believed to be ubiquitous, changes in angular position of one branchpoint relative to its neighbors are expected to be rather scarce. This is because in a flawless rigid network each such change requires the deformation of several adjoining stiff segments with large investment of energy. The primary mechanism of stiff aromatic segment motion leading to significant translations of rigid branchpoints is rotations around single bonds in the chain backbone which are governed by relatively soft torsional potentials.

In the limit of very long chains, many qualitative properties are practically independent of the details of chain structure [788–790], but in the aromatic networks with, generally, very short segments, the structural details become extremely important. Thus, in rigid branchpoint networks, the deformation of rodlike segments is expected to involve fundamentally different modes of deformation than stiff worm-like segments. Furthermore, in networks where junction points of varying rigidity are inserted by either reacted end-caps or less than fully-rigid branchpoints, the overall network rigidity may be affected by the interplay of segment and junction rigidity. In fact, with increases in the volume fraction of the junction points, and with increased junction flexibility, the nature and characteristics of the junction points may start dominating and the stiff segments begin contributing less and less to the properties of the network.

We may state, then, that for the purpose of describing differences in modes of segmental deformation, the structural details of stiff aromatic network segments gain in importance as the length of these segments becomes smaller and as the defect concentration in the network decreases. The relationship between chain structural features and possible deformation modes of stiff segments in flawless rigid aromatic networks will be explored below.

4.4.2 Deformation of Aromatic Segments with Bridges Consisting Exclusively of Coaxial Single Bonds

Typical of such segments are the poly(p-phenylenes) discussed in Sect. 3.1.1.1 and probably 3.1.1.2, the stiff aliphatic segments of Sect. 3.1.1.3, some of the triazine branchpoint networks in Sect. 3.1.5 and some end-capped polyimides, polybenzimidazoles and similar highly rigid segments mentioned in Sect. 3.1.6.2. Their common structural feature is that their aromatic phenylene or condensed rings are connected to each other by a single bond at each location and all these single bonds are coaxial with the segmental axis:

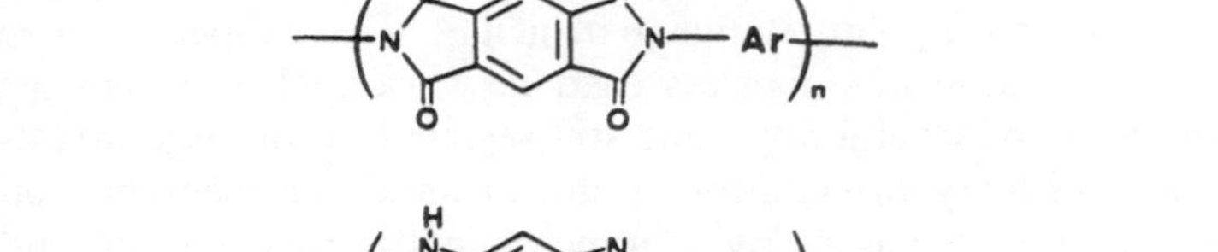

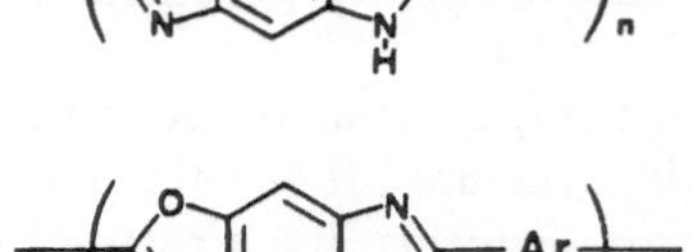

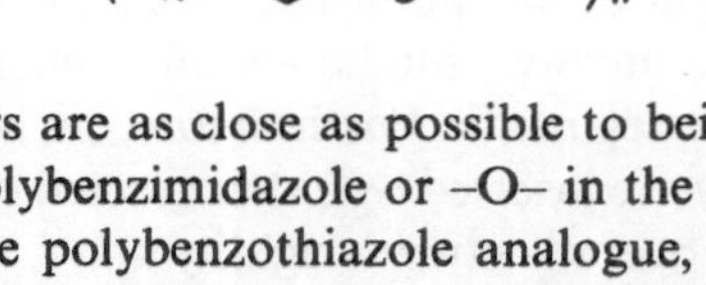

These polymers are as close as possible to being truly rodlike. If, however, the N–H in the polybenzimidazole or –O– in the polybenzoxazole are replaced by –S– to give the polybenzothiazole analogue, then the single bonds along the chain cease being fully coaxial and the polybenzothiazole chain adopts an undulating configuration [791]. The loss of a truly rodlike configuration results in a dramatic decrease in persistence length, dropping from about 650 Å for the polybenzoxazole to about 270 Å for the polybenzothiazole [791]. The slightly larger size of the sulfur atom and the longer single bonds connecting it to the rest of the heteroaromatic group, relative to those of the $= N-H$ and $-O-$ moieties, are the reasons for the loss of coaxial conformation in polybenzothiazoles. These differences also change the orientation of the phenylene rings from being coplanar with the heteroaromatic groups in polybenzimidazole and polyben-zoxazole, to being oriented at angles of 23° [792] and 55° [228, 229] in polybenzothiazole.

In the coaxial rodlike polyphenylenes the aromatic rings along the chain are not coplanar. Spectroscopic studies [208, 793] showed that the dihedral angle between the two rings in biphenyl is about 20 degrees in solution and about 45° in the gas phase [794, 795]. X-ray studies of the crystalline phase of a polyimide containing biphenylene residues along the chain revealed that the dihedral angle in the biphenylenes is close to 45° [795]. In light of the above, torsional angles between rings of poly((p-phenylenes) residing in the interval of 20–45° appear to be highly reasonable. Crystal structure studies on the polyimide

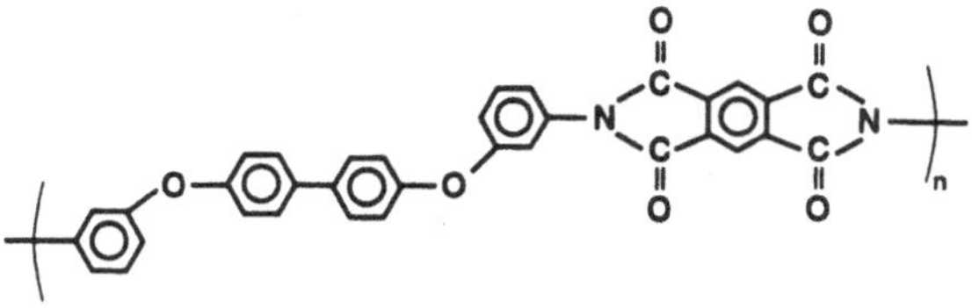

indicate that in this polymer the phenylene rings and the adjacent condensed heteroaromatic groups are not coplanar. In the polyimide the mutual orienta-tion of the planes of the phenylene rings are close to 45° and 135° relative to the plane of the bisimide in between [795]. In crystalline polybenzoxazole, the planes of the phenylene rings are nearly coplanar with the bisbenzoxazole groups [796, 797], but occasional chain-twists of about 90° may be present [797]. Here, at frequencies of about once every four motifs, the plane of the phenylene rings twist to be oriented at about normal to the planes of the adjacent bisbenzoxazoles [797]. A more or less coplanar orientation of the phenylenes and heteroaromatic groups is expected by analogy in the case of polybenzimidazole too, and here again an occasional chain twist is likely. We conclude that in rodlike segments both conditions of coplanarity and of angles relative to one another arise in the mutual orientation of the coaxial aromatic groups. In the crystal state, for the sake of packing symmetry, the angles along the chain usually alternate. In a network, especially in the gel state where segmental packing practically ceases to exist and intersegmental interactions fade away, the angles along the chain need not alternate regularly and random sequences may occur.

Upon reflection we find that when rodlike segments are bent, the coplanar or out-of-plane orientations of the aromatic groups lose their distinction. For a rodlike segment to bend, the valence angles around the single coaxial bonds must change, or the aromatic residue pucker somewhat, or covalent bonds alter their length. It has been shown in the literature [798] that when chains of both poly(p-phenyleneterephthalamide) and poly(m-phenyleneisophthalamide) are stretched, most of the deformation is caused by valence angle changes with only a very small contribution coming from changes in bond lengths. We will therefore neglect changes in bond lengths in the following discussion. When a rodlike aromatic segment is bent under external force, changes in the valence angles of the single bonds and puckering of the aromatic groups can take place. The changes in valence angles can occur in the plane of the adjoining aromatic group, or perpendicular to this plane. Changes in the plane are more likely to be hindered by van-der-Waals-type interferences by adjacent aromatic units. On the other hand, changes perpendicular to the aromatic plane may be enhanced by a concerted puckering of the aromatic unit. Thus, the most efficient mode of segment bending involves the combination of several movements. In the first, one or more aromatic units assume, by torsional movement around the single coaxial bonds, a position in which their planes are normal to the plane of the applied stress. This torsional movement is a low energy one, requiring only a few kcal/mol, and takes place when the rings' initial orientation contains a significant vectorial contribution from the plane of the applied stress. In the second movement, one or more single bonds alter a little the valence angles around the connections to the aromatic group, in the direction of the applied stress. Such changes have much higher energy barriers. At the same time puckering of the aromatic groups normal to the plane of the applied stress may take place to relieve excessive stresses on the single bonds. The resulting deformed segments may appear as in Fig. 35. It is apparent from this figure that, when rodlike segments bend, they do so in small increments, making the segmental curvature close to uniform. Because the energy barriers to the bending motions are very high, the modulus of flawless network gels made from such segments is expected to be far higher than the modulus of worm-like segments where the bending may be accomodated by lower energy torsional motions.

4.4.3 Deformation of Stiff Aromatic Segments with Bridges Consisting of a Single Non-Coaxial Bond Connecting Two Coaxial Bonds Between Aromatic Groups

In each bridge between two adjacent aromatic groups, these stiff segments contain two coaxial bonds on both sides of a single non-coaxial bond. The presence of the middle non-coaxial bond allows for relatively low-energy torsional rotations, changing the direction of the segmental axis and introducing a bend in the segment. Polymers typical of such structure are

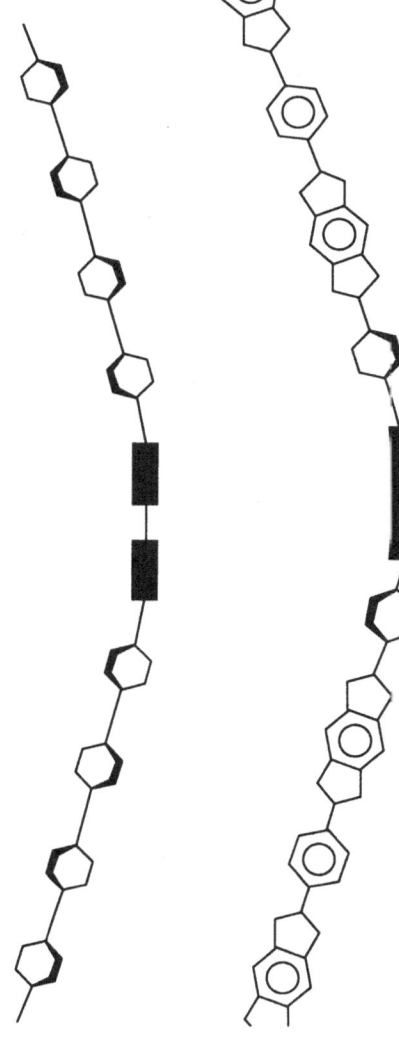

Fig. 35. Deformed stiff chains with single coaxial bonds between aromatic and heteroaromatic units. Twisting of the aromatic groups out of the bending plane may depend on their specific structure and their position along the deforming chain

polyamides

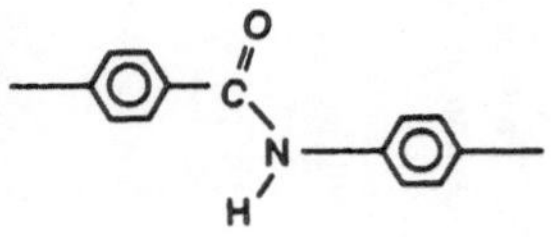

and polyesters

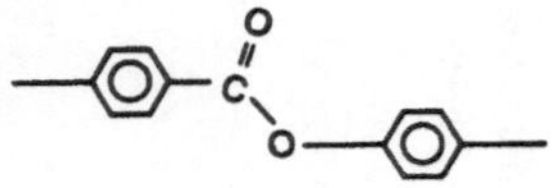

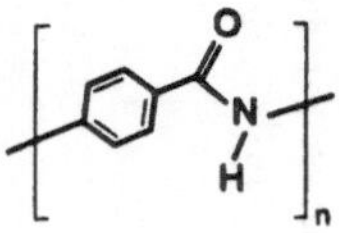

p–BA

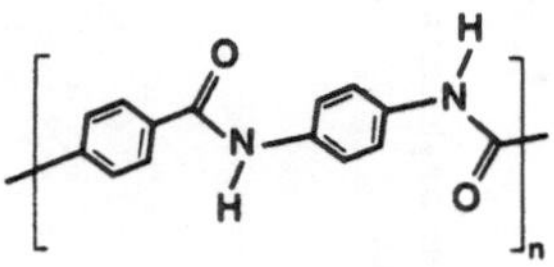

p–PT

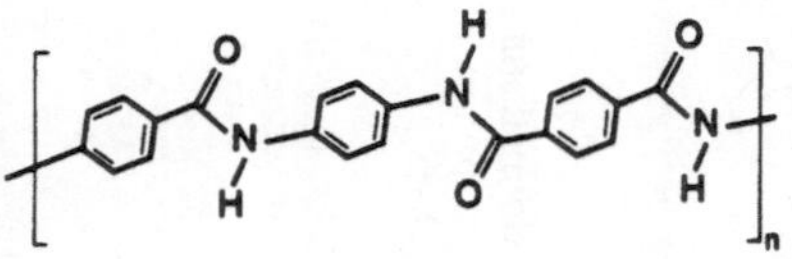

p–BT

the torsional and valence bond angles being defined in the scheme below:

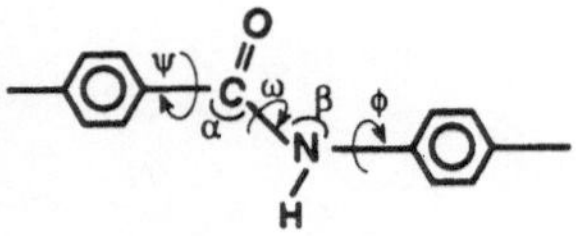

Because neither angle α nor β equals $120°$ and $|\alpha - \beta| > 0$, there is a curvature imparted to the overall chain direction, a curvature reflected in the magnitude of the persistence length, a. The magnitude of a is, hence, inversely proportional to $|\alpha - \beta|$ [663]. It is well accepted that in p-BA p-PT and p-BT the amide group exists in the planar trans conformation. Relative to the aromatic ring, the amide groups can adopt either anti- or syn- conformations:

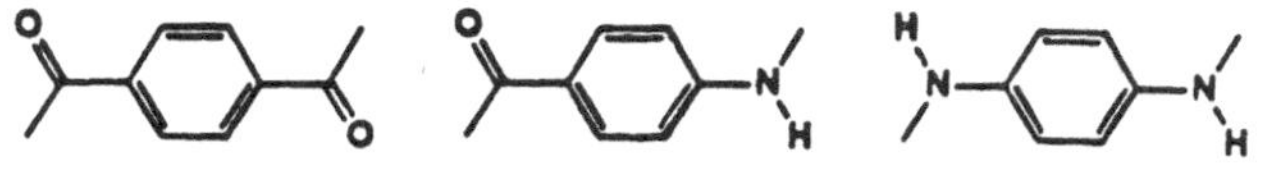

anti

syn

but even this assumes coplanarity of the amide and aromatic groups. In fact, the planes of the aromatic rings are oriented at $+30°$ and $-30°$ relative to the planes of the amide groups [798–800]. Therefore, the para-substituents can adopt two syn- and two anti-conformations. We illustrate this for the benzamide group. For the anti-conformation, one group can be above and the second group below the plane of the phenylene. In the first case, which we designate as $(-)$anti-, the dihedral angle between the amide carbonyl and amide N–H groups is 180°. In the second case, designated as $(+)$anti-, the carbonyl oxygen and the N–H hydrogen are on the same side of the phenylene group and the dihedral angle between them is 120°:

(–)anti-(180°)   (+)anti-(120°)

In a similar fashion, we represent the syn-conformers as $(+)$syn- and $(-)$syn-, with torsional angles between the carbonyl and N–H groups of 0° and 60°, respectively:

(+)syn-(0°)   (–)syn-(60°)

A similar analysis was applied to the carbonyl pairs in terephthalamide groups [663] and can easily be applied to the diamines in 1,4-phenylenediamides.

We have previously observed [412, 664] that in a fully anti-chain, a single interconversion from anti- to syn- placement creates a bend of about 20° in the chain direction. The same bend occurs with a single anti-placement in an otherwise fully syn- chain. Very recently these observations were expanded to cover a broader series of aromatic polyamides [801]. In modeling studies of aromatic polyamide chain bending [663], we have found that two other chain bending modes appear in higher frequency than the syn-anti and anti-syn interconversions. These two more frequent modes of deformation were found to be [663] a concerted one-direction departure of the torsional angle ω from its energy minimum, and 60° phenylene ring flips from (+)syn- to (−)syn- and vice versa, or from (+)anti- to (−)anti- and vice versa. Importantly, the three modes of deformation were found to coexist in the same free chain and the shorter segments even when firmly anchored in rigid network. The frequency of appearance of the three deformation modes diminished from the concerted departures from energy minima, through the 60° ring flips to the syn-anti interconversions. The weight of each mode could be altered by changing the height of the energy barriers to torsional rotations around the ring-to-carbonyl and ring-to-nitrogen bonds, the higher the barriers the lower the weight of the concerted departures from energy minima and the larger the weight of the two other modes. Representing the polyamide chain or segment as a ribbon or a long thick strip, the three modes of deformation are shown pictorially in Fig. 36. Additional, less frequent modes of deformation of stiff aromatic polyamide segments may exist but were not detected in our simulations studies [663].

It is important to recognize that, because the mechanism of the above deformations involves torsional rotations and not valence angle changes, all the deformations are twist-bends and not simple breaks in the chain direction. Furthermore, because the rotations are all torsional, the modulus of the network is far lower than the modulus of rigid rodlike networks of comparable l_o, f and concentration.

During the simulation studies [663], we investigated what will happen to a stiff polyamide segment whose rigid branchpoints are firmly held by a rigid network, when these branchpoints are incrementally brought closer together without changing their relative orientation. In addition to manifesting the above three modes of deformation, the negatively strained segments adopted different configurations depending on their axial ratio. When $x \leq 6$, the segments appeared to prefer an arched shape. For longer segments, with $9 \leq x \leq 15$, the segments adopted a corkscrew-shaped configuration. It remains to be seen how many configurations exist in real networks with stiff polyamide segments.

In Sect. 4.1.1 it was indicated that, because the difference in the energy of the anti- and syn- conformers is only about 0.3 kcal/mol [665], there is very little preference to the anti- over the syn-placements during the polycondensation reaction, provided that intersegemental H-bonds and parallel packing are totally or mostly absent. Combining all the above we may state that, in flawless rigid polyamide networks with average axial ratio x, the stiff segments may deform by several twist-bend modes and the deformed segments may adopt several length-dependent configurations. Because of the plurality of syn-anti placements and deformation modes coexisting along the segments, a full theoretical treatment capable of correctly predicting the behavior of stiff polyamide segments is still far out of reach.

Aromatic polyester segments are similar in many respects to the aromatic polyamides. An important difference is the inability of polyesters to participate

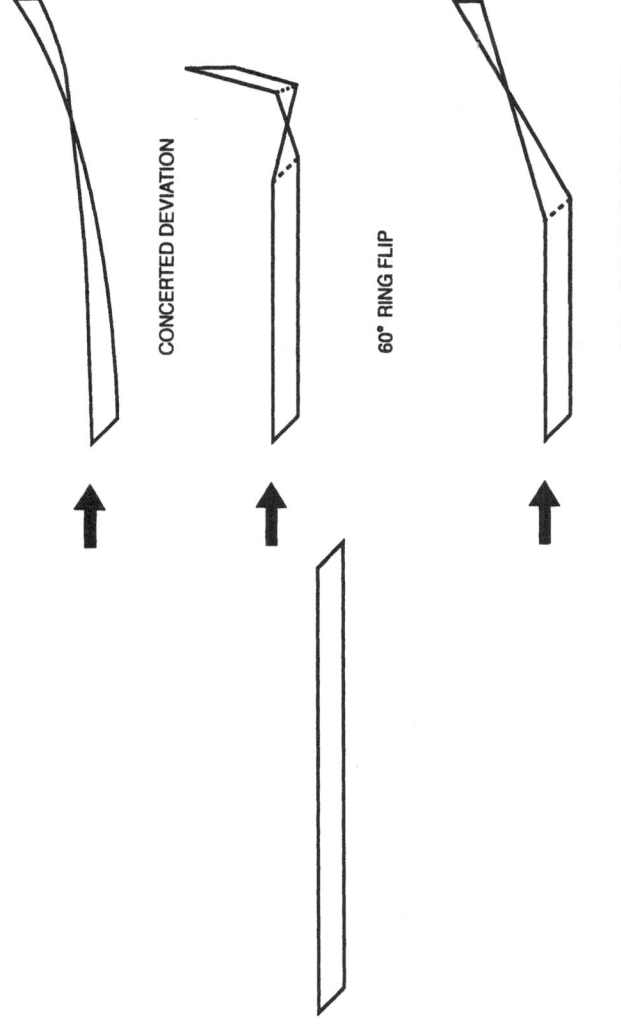

Fig. 36. Three deformation modes of aromatic segments containing single non-coaxial bonds between stiff coaxial units

in intersegmental H-bonds, a fact that greatly relaxes the demand on parallel polyamide chains to be in register in order to maximize the number of intersegmental H-bonds. Therefore, stiff aromatic polyester chains can be rather nicely packed in parallel arrays without being in register [802–806].

In a fashion similar to polyamides, the planes of the aromatic rings in polyesters are not coplanar with the planes passing through the ester groups. The ring near the carbonyl is oriented at an angle of 6–10° relative to the flat plane through the ester [665, 807, 808a], making them very close to coplanar. The ring attached to the oxygen, however, is oriented at an angle in the range of 65–70° relative to the ester [665, 807–811]. Although the ester group is flat, it may exist in trans- or cis- configurations, adding to the number of possible chain configurations and contributing to the reduction of the chain persistence length. The lowest energy barriers between rotational isomeric states of fully aromatic polyesters are those associated with torsional motions in the ester group. They are all in the range of 5 kcal/mol or lower [807], placing them lower than the comparable energies in aromatic polyamides. The difference is most dramatic for the central bond: while the activation energy for cis-trans interconversions in aromatic polyamides is of the order of 20 kcal/mol [664], it is only about 6 kcal/mol in the aromatic polyester analogue [807]. This surprisingly low energy barrier facilitates motions in transverse direction to the chain axis and, as a result, shorter persistence length for polyesters than for polyamides.

Furthermore, the angles α and β in aromatic esters are respectively 111.7° and 118° [807] or very close thereabouts [665, 808]. This means that $|\alpha - \beta| \geq 7.1°$ and that the axes of the aromatic rings are oriented at an average angle of $\geq 7.1°$ relative to the chain axis. This imparts a substantial curvature to the chain, in a fashion similar to that present in aromatic polyamides [663]. Ring flipping through 180° is rather common in fully aromatic polyester chains even at ambient temperature [812, 813]. The onset temperatures for ring flipping are substantially below the T_g of the respective polymers [813]. Because there is practically no dipolar interaction between adjacent ester groups [665], the frequency and ease of ring flipping are expected to be similar to those of ester group flipping by 180°. This mode of deformation was previously found in the simulations of polyamides [663] and is expected to be present in polyesters. Because the energy barriers to torsional rotations of polyesters are lower than those of polyamides, concerted departures from energy minima in the same direction are expected to be the commonest mode of deformation of polyester chains and segments. Less frequent modes will be those associated with 180° ester group flips (syn-anti interconversions). At present we have no estimate of the frequency of segmental bending due to ring-flips and what are the angular changes in ring orientation relative to the planes of the ester groups. We believe, however, that a mode of chain bending associated with ring flips does exist. A very recent brief computer simulation study [814] leads to a similar conclusion.

4.4.4 Deformation of Aromatic Segments with Swivels Consisting of Two Single Bonds Connected by a Single Atom

The lowest energy deformation mode of such segments is twist-bends at their swivel points. Typical of such swivels are the following structures:

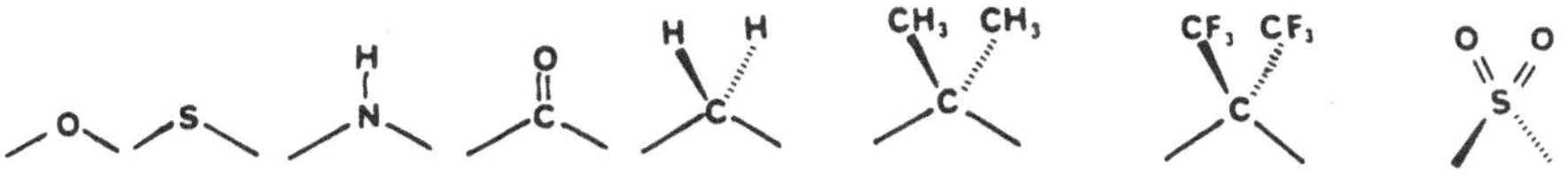

The rigid aromatic residues between swivels and junctions have the general shape of stubby flat bars. When the segments deform, the aromatic flat sections change their directional orientation and, concomitantly, perform torsional twists by rotating around one or both of the single bonds leading to the central atom in each swivel [815]. Full 360° rotations are possible, but as a rule the rotations settle in energy minima at several angles, the number of which depends on the specific structure of the swivel and the aromatic groups connected by it. While the coplanar conformation of the two p-phenylene groups on both sides of

and  are all associated with

relatively very high energy [815], a skew of the rings out of coplanarity rapidly brings them to stable torsional angle energy minima. Thus, for instance, the planes of the p-phenylene rings are at $\pm 40°$ relative to the plane containing –O– or –S– in the respective poly (p-phenylene oxide) or poly (p-phenylene sulfide) [815], but only about 30–34° in PEK and PEEKK [816]. In the case of the ketone bridge the conformational energy minimum is found for the skewed form, where the torsional angles are both at $\pm 30°$ [617]. In a crystalline polyimide containing ketone swivels the torsional angles were calculated from X-ray data to be at $\pm 33°$ [817], in good agreement with computer simulation results. In the crystal form of diphenylmethane the lowest conformational energy was found to be in the skewed $\pm 52°$ [818] state, and in structures containing the 2, 2-bis (p-phenylene)propene the p-phenylene rings are known not to be coplanar [819, 820].

The salient point of all the above is that in each swivel the aromatic segment changes its direction by an angle equivalent to the valence angle of the swivel, and deviates from coplanarity by an angle equal to $\pm$ the minimum energy torsional angle. If in a given segment there exists more than one swivel point, such as in

then the number of minimum energy conformations rapidly increases. If, in addition, not all phenylene rings are substituted in para positions, but some are meta- or ortho-substituted instead, then the number of geometrical isomers rapidly increases and the probability of colinear segments rapidly diminishes [815].

Segments containing swivels are not straight. The easiest deformation mode of such segments is a change in the torsional angle of the bent swivel. This is somewhat similar to the model offered by Yannas and Luise a decade ago [821, 822]. Because the energy barriers to such changes are very low [815], they take place before any other changes may occur, such as changes in valence angles. For a stiff segment attached to the network at only one end and containing a single swivel in the middle, a 360° rotation around one bond of the swivel drives the unattached half of the segment through a "cone of rotation" [664] as is shown in Fig. 37. In this case the attached half of the segment and the central atom of the swivel need not change their position. When both ends of the segment containing a swivel are rigidly connected to a rigid network, then concerted torsional rotations around several single bonds in both portions of the segment are required. When these rotations are insufficient, puckering of aromatic and heteroaromatic groups, and twisting and deformations of branch-points may participate in facilitating the segmental bend and relieving the stress.

When a stiff segment contains more than only one swivel, the mode of deformation consuming the least amount of energy is one in which a series of

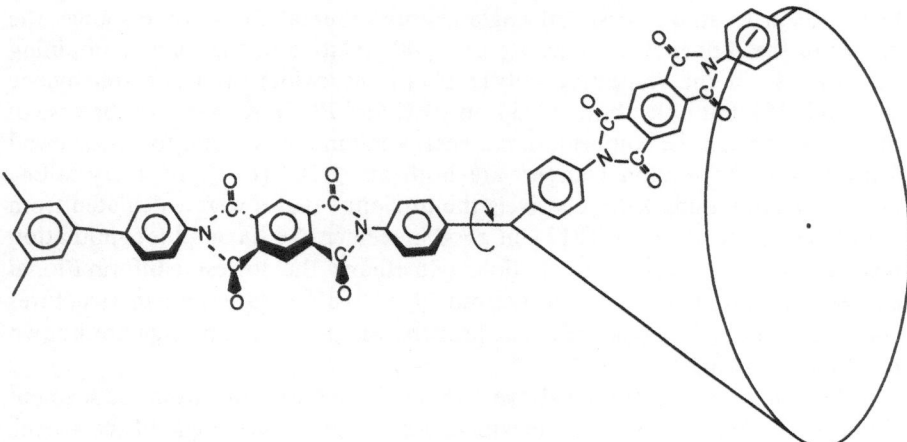

Fig. 37. "Cone of rotation" of stiff chain with a single swivel in it

torsions are executed around several single bonds. The torsions around bonds belonging to the swivels in the segment are those capable of producing the largest deviations of the segment from its initial conformation, but other single bonds may contribute to the deformations. As the number per segment increases of single bonds capable of substantial changes in direction, by moving from one location to another in their "cones of rotation", the deformation of the whole segment becomes easier: large directional changes of sections between swivels are facilitated by torsions encountering relatively low energy barriers. The rotations may be concerted and concomitant, or in sequence, each releasing stress buildup caused by the applied external force and by previous rotations. The great variety of deformations in segments containing several swivels each, is schematically shown in Fig. 38. Here, two cases are depicted. In the right panel the segment deforms by a change in the position and direction of its middle sections, allowing only a small change in the position and no change in the orientation of the branchpoint. In the left panel the same segment is deformed by rotation of other swivels. Here a large change in the position of a branchpoint is allowed but not in its orientation in the plane of the page.

4.4.5 Deformation of Networks with Stiff Aromatic Segments and Relatively Flexible Aliphatic Junctions Created from Reactive End-Caps

The point that the deformability of stiff segments rigidly connected at both ends to a rigid network is severely restricted cannot be over-emphasized. This is

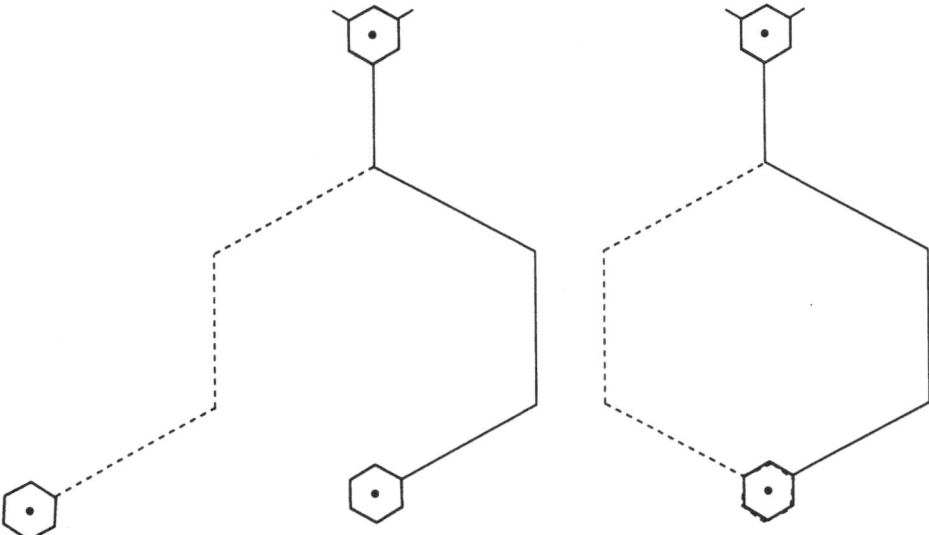

Fig. 38. Deformation of stiff segments with three swivels in each. *Solid line*: initial position. *Dashed line*: final position

because the stiff segments are generally very short and have only a few locations each at which significant deformations can take place. Each of these deformations must overcome relatively high energy barriers and has a small number of rotational isomeric energy minima to settle into. Such systems may be collectively called enthalpic because their deformability is not directly dependent on the absolute temperature. In fact we believe that, in stiff segments, changes in the ease of deformation may take place over narrow temperature intervals, in which the kinetic energy of the segment surpasses one or more energy barriers between rotational isomers.

This is dramatically different from flexible segments, each of which contains many locations at which significant changes in chain direction may take place, and where each of these reorientations involves rotational isomerization over relatively low energy barriers. Because the energy barriers for these rotations are so low, they are in a range comparable to kT and therefore a large number of these rotations takes place at ambient temperatures or thereabouts. The number of these rotations increases with temperature, making networks of flexible segments behave in a heat-activated, entropic manner [5–16, 38–46].

In Sect. 3.1.5, rigid networks with triazine branchpoints, and in Sect. 3.1.1.2, aromatic hydrocarbon networks were described. These are similar to the networks described here in Sects. 4.4.2, 4.4.3 and 4.4.4. In Sect. 3.1.4, systems were described with multiple swivels connecting small aromatic units strung along very short segments. They are conceptually similar to the systems discussed in Sect. 4.4.4. These networks have high modulus and substantial strength below their respective glass transition temperature, and soften rapidly above T_g. In Sect. 3.1.6, networks prepared from end-capped stiff segments were described. When the reactive end-caps are unsaturated imides or strained rings, the junction point created by their reaction with one another contains a significant, generally poorly defined, number of aliphatic carbon atoms with sp^3 orbitals, usually connected to one another by single bonds. The junctions will retain varying levels of mobility and deformability when not too large a number of the carbon atoms orbitals are involved in direct carbon-carbon bond formation to the exclusion of carbon-hydrogen bonds. The energy barriers to the rotational motions necessary to bring about positional translations of atoms and single-bond directional reorientation are relatively low, typical of barriers usually associated with rotations around single bonds with sp^3 geometry. Because of this, the aliphatic junction points are entropic in nature, in contradistinction with the enthalpic behavior of rigid aromatic branchpoints. It should be recalled in this context that the T_g of aliphatic structures, such as those present in the junction points, is usually substantially lower than the T_g of the aromatic structures of the stiff segments.

Therefore, when an aromatic network with aliphatic junctions is macroscopically deformed, the molecular-scale sites accomodating most of the deformation are the aliphatic junctions, provided these are not internally crosslinked to such a level that the T_g of the aliphatic junction equals or transcends that of the aromatic segment material. In general, the aliphatic junctions are not geometrically well-defined and may be taken to be isotropic in nature. Applied stresses are dissipated by the junction performing a multiplicity of torsional motions, in different directions and to differing extents, resulting in the junction itself adopting internal bond alignments different from the starting point, and the stiff segments emanating from this junction pointing in new directions. The torsional motions within the junction and those of the bonds connecting the junction with the stiff aromatic segments may occur all at the same time or, more probably, in a cascade-like sequence. Thus, only very large macrodeformations that cannot be accomodated by deformations of the junctions and reorientations of the stiff aromatic segments may result in deformations of these segments. These deformations may vary according to the nature of the stiff segments, but they are all variations on the themes discussed above of bending, twist-bending and torsions around swivels.

Because the chemical nature of the aliphatic junction points and the stiff aromatic segments is so different, their solubility parameters and affinity to solvents are expected to be different too. Therefore, it is expected that upon swelling in a given solvent, locally selective swelling will occur. In solvents better for the aromatic segments, segment-rich domains will preferrentially absorb more solvent and swell, and in solvents better for aliphatic units, the aliphatic junctions will preferentially swell. Large internal stresses are, hence, expected to appear in such systems. At present, however, not much is known about such systems.

4.5 Questions of Elastic Constants

The general relationship between stress and strain is

$$\sigma_{ij} = p_{ijkl}\varepsilon_{kl} \tag{52}$$

but for homogeneous material this reduces to

$$\sigma_{ij} = \left(\frac{E}{1+v}\right)\left(E_{ij} + \frac{v}{1-2v}\delta_{ij}(\varepsilon_{11} + \varepsilon_{22} + \varepsilon_{33})\right) \tag{53}$$

where s_{ij} is the unit matrix, E is Young's modulus and v is Poisson's ratio. If we assume that our network solid is homogeneous and isotropic, which is true on the macroscale but not necessarily true at segmental dimensions, we can expect to derive from Eq. (53) expressions for E and v in terms of the constants of a single rodlike segment and the nature of the junction points.

The strength of the network is measured by the force required to fracture it, or to make it flow in the sense that a failure can be catastrophic or many microscopic failures. If the material shows some non-recoverable strain prior to failure in the form of ductile deformation, then its strength reflects contributions from the shear modulus, G, in addition to contributions from E. Because of the complexity of the failure process [823] and its dependence on the number and size of sample imperfections, in addition to the characteristic moduli, we shall not discuss below the macroscopic strength of polymeric networks and gels except for its kinetic dependence. We shall start, then, with a brief discussion of segmental strength, follow with a short exposition of the kinetics of failure of bulk specimens, and then concentrate on the strain modulus of individual stiff segments.

The potential energy of a bond in a diatomic molecule in the absence of stress is usually approximated with the Morse function [824]:

$$V(l) = D(1 - \exp[-a(l - l_e)])^2 \tag{54}$$

where l is bond length, l_e is the equilibrium separation distance of the atoms, D is the bond dissociation energy, $a = (K/2D)^{1/2}$ is a parameter which defines the width of the minimum and K is the force constant of the bond in the neighborhood of the equilibrium separation [758]. When an external axial force is applied to the bond, it is counteracted by the binding force which is the derivative of the Morse potential:

$$F_1 = -dV(l)/dl = -2Da(\exp[-2a(l - l_e)] - \exp[-a(l - l_e)]). \tag{55}$$

This function has a maximum at

$$l_b = l_e + \ln(2/a). \tag{56}$$

The maximum corresponds to the critical elongation at break. The force producing this elongation is known as the breaking strength of the bond:

$$F_b = F(l_b) = \tfrac{1}{2}Da. \tag{57}$$

It is obvious that in covalently bonded systems more complex that diatomic molecules, more parameters are needed in order to approximate the potential energy of the various bonds [758] and estimate the macroscopic breaking strength of the system. For a single segment of a rigid network fully oriented in the draw direction, we may estimate the breaking strength, σ_b, as the breaking force for a single bond in the direction of the segment axis applied to it in the axial direction, F_b, divided by the effective cross sectional area of the segment, d^2:

$$\sigma_b = F_b/d^2. \tag{58}$$

In the linear response regime the stress, σ, carried by the axially strained segment, containing several single bonds is

$$\sigma = F/d^2 \tag{59}$$

with the provison that for no single bond will the stress σ surpass σ_b. The modulus for small deformations of a single chain or segment is, then, given by $\sigma = E\varepsilon$, but in this case ε is the segment strain in the axial direction of the applied stress. It is commonly accepted that, instead of E, the force Φ necessary to stretch the chain by 1% strain is used:

$$\Phi = 0.01(Ed^2) \tag{60}$$

where Φ is a measured of the rigidity of the polymer segment. In samples where all the chains are fully aligned in the draw direction, one can obtain Φ from the Young's modulus of the specimen and knowledge of the effective cross-sectional area of each chain. Of course, when the segments are not all aligned in the draw direction, larger forces must be applied to the sample in order to reach σ_b, E or Φ of the individual misaligned segment.

Segmental bond rupture under stress can be considered as a chemical reaction whose rate is increased by mechanical stress. It can therefore be treated in the framework of the transition state theory:

$$k_c = A \cdot \exp(- E_A/RT) \tag{61}$$

where k_c is the kinetic rate constant, E_A is the Arrhenius activation energy $(\Delta H^* + RT)$, ΔH^* is the enthalpy of activation and the pre-exponential factor A is exponentially linked to the entropy change, ΔS^*, occurring during the activation process. By combining the transition state theory with De Boer's [825] thermodynamic formulation, Tobolsky and Eyring [826] developed the rate theory for thermally activated failure according to which

$$k_c = A \cdot \exp[- U_o - \gamma\sigma)/RT] \tag{62}$$

were U_o is the thermal activation energy for bond rupture $(= D - \frac{1}{2}RT)$ corresponding to ΔH^* in the unstressed state, σ is the applied stress and γ is the stress concentration factor for brittle failure. Zhurkov and associates deduced that this very same relationship holds true for stress activated bond scission [827–829] in the brittle state. In this relationship γ has the dimensions of volume and is identified as the activation volume for the reaction [758]. The time to failure, τ, of brittle samples below T_g is, then,

$$\ln\tau = \ln\tau_o + (U_o - \gamma\sigma)/RT \tag{63}$$

where τ_o is a time constant of about 10^{-13} seconds. In the case of ductile samples, where motions of whole segments between entanglements or crosslinks may take place, the activation energy for brittle rupture, U_o, is replaced by the activation energy for the glass transition process, ΔH_a, and the symbol ω is used instead of γ to emphasize the fact that non-brittle failure is being described [714, 829–835]:

$$\ln\tau = \ln\tau_o + (\Delta H_a - \omega\sigma)/RT. \tag{64}$$

The transition from ductile to brittle failure occurs when

$$\gamma + (\Delta H_a - U_o)/\sigma = \omega, \tag{65}$$

brittle failure occurs when

$$\gamma + (\Delta H_a - U_o)/\sigma > \omega \tag{66}$$

and ductile failure takes place when [835]

$$\gamma + (\Delta H_a - U_o)/\sigma < \omega. \tag{67}$$

Because γ is relatively small and constant [832], the tendency of a polymer to fail in a ductile fashion will increase with an increase in $(\Delta H_a - U_o)/\sigma$.

Even though the mode of failure depends on the relationship between ΔH_a and U_o, we have observed in the past that the strength of both brittle and ductile polymers is related only to their respective ΔH_a values [834, 835]. These are obtainable from the temperature dependence of dynamic mechanical or dielectric loss determinations of T_g [836]. The magnitude of the activation volume ω can be calculated from the empirical equations [170]

$$\omega\sigma = \Delta H_a \tag{68}$$

and

$$\omega(\sigma - 7) = \Delta H_a - 70 \tag{69}$$

found in [834] and [835] to describe the relationship between the applied stress and ΔH_a for brittle and ductile polymers respectively. The magnitude of ω was calculated by Aharoni [170] for a large number of polymers. The volume of the polymeric segments between entanglements, $L_c d^2$, and the volume of Kuhn segments, $A d^2$, were also calculated for the same polymers. While no obvious correlation between ω and $A d^2$ was observed, a strong correlation between ω and $L_c d^2$ was found. In all cases where

$$\omega > L_c d^2 \tag{70}$$

the failure was ductile and whenever

$$\omega < L_c d^2 \tag{71}$$

the failure was brittle [170]. We believe that the same reasoning should hold for crosslinked networks except that, instead of the volume between entanglements, the segmental volume between branchpoints or crosslinks should be evaluated and compared with the activation volume for the glass transition of the network. In the case of gelled rigid networks, the reduced polymer concentration is factored in as an increased in d^2 which in this case becomes the mesh size of the network. Equations (70) and (71) instruct us that when a network characterized by a large ω and ductile failure is swelled with solvent to become a gel, the mode of failure may change from ductile to brittle once the dilution, measured by d^2, is large enough.

For the above, detailed knowledge of chain architecture is not necessary. It is well known, however, that the modulus or Φ of each polymer are strongly dependent on the detailed structure, actual and virtual bond lengths, rotational and valence bond angles, and the conformational isomeric states of the individual chains or segments. In a crystalline polymer, the highest elastic modulus is one parallel to the chain axis. The lowest modulus is in the plane transverse to the chain axis. In this plane the interactions are exclusively intermolecular in character and contain no intramolecular, covalent bonds. The intermolecular interactions may be common van der Waals dispersive forces, the somewhat stronger dipole-dipole interactions and the even stronger hydrogen-bond inter-actions. These are all far smaller than interatomic bond forces along the chain, resulting in transverse modulus, E_t, maybe an order of magnitude smaller than the longitudinal modulus, E_l, along the chains, provided these do not have an exceptionally large cross-sectional area [837]. The transverse moduli of several polymers were found to fall in a relatively narrow interval [837], from about 3 GPa for polymers with only van der Waals interactions between the chains in a crystal, through about 8 GPa for polymers with strong dipole-dipole interac-tions, to 11 GPa in the H-bonded planes in crystals of nylon-6 [837, 838]. Along the chain direction the moduli E_l and the extensibilities Φ are strongly depen-dent on the chain structure and conformation, ranging from very low values for aliphatic helical chains to very high values for fully aromatic coaxial chains.

Aliphatic polymers may be present in the crystal exclusively in the planar zigzag form, such as the T 2/1 structures of polyethylene or polyvinyl alcohol. When such crytals are longitudinally strained, the individual chains strain by the opening up of the valence angles and the elongation of chain backbone bonds. No torsional motions are involved since none can take place. Because all bonds are actual and short, backbone atoms undergo only small translations towards the chains' axes of symmetry. The combination of these structural features results in the inability of the chain to strain significantly without encountering strong resistance caused by the very high energy necessary in order to alter covalent bonds and valence angles. This very high resistance to strain of each chain combines with lack of side-groups that do not contribute to chain strength but increase its cross-sectional area, to produce polymers with very high E_l values. When groups containing a fixed number of actual bonds are replaced by a single virtual bond, atoms of the planar zigzag chains farthest removed from the axis of symmetry serve as fulcrums for easier straining of the chains. Because of the length of each virtual bond, the opening up of each valence angle at the fulcrum results in larger translations toward the axis of symmetry and larger strains in the overall chain axis direction. A force capable of opening the valence angle and straining the covalent bonds that produced only very small strains in planar zigzag chains with only short actual bonds will generate large strains when these bonds are replaced by long virtual bonds, and lower E_l modulus. When, in addition to the trans-conformers, gauche-conformations are present in the crystalline polymer, helical structures appear. Depending on the specifics of the helix, the average distance of backbone atoms from the axis of symmetry may vary greatly. When such helices are longitudinally strained, torsional motions take place in addition to valence angle opening and some bond lengthening. The average distance of backbone atoms from the axis of symmetry decreases with strain but now, with contributions from relatively low activation energy rotational motions, a lower force is necessary in order to obtain a given degree of strain. Hence, polymers with crystalline helical geometry are charac-terized by E_l lower than polymers with only planar zigzag conformations, and the larger the inner diameter of the helices the lower is E_l. The above is consistent with the results obtained by Sakurada and co-workers on aliphatic and largely-aliphatic crystalline polymers [837–839].

Depending on the deviation or lack of deviation of aromatic groups from coplanarity with the planes defining the single-atom swivels, as described in Sect. 4.4.4, and on the magnitude of the difference between different valence angles present in a given aromatic chain with more than one kind of swivel, such chains may exist in planar zigzag or helical forms. Their E_t modulus is expected to be rather small and insensitive to structural details while E_l is expected to be large. E_l is also expected to change according to whether the lowest energy conformation is planar zigzag or helical, and the length of the virtual bonds traversing the aromatic groups between the swivel points. Analogous with the zigzag and helical aliphatic polymers, the longer the virtual bonds between the swivels the farther the lateral reach of the backbone atoms away from the axis of symmetry and the lower the force needed to reach a given strain. That is, the E_l modulus of chains with swivels is expected to decrease with increased l, in a fashion similar to that encountered in comparable aliphatic polymers endowed with similar modes of chain extensibility.

It has been repeatedly shown [840–843] that, in polymers containing single-atom swivels, the aromatic rings perform 180° ring flips at rather high frequency. It has further been demonstrated by NMR studies [844] that the reorientation angle is not exactly 180°, but could be fitted to a distribution of reorientation angles having a width of about 20°. The rapid ring flips and the distribution of reorientation angles both contribute to a reduction of E_l, relative to its calculated values in the absence of such motions, by easing the change from one torsional angle minimum to another caused by the applied stress.

Aromatic chains with a single non-coaxial bond in the bridge between aromatic groups, as described in Sect. 4.4.3, deviate slightly from the chain axis of symmetry. Two factors contribute to this deviation. One is the fact that the virtual bonds traversing the para-substituted aromatic rings are oriented in an angle relative to the overall chain axis. Another is the fact that the valence angles α and β at both ends of the actual central bond in the amide or ester groups, each placed between two virtual bonds, are not 120° each and $|\alpha - \beta| > 0$. A schematic description of the above, neglecting the contributions of $|\alpha - \beta| > 0$, is:

When the chain is stressed along its axis, the atomic displacements toward the axis of symmetry are rather small, being limited by the length and valence angles of the non-coaxial single bonds. Because in the crystalline state the chains are fully extended and syn-anti-interconversions or 60° ring flips are not present, the strain of aromatic chains with single non-coaxial bonds in the bridge, such as polyamides and polyesters, is accommodated by atomic translations towards the chain axis, opening of valence angles (mostly the α and β angles) and some bond elongation [798, 808, 845]. Deviations of ring orientation from the minimum energy of the torsional angles in the relaxed state are expected to occur, contributing to the lowering of the forces required to strain the chains. Unlike aromatic polymers with single-atom swivels, in polymers with one non-coaxial bond the virtual bonds traversing the aromatic groups are placed in a rather acute angle relative to the axis of symmetry. Because the virtual bonds are connected by a single non-coaxial bond, the longer the virtual bond the more acute its angle with the axis of symmetry. The longer virtual bonds lead, then. to decreasing the average lateral displacement of backbone atoms from the axis of symmetry and a decreasing number of non-coaxial bond and their associated valence angles. An applied axial stress can now be relieved by only small atomic translations toward the axis of symmetry, and the opening up of a diminishing number of valence angles. A higher force is now needed in order to cause a given strain. We conclude, hence, that in their crystalline form, aromatic polyamides and polyesters with longer virtual bonds are expected to possess higher E_l than their analogues with shorter virtual bonds. The presence of the non-coaxial bond, however, limits the E_l of these polymers to only about 35–40% of the longitudinal modulus of poly-p-phenylene chains in the crystal, perfectly aligned with their axis of symmetry [808].

In the non-crystalline state polyamide and polyester chains do not contain exclusively anti- or syn- conformers, but a mixture of both placements. Under strain these are easily allowed to interconvert in order to release the stress [412, 663, 664]. In the case of the aromatic polyamides at least, 60° ring flips also contribute to stress release [663]. These isomeric interconversions involve relatively low energy-barrier rotational motions, resulting in substantial reduction in the forces needed to strain such chains and a commensurate drop in modulus. Isolated aromatic polyamide or polyester chains, each free to adopt several conformational isomers, have, then, lower modulus than when present in the crystal, fully extended in a single conformational isomeric state.

When aromatic chains with bridges consisting exclusively of coaxial single bonds, as described in Sect. 4.4.2, are longitudinally pulled, their ability to strain is fully dependent on straining of individual bonds from their equilibrium bond lengths, and on opening of valence angles beyond their equilibrium values. Because of the symmetrical structure of these polymers and the correspondence of their actual and virtual bonds with the axis of symmetry, no meaningful atomic translations in directions perpendicular to the chain axis take place when the chains are strained. The major contribution to the strain of poly-p-phenylene, poly-p-phenylene-bisbenzimidazole and similar such polymers are the stretching of the single coaxial bonds between aromatic and heteroaromatic groups, and changes in the valence angles between these single bonds and aromatic units [808]. Because there are practically no other modes of deformation to accomodate longitudinal strains and because these modes require passing over very high energy barriers while contributing only very little to chain elongation, very large forces are needed in order to effect very small longitudinal chain strains. As a family, then, the aromatic and heteroaromatic polymers with bridges consisting of coaxial single bonds are endowed with the highest values of E_l and Φ; two to three times higher than aromatic polyamides and polyesters [808].

Based on the above, the longitudinal modulus E_l of individual aromatic chains should scale as: chains with coaxial single bonds > chains with non-coaxial bonds in bridge > chains with single-atom swivels > chains with flexible multiple-atom bridges.

The Young's modulus can be calculated with considerable accuracy, once a model is chosen, by using all the information listed above about the properties of individual segments and their linkages .If one assumes the system to be homogeneous down to segment levels, there are established papers by Treloar [846–848], Tashiro and associates [849] and Tashiro and Kobayashi [808] adding translations due to rotational motions and the effects of such motions on the deformation and modulus of the segments. Figure 39 is for aromatic polyamide segments. It is easily translated to polyester by the replacement of all N–H groups with –O–. It may be applicable to structures with swivels by omitting the carbonyl or N–H in each repeat unit, and to coaxial polymers by removing the amide group in its entirety. By judicious selection of the factors in the following equations, the mathematical description of the polyamide may be reduced to that of polymers with swivels or coaxial polymers. From Fig. 39 we define:

R_1 is the non-coaxial, real bond along the segment. R_1 may have an average length, depending on the exact nature of the groups in which it resides;

R_2 is the virtual bond along the segment, tranversing the length of the aromatic groups from one substituent atom to the other. R_2 may also have an average length, depending on the particular structure of the segment;

I is the projection of R_1 and R_2 on the segment axis. Depending on the structure, there may exist in the segment more than one I and an average length should be used;

J_i is the projection on the segment axis of each portion of the segment between branchpoints and bends and/or twists. There may exist several J units in a segment, but because in our networks the segments are rather short, a rather small number, $i = 1, 2, 3, \ldots$ of J portions is expected to be present;

α is the valence angle of the   C = O group, and

β is the valence angle of the   N–H group. Because of their similar high energy barriers and for convenience we average them as $(\alpha + \beta)/2$ and use the averaged value in the calculations;

ω is the torsional angle around the central bond in the amide group. Usually, $\omega = 180°$ with rather narrow distribution and intermediate rotational energy barriers;

ψ is the torsional angle around the bond connecting the aromatic ring to the amide   C = O moiety. It is characterized by a rather low energy barrier, allowing for rather easy rotations;

ϕ is the torsional angle around the bond connecting the aromatic ring to the N–H moiety in the amide residue. It is typified by a rather low energy barrier, allowing for easy rotations. Because ψ and ϕ have low activation energies and both contribute to translations in a similar fashion, we felt that for the

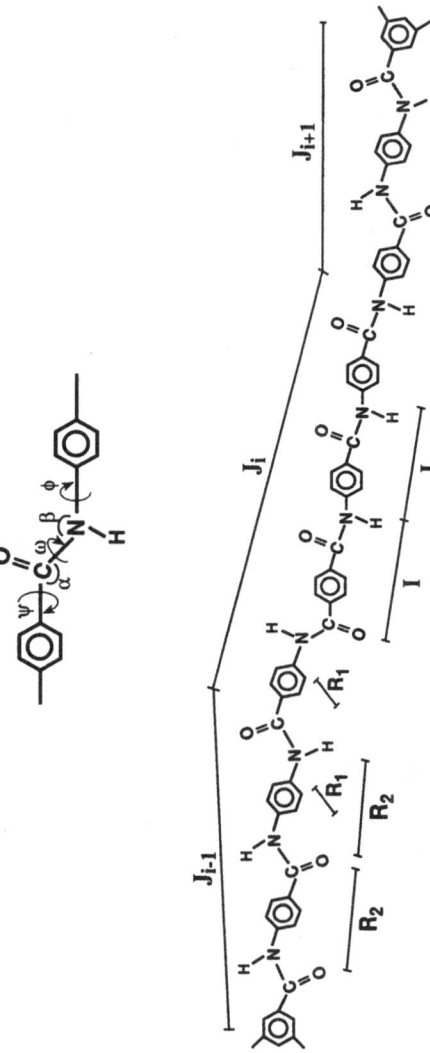

Fig. 39. Aromatic polyamide segment and its J subdivisions

zero-order estimates in the following equations they could be added together and treated as the sum of $\psi + \phi$;

K_1, K_2, K_3, K_4 and K_5 are force constants;

ΔR_1 gives the internal displacement of R_1 (force constants K_1 and K_3);

ΔR_2 is the internal displacement of R_2 (force constants K_2 and K_4);

$\Delta[(\alpha + \beta)/2]$ is the deviation from the energy minimum of the average angles $(\alpha + \beta)/2$, (force constant K_5);

$\Delta\omega$ is the deviation of the angle ω from $\omega = 180°$, (force constant K_4, (large));

$\Delta(\psi + \phi)$ is the sum of the deviations of $\psi + \phi$, (force constant K_3, (small));

d^2 is the effective segment cross sectional area;

f is the tensile force applied to the segment; and

E_l is the longitudinal tensile modulus.

We use straightforward geometry to obtain

$$\Delta R_1 = f\left[(R_1 \cos(\psi + \phi))/K_3 J + \left(R_1 - R_2 \cos\left(\frac{\alpha + \beta}{2}\right)\right)\bigg/ K_1 I\right] \quad (72)$$

$$\Delta R_2 = f\left[R_2 \cos\omega/K_4 J + \left(R_2 - R_1 \cos\left(\frac{\alpha + \beta}{2}\right)\right)\bigg/ K_2 I\right] \quad (73)$$

$$\Delta\left(\frac{\alpha + \beta}{2}\right) = f R_1 R_2 \sin\left(\frac{\alpha + \beta}{2}\right)\bigg/ 2K_5 I \quad (74)$$

$$\Delta\omega = f R_1 R_2 \sin\omega/K_4 J \quad (75)$$

$$\Delta(\psi + \phi) = f R_1 R_2 \sin(\psi + \phi)/K_3 J. \quad (76)$$

The spatial translations of branchpoints and bends are most influenced by rotational motions and their ease. We sum up over the appropriate J_i to obtain the relevant translation

$$\sum_i \sum_R \Delta R_1 \Delta R_2 = f\left\{ \sum_R \left[\left(R_1 - R_2 \cos\left(\frac{\alpha + \beta}{2}\right)\right)\bigg/ K_1 I \right.\right.$$
$$+ \left(R_2 - R_1 \cos\left(\frac{\alpha + \beta}{2}\right)\right)\bigg/ K_2 I$$
$$\left. + \left(R_1 R_2 \sin\left(\frac{\alpha + \beta}{2}\right)\right)\bigg/ 2K_5 \right]$$
$$+ \sum_c [(R_1 \cos(\psi + \phi))/K_3 J_i + (R_1 R_2 \sin(\psi + \phi))/K_3 J_i$$
$$\left. + (R_2 \cos\omega)/K_4 J_i + (R_1 R_2 \sin\omega/K_4 J_i]\right\} \quad (77)$$

where the summation over all R adds up the positive and/or negative contributions of all actual and virtual bonds along the segment, and the summation over i adds up all the contributions due to reorientations of all segment portions J.

When the network segment is fully extended prior to the application of tensile force, then all contributions of torsional motions reduce to zero and we obtain a Young's modulus

$$1/E_1 = (d^2/I)\left[K_1(\Delta R_1/f)^2 + K_2(\Delta R_2/f)^2 + 2K_5\left(\Delta\left(\frac{\alpha+\beta}{2}\right)\Big/f\right)^2 \right]$$

(78)

which is identical with the results obtained by Tashiro and Kobayashi [808]. When the segment is bent, the tensile modulus is lower. When bent, pulling on the segment straightens it, a motion involving translations of branchpoints and bends. When this happens, the contributions of the torsional angles have to be factored in and we obtain

$$1/E_1 = (d^2/I)\Big\{\left[K_1(\Delta R_1/f)^2 + K_2(\Delta R_2/f)^2 + 2K_5\left(\Delta\left(\frac{\alpha+\beta}{2}\right)\Big/f\right)^2 \right]$$

$$+ \left[K_3(R_1\cos(\psi+\phi)/f)^2 + K_3(R_1 R_2\sin(\psi+\phi)/f)^2\right.$$

$$\left. + K_4(R_2\cos\omega/f)^2 + K_4(R_1 R_2\sin\omega/f)^2\right]\Big\}$$

(79)

in which the summation over i adds up all the reductions of the Young's modulus E_1 due to the deviations of the J_i portions of the segment from colinearity. We find, hence, that large bends or twist-bends in a stiff segment reduce the Young's modulus relative to the fully extended segment, just the same as the presence of single non-coaxial bonds in polymers such as polyamides and polyesters reduces their modulus relative to fully coaxial aromatic polymers.

The equations above may not give us exact translations and modulus but, we believe, are sufficient for comparison between stiff polymers belonging to different classes, provided correct values are available for all the necessary parameters and constants. We must note that the equations above describe the tensile behavior of individual stiff segments. The corresponding bulk properties require the chains or segments to be fully or mostly oriented in the applied stress direction. This, however, is not the case in isotropic networks and gels. In these systems the situation becomes more complicated because the vast majority of aromatic segments are aligned neither fully parallel nor exactly perpendicular to the draw direction. Here, in addition to the longitudinal modulus E_1 characteristic of each polymer, the angle the segment makes with the network's draw direction and the presence or absence of bends in the segments are both of importance in evaluating the modulus of the tested network. Accepting as fact that homogeneous stress distribution [798, 808, 845] exists in flawless aromatic networks, one may apply a "parallelogram of forces" to determine what fraction of the force applied to each segment is transmitted along the segment and what fraction of the force is laterally transmitted through the branchpoints and entanglements.

Recent work has extended such calculations to fractal and tree networks with considerable subtlety [311, 400, 690, 732, 781, 850–864]. From amongst the extensions to tree and related network, the work of Jones and Marques [732] will be discussed in the Theory section, but that of Jones and Ball [781] offers some results of immediate value to this section. These authors consider the network by completely ignoring the circles formed when it is possible to find a route along rods returning to the starting point. As will be argued in the Theory section, this can often be highly unlikely so that the strength of the network stems from its entanglements within a single tree or across trees. Without circuits the geometry of the network is the geometry of a tree, and the mathematics suddenly becomes very simple because one can "renormalize" the network, i.e., the mathematics of structure at Fig. 40a is equivalent to structure at Fig. 40b which is equivalent to the structure at Fig. 40c.

Starting from the simple structure in Fig. 41, depicting a rod rigidly clamped at its left-hand end (a) and a similar rod that can translate and rotate in space (b), Jones and Ball [781] show that

$$\sigma_{\bar{y}} = \frac{nY\gamma}{15s}[(5 - 3D)\varepsilon ij + D\delta ij(\varepsilon_{11} + \varepsilon_{22} + \varepsilon_{33})] \tag{80}$$

where Y is a limiting number (the 'fixed point' as the network becomes infinitely large), n the number density of rods, s their length and γ their bending compliance per unit length . The quantity D is related to the force constants m_1, k_1 which mechanically couple one end of the rod to the rest of the network, and k_{11} which measures the stretching modulus

$$D = \frac{k_1(1/m_1 + 1/3) - k_1/k_{11}}{1 + k_1(1/m_1 + 1/3)}$$

$k_{11} = K_{11}s^3/\gamma$ where K_{11} is the stretch modulus.

The Jones and Ball paper [781] raises and answers other questions which are taken in the Theory section below.

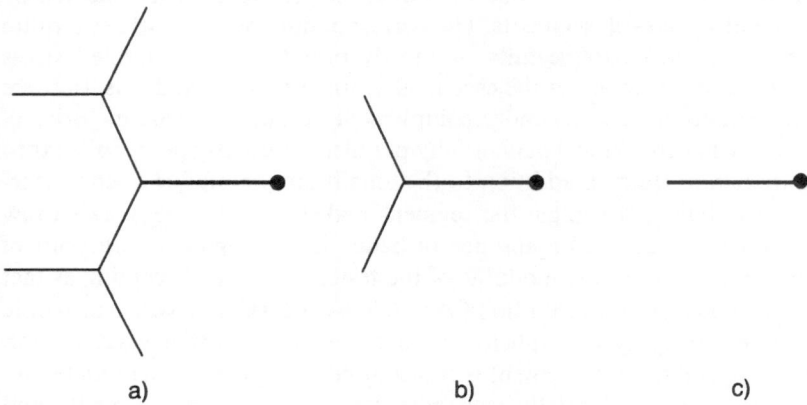

a)   b)   c)

Fig. 40. Renormalization schematic of (a) to (b) to (c)

a)

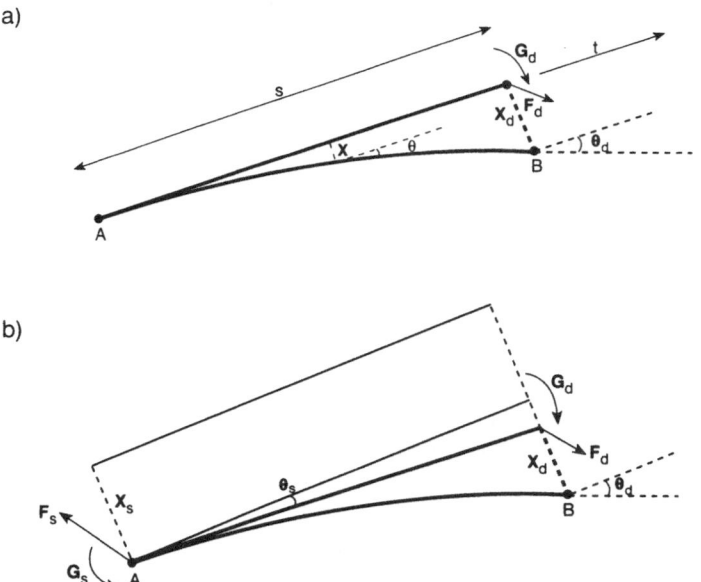

b)

Fig. 41a, b. Rod AB with projection length s: **a** rigidly clamped at A; **b** free to translate and rotate. Translations are **X**, forces are **F**, couples are **G**, and angles are **θ**. Sub s and d stand for left and right, respectively.

5 Theory

Networks of rigid polymers are new and their theoretical description has nothing like the maturity of the description of flexible networks [8–46, 865, 866]. The all-pervading difficulty of all network theories is the inadequacy of characterization. It is extremely difficult to make theories of how a network is formed [17, 38, 865] and the results are not always reliable. Hence, the ideal situation is to discover experimentally what is there and then predict the consequences. It is, however, very difficult to give an experimentally based characterization. For example, when a network of flexible polymers is crosslinked it is not easy to say whether the links are at random or they clump into densely populated regions with rather thinly populated regions around them, and there is indeed evidence in some materials that clumping takes place [722–728]. However, the anomalous neutron scattering [716–718, 720, 867–869] found in materials which do not show obvious signs of clumping, suggests a rather different picture to the conventional uniform behavior. Nevertheless, if a flexible network is created from a melt of long linear chains by, say, sudden irradiation, the distribution of crosslinks will be close to uniform and the question is then better posed by asking how close a vulcanized rubber is to this ideal. There are many theories of flexible chain networks [8–46, 862, 866, 870–876], some of them with non-classical connectivity [782–784, 877], and the difficulty faced by the investigator is that even quite naive theories give, by the standards of the constitutive equations of other complex materials, quite good answers. For example, the assumption that a segment between crosslinks moves with complete freedom, generates a free energy which is entirely entropic, and whose free energy change on deformation can be attributed solely to the affine movements of its endpoints, gives

$$kT\sum\lambda_i^2$$

per segment. For an incompressible material this is

$$kT(\lambda^2 + 2/\lambda)$$

per segment. The expression $(\lambda^2 + 2/\lambda)$ is not perfect and the kT is also correct to a few percent, but since the number of "effective" segments is not well known, the results obtained are sufficiently close to experimental values to be accepted by technologists. Modern theories improve on all these aspects. For example, by putting the polymer segment in a tube one allows for topological constraints and $\sum\lambda_i^2$ is replaced by $\sum 1/\lambda_i^2$, as has been shown by Gaylord [878]. A combination of end-point restrictions plus topological restrictions gives expressions containing aspects of both of these forms, albeit in a more complex form than a simple addition of these terms. Such an addition was put forward by Mooney [879] and Rivlin [880] as a description of the experimental results long before the more complex modern theories were offered. It is fair to say that the theory of flexible networks now languishes because it is not worth a theorists's while to work hard on the consequences of a physical structure when experiments have not reached the point of specifying the situation any better than the current theories can handle. A good general account of the situation is given in the book of Mark and Erman [46].

When one turns to rigid networks, these difficulties are still present. One can build up a very reasonable intuitive picture of how the networks described in earlier sections must look. Experiments bear out these pictures in general terms. The difficulty comes in that more refined theories will need characterizations of the network materials, the pathway to which is not easy to conceive, let alone perform. Let us enumerate the problems.

5.1 The Problem

5.1.1 Responses of Flexible and Rigid Networks to Stress Fields

Firstly, what is not really a problem any longer given the power of molecular simulation by computers? Given the appropriate force-field parameters and structural details of a polymer chain, the crosslinking mechanism and the specifics of the crosslinking junction, one can work out how this polymer deforms under stress, i.e., the energy function of the molecule under elongation, bending, torsion and twisting can all be found reliably; similarly for the crosslink behavior. These studies give the size required for a polymer to be worm-like, i.e., give the peristence length, and discriminate between those chains which cannot bend, and those capable of changing direction abruptly at one or many points by overcoming a potential barrier by the application of external stress of accessible magnitude [663]. Thus one can with confidence compare structures dominated by entropy, e.g., in rubber, as against rigid networks where thermal motions are secondary, and situations intermediate to these two extremes. Likewise one can decide whether the effect of external stress is to alter the number of configurations and hence the entropy, or directly approach the molecular configurations at the microscopic level so that the molecular network is like a steel girder frame under stress. One naturally thinks of bridgework with rigid joints, but one can conceive a bridge with flexible joints, e.g., the universal joint of mechanical engineering. As was demonstrated by Maxwell [881] a network of rodlike segments and flexible joints is only rigid if there are six or more segments emanating from each flexible joint. If there are fewer, the network rigidity has to be sustained by osmotic pressure.

5.1.2 Complexities Uniquely Linked to Network Rigidity

Flexible networks are most commonly prepared by one of two methods: poly-merizing an appropriate mixture of monomers in one step, as in the case of styrene-divinylbenzene or polyurethane networks, or by first preparing long chains in a first step and then crosslinking them in a second, chemical or irradiation step. Weaker but similar crosslinks can be formed by hydrogen bonding [882–885] and are reversible. A less frequent pathway is by end-crosslinking [886–887], which has considerable technical importance [888]. Such end-links are often weak and can break and reform thermally or under stress [888]. A general characteristic is that, although the specification of the final network can be complicated and difficult to characterize with confidence, the processes of crosslinking of flexible networks are easy, accessible, and predictable. Not so with rigid networks: the extra constraints of end-to-end length and angles at the junction points means that *topology* is not enough, one has to satisfy detailed *geometry* as well. This implies that, given a set of rods with no additional crosslinking moieties, one cannot in general make a network out of them as the lengths and angles are all wrong. What one can always make are *trees*, and these then make up a central part of the analysis of networks, as we have seen above. The two limiting cases of dilute formation followed by concentration, e.g., at $C_o \leq C_o^*$ (see Sect. 4.1.2), leading to Voronoi-type structures, and concentrated formation at $C_o > C_o^*$, possibly leading to DLA problems, require different theoretical treatments and present quite different problems on straining.

All these matters have arisen in the appropriate positions in the previous Sections. We now describe them as *problems* in greater detail in the hope that they may stimulate further interest.

5.2 Theoretical Description of Rigid Networks

5.2.1 One-Step Networks

Figures 42 and 43 represent two extreme cases of one-step networks, one formed in a concentrated environment at $C_o \gg C_o^*$, and the other in a dilute solution at $C_o < C_o^*$ and subsequently concentrated. In Fig. 42 we draw one tree which, for simplicity, is in the plane. The other trees represented by dots where they cross the plane. These dots do not represent unreacted monomers.

It is clear that there is low probability of circuits in this kind of network, i.e., configurations like the one in Fig. 44 are very unusual when each belongs to a single tree. If the junctions between segments are flexible there is a much higher probability for the formation of circuits than if the junctions are rigid, but in both cases the tree-like configuration is the most likely. When the trees are concentrated as in Fig. 42 the topological constraints on the system must mean that the deformations are roughly affine. For the dilute system subsequently concentrated one can expect some "surface" interactions if crosslinking continues, as is shown at one point in Fig. 43, and possibly the surface fractal structure described earlier, with whole trees taking to some extent the role of Voronoi polyhedra.

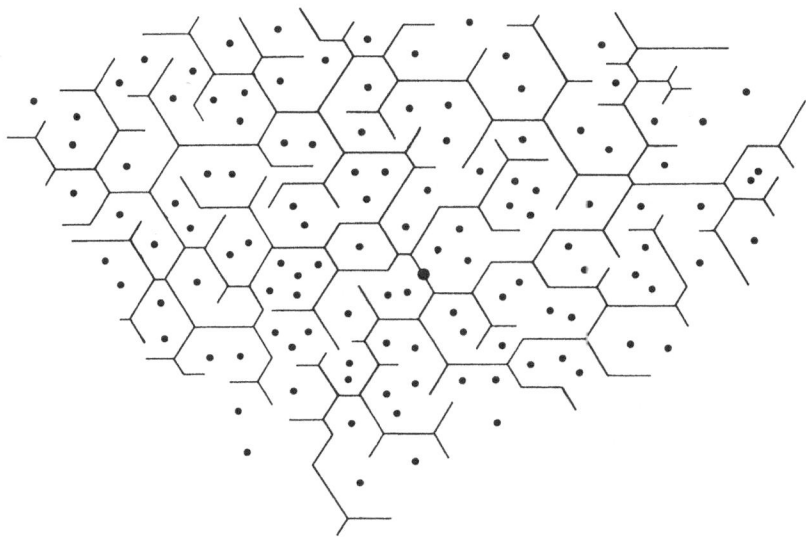

Fig. 42. One-step network prepared at high concentration, $C_0 \gg C_0^*$. Each tree (fractal polymer) is deeply interpenetrated with others resulting in rather homogeneous network structure and low probability of large voids

In more detail, the crudest version of the affine deformation problem of flexible junction-point networks will lead to an energy

$$\int V(\Lambda R)d\Omega$$

where V is the potential energy of a rod whose ends are separated by R. Λ is the deformation matrix

$$\Lambda_{ij} = \delta_{ij} + e_{ij} \tag{81}$$

where e_{ij} is the strain. The integral over Ω is over the orientation of R relative to e_{ij}. This leads to

$$dE = \sum_{ij} \frac{\partial V}{\partial R_i} e_{ij} R_j + \frac{1}{2} \sum_{ijkm} \frac{\partial^2 V}{\partial R_i \partial R_j} e_{ik} R_k e_{jm} R_m \tag{82}$$

$$d\bar{E} = e\left(R \frac{\overline{\partial V}}{\partial R} \right) + \frac{1}{2} ee R R \frac{\partial^2 V}{\partial R \partial R}. \tag{83}$$

The average will be isotropic, so

$$dE = v_1 \sum_k e_{kk} + v_2 \sum_i \sum_k e_{kk} e_{ii} + v_3 \sum_i \sum_k e_{ik} e_{ki}. \tag{84}$$

These are standard forms with $v_1 = 0$ when one is deforming an equilibrium system. To improve on such expressions one must review the physical circumstances. The problem arises that it is far easier to compress a rod by the Euler

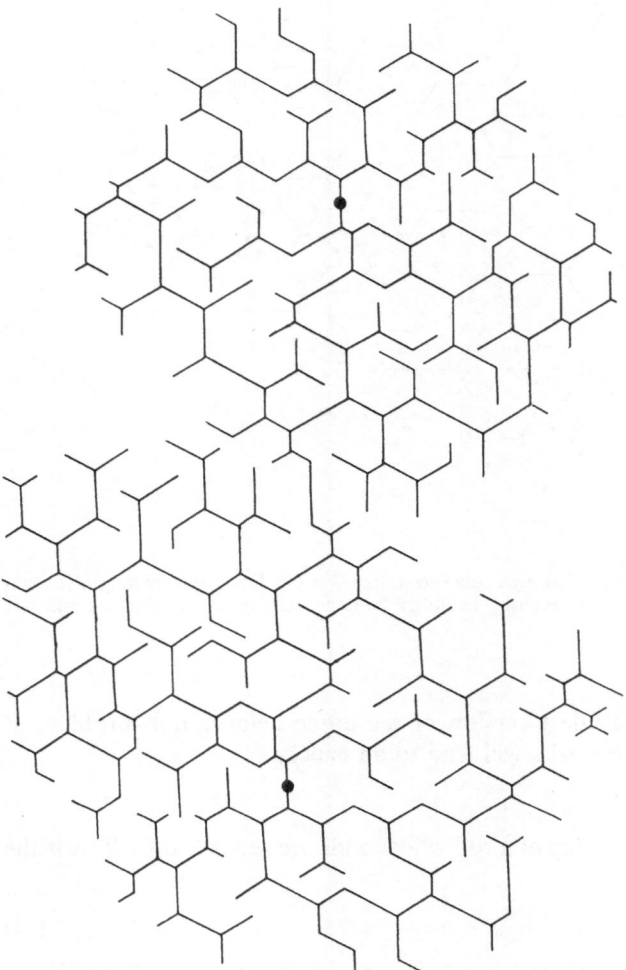

Fig. 43. One-step network prepared by first polymerizing trees (fractal polymers) at high dilution, $C_0 \ll C_0^*$, and then concentrating the solution to C_0^* to bring the trees in contact with each other. A small number of interfractal covalent bonds may be formed if polycondensation is allowed to proceed, and the network is inhomogeneous with high probability of large interfractal voids

instability than to extend it. A crude representation of this is to argue that $\partial V/\partial R$ is very large under extension and under deformation the rod can be regarded as inextensible. Thus, considering for simplicity elongation of the material, as in Fig. 45a, rods perpendicular to the axis of elongation buckle, as in Fig. 45b, whereas rods parallel to the elongation cause a deformation of their surroundings. Crudely speaking, the modulus comes from the compression of these rods whose angle to the strain leads to them being compressed and the general deformation of the system, which is taken to be an elastic continuum

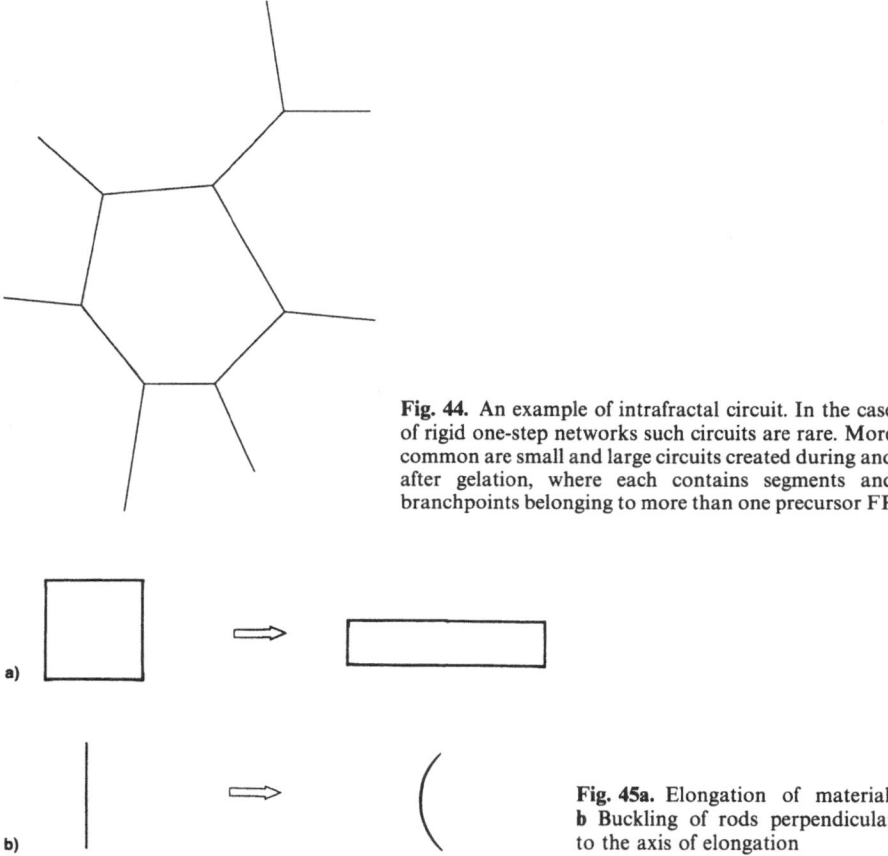

Fig. 44. An example of intrafractal circuit. In the case of rigid one-step networks such circuits are rare. More common are small and large circuits created during and after gelation, where each contains segments and branchpoints belonging to more than one precursor FP

a)

b)

Fig. 45a. Elongation of material. **b** Buckling of rods perpendicular to the axis of elongation

with some mean modulus, from those rods whose angle leads to forces trying to extend the rod. There is a considerable literature of the problem of materials reinforced by fibers, in particular fibers which can bend but not extend [889–903], a problem related to the mean hydrodynamics of such systems [904–907] and the electronic properties of random systems [908].

To give a rough derivation of a shear modulus of an array of rods, we divide the rods into those in compression, fraction α of the whole, and those in tension, fraction $(1 - \alpha)$. Those in compression can be considered to generate an elastic solid of modulus E_m, and those in tension do not extend. They contribute to the final modulus E by causing a deformation of the surrounding matrix, which is of course just the compressible rods. The references cited [889–903] show that if a volume fraction V_f of rods of length l, radius r_o, Young's modulus E_f, are placed in a matrix of Young's modulus E_m, the net modulus is

$$E'_m = E_m + (\alpha/5)(E_f - E_m)V_f\eta_l \tag{85}$$

where

$$\eta_l = 1 - \tanh\tfrac{1}{2}\,\beta l/\tfrac{1}{2}\,\beta l$$

$$\beta = [2\mu\pi/E'_m a \, \log(\xi/r_o\gamma')]$$

$\gamma' = $ numerical factor ($\tfrac{1}{2}e^\gamma$, where γ is the Euler constant $= 0.57721...$)

$\mu = $ shear modulus ($= E/3$ if incompressible)

$\alpha = $ fraction of rods compressed

and

$a = $ cross sectional area of a rod.

Since the compressible rods make up the matrix, we put

$$E_m = 0$$

to get

$$E'_m = (E_f/5)\,V_f\,\eta_l. \tag{86}$$

We now apply Eq. (85) again with E_m replaced by E'_m and $E_f \to \infty$, whereupon

$$E_f\,V_f\,\eta_l \to \frac{(1-\alpha)}{5}\,\frac{Nl^3}{\log(\xi/r_o\gamma')}$$

as is shown in [903]. Here N is the number density of rods, so using this to write a uniform form we obtain the final modulus E to be

$$E = \frac{\alpha N l r_o^2 \eta_l E'_m}{5} + \frac{Nl^3(1-\alpha)}{5\log(\xi/r_o\gamma')} \tag{87}$$

where E'_m is the compression modulus.

Crude as this model is, it shows that one can derive an effective modulus, and such models can be refined if experimental situations warrant it. However, these models depend on a uniformity which does not in general occur unless some agency (like a bridge building engineer) forces it. In general, more uniform structures result from tree type growth as we will see below.

More realistic calculations need to include features which are intuitively obvious, but not easy to handle. The affine assumption can be extended to rigid junction-points, so that if there is an angular energy $\cos\theta$ where

$$\cos\theta = n_1 \cdot n_2 \tag{88}$$

n_1, n_2 being the directions of rods at a junction, the new energy is $\cos\theta'$:

$$\cos\theta' = n'_1 \cdot n'_2 = \frac{n\cdot\Lambda\cdot\Lambda\cdot n}{|n\cdot\Lambda||n\cdot\Lambda|}. \tag{89}$$

If the rods buckle there has to be a recalculation of n allowing for this, and if the junction point is truly rigid, i.e., $\cos\theta$ remains the same and extra buckling has to take place, a calculation analogous to that of extensive fiber reinforcement (e.g., as in [903]) is needed. We do not know of any such calculations.

A further complexity arises when the twisting of the rods has an influence on the energy and when this twisting is coupled to the bending. Again, these calculations do not exist at present, but although the problem is by now of considerable algebraic complexity, it is still a straightforward extension of the work cited.

A more difficult class of mathematical problems arises when the rod has an internal potential barrier so that on experiencing a stress threshold it buckles at that barrier. Although again it is intuitively obvious what happens, indeed an everyday occurrence in compressing, say, straw, it is not easy to handle because the local stress will reach the threshold in different places at different external stress since the material is not locally homogeneous. If one attempts to use average methods, the whole system will buckle together. Thus, the distribution of stress is needed in order to make a sensible theory. (There is an analogy here in the theory of dislocations; without them colossal stresses could be sustained by crystalline materials until they would suffer catastrophic failure.) Again, this problem remains to be developed.

So much for simple homogeneous pictures. Several authors have suggested that, although one can not get simpler pictures than those above, there are crucial features missing from the "mean field" pictures which dominate the real networks and which, albeit at a crude level, can be usefully described and employed to give good predictions. Since these issues arise in the two-step networks also, we consider them below.

As an example of the difficulty of averaging, even under uniform conditions, we can offer the case of auxetic materials, i.e., materials with negative Poisson ratio. The Ashby model [864] of such a system is depicted in Fig. 46. With freely hinged junctions, we see that when extended in one direction this material will also extend in the perpendicular direction. The molecules of Sect. 3.1.1 may make a molecular version of this if (a) a method is found to have all symmetrical elements of identical lengths, and (b) a way of defect-free, or almost flawless, assembly could be found [909]. As we mentioned at the begining of Sect. 3, such a system, if made, is expected to behave as a macroscopic single crystal. Alternatively, ordered microcrystalline domains embedded in amorphous "connective tissue" may emerge as the preferred structure. It has also been suggested by Baughman and Galvão [660c] that certain carbon and hydrocarbon networks possessing the correct symmetry may be constructed to behave in an auxetic manner. The synthetic efforts of Moore [386a] and those described by Diederich and Rubin [910], may result in rigid networks having the appropriate symmetry for auxetic behavior.

Clearly, the mathematical analysis of materials based on this kind of structure, which is likely to appear in small domains, cannot be handled by the overall averaging procedures.

We can now approach the theoretical problem from the other end. Is there an all-embracing mathematical formulation from which a sequence of approxi-

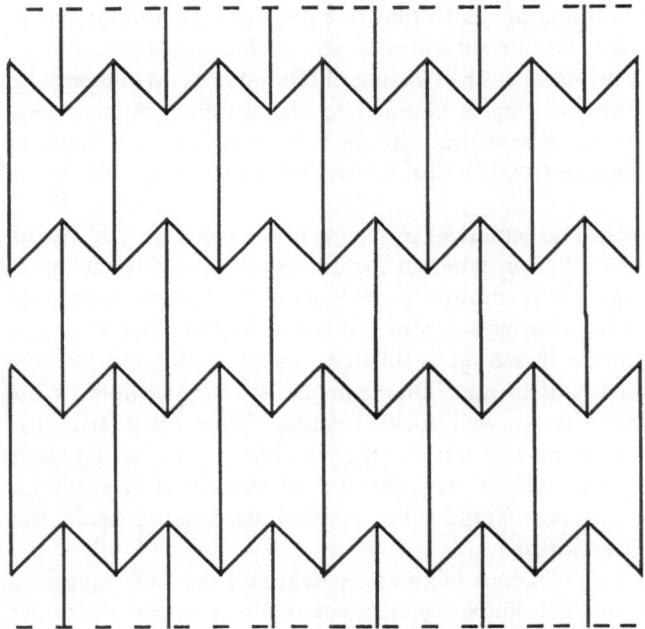

Fig. 46. A model of an auxetic material, i.e., material with negative Poisson ratio

mations can be developed? For example, in statistical mechanics and in many-body theories one knows the answer to this question. It turns out that there is a formulation available but it has deficiencies in that, in spite of progress in the pure mathematics of the topology of curves, there still does not exist a simple way to describe the entanglement of curves. For rods it is possible, in forms which again are obvious, but still algebraically too complex for easy manipulation.

5.2.1.1 General Mathematical Description

The trick of handling units which can combine or separate was learned in the second quantization formulation on quantum mechanics and was first introduced into polymer science by Fixman [911]. An easy way to think of this, avoiding operator calculus, is to use integration methods. The key formula is found in the extensibility of the integral

$$\int x^2 e^{-x^2-y^2} \, dx \, dy = \tfrac{1}{2} \int e^{-x^2-y^2} \tag{90}$$

$$\int y^2 e^{-x^2-y^2} \, dx \, dy = \tfrac{1}{2} \int e^{-x^2-y^2} \tag{91}$$

$$\int xy \, e^{-x^2-y^2} \, dx \, dy = 0 \tag{92}$$

to steadily more general forms

$$\int x_i x_j e^{-\sum_a x_a^2} \Pi \, dx = \tfrac{1}{2} \delta_{ij} \int e^{-\sum_a x_a^2} \Pi \, dx \tag{93}$$

and

$$\int x_i x_j e^{-\sum_{ab} x_a x_b A_{ab}} \Pi \, dx = \tfrac{1}{2} B_{ij} \int e^{-\sum_{ab} x_a x_b A_{ab}} \Pi \, dx \tag{94}$$

where B is the inverse matrix of A:

$$\sum_c A_{ac} B_{ce} = \delta_{ae} \tag{95}$$

or

$$B = A^{-1}. \tag{96}$$

It can be shown that these formulae will extend to the case of the indices i, j, a, b, etc., being in the continuum. There δ_{ij} is replaced by $\delta(r_1 - r_2)$ and Eq. (94) becomes

$$\int \cdots \int \phi(r_1)\phi(r_2) e^{-\iint \phi(r')\phi(r'')A(r'r'')d^3r'\,d^3r''} \Pi \, d\phi$$
$$= \tfrac{1}{2} B(r_1 \, r_2) \int \cdots \int e^{-\iint \phi(r')\phi(r'')A(r'r'')d^3r'\,d^3r''} \Pi \, d\phi \tag{97}$$

where

$$\int A(rr'') B(r''r') d^3 r'' = \delta(r - r') \tag{98}$$

or again we can write $B = A^{-1}$.

A more useful form is to think of ϕ as having two components and extend the formula to complex ϕ (in an abbreviated notation):

$$\int \phi\phi^* e^{-\iint \phi A \phi^*} (\delta\phi) = \tfrac{1}{2} A^{-1} \int e^{-\int \phi A \phi^*} (\delta\phi) \tag{99}$$

and

$$\int \phi\phi \, e^{-\iint \phi A \phi^*} (\delta\phi) = 0 \tag{100}$$

$$\int \phi^* \phi^* e^{-\iint \phi A \phi^*} (\delta\phi) = 0 \tag{101}$$

where $(\delta\phi)$ means integrate over every $d\phi(r)$, all r. Now think firstly of a flexible chain, with the monomer described by $g(r_1 r_2)$: $r_1 \bullet\!\!-\!\!-\!\!-\!\!-\!\!\bullet r_2$. If we consider $\iint \phi g \phi$ to represent an unreacted monomer and $\iint \phi^{*^2}$ a a difunctional linker, then

$$\left(\int\!\!\int \phi g \phi \right)^N \left(\int\!\!\int \phi^{*^2} \right)^M \exp\left(-\tfrac{1}{2} \int \phi\phi^* \right) \delta\phi \, \delta\phi^* = \text{all ways of joining up} \tag{102}$$

e.g.,

$$\left(\iint \emptyset g \emptyset\right)^3 \left(\iint \emptyset *^2\right)^3 = 1 + \bigcirc + 1/4[\bigcirc]^2$$

$$+ 1/36 \left\{ [\bigcirc]^3 + 6 \bigcirc\!\!\!\bigcirc\!\!\!\bigcirc + 4 \triangle \right\}$$

$$+ \cdots \tag{103}$$

If $g(r_1 r_2) = 0$ for $r_1 = r_2$, e.g., if g represents a rigid object of length l,

$$g = (1/4\pi l^2)\delta(|r - r'| - 1) \tag{104}$$

then only the triangles survive. Clearly our integral is zero unless $N = M$.

Now consider a trifunctional crosslinking agent $\int \phi *^3$ (or if a shape to the crosslinker is required

$$\iiint \emptyset*(r_1)\emptyset*(r_2)\emptyset*(r_3) \, C(r_1 r_2 r_3) d^3r_1 d^3r_2 d^3r_3 \text{ representing } r_2\overset{r_1}{\underset{}{\diagup C \diagdown}} r_3).$$

Now the addition of $(\int \emptyset*^3)^L$ gives a collection of chains with L links as in Fig. 47. These expressions can be more easily handled by fugacities with the result that the soup of monomers, polymers and network is contained in the expression

$$\exp\left(-\mu_m \int \emptyset g \emptyset - \mu_1 \int \emptyset*^2 - \mu_3 \int \emptyset*^3 - \int \emptyset \emptyset* \right). \tag{105}$$

| monomers | difunctional linear joints | trifunctional junctions | performs the linkage mathematically |

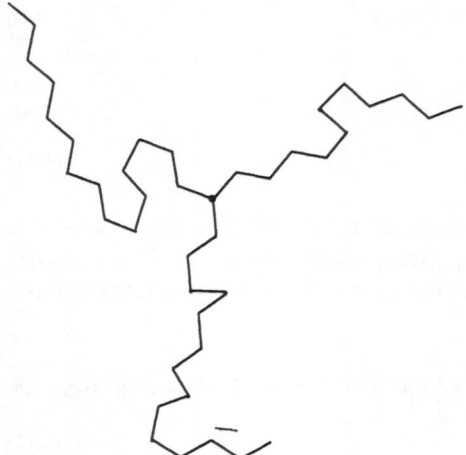

Fig. 47. Three chains connected by a single trifunctional crosslinking agent, $L = 1$

Such expressions are found in field theories so the problem of the one-step polymerization in the sense of a mathematical description of the consequences is expressed in this kind of function. These ideas can be extended to the actual dynamics of the creation of networks, but it is inappropriate to include that here.

Models like the affine deformation emerge as simple approximations to the evaluation of expressions like Eq. (105), but the reader can sense that integrating over all functions is not easy and most field theories end up with versions of perturbation theory in the constants like μ_3 or the dimensionality of the problem, the renormalization group, both of which are too delicate to handle real physical problems and also work perfectly in physically unusual situations such as very weak conditions or criticality. In order to get an orderly progress with these methods, which in principle are complete but still largely unexplored, one should be matched by well defined experiments which progressively explore the same territory as the mathematics. Alas, the theory and experiment do not go hand-in-hand and things like "homogeneous" and "affine" are very hard to create.

5.2.1.2 Field Formalism for Deformation

Our analysis applies equally to flexible and rigid polymers and to rigid or flexible junctions. However, it gives the distribution of species if equilibrium is maintained. It is therefore correct for a hydrogen-bonded network if the system is at a sufficiently high temperature to break the H-bonds occasionally, but not so high as to make them very short lived. The general problem of building the macromolecules and the lattice irreversibly is much more difficult to handle and we do not attempt it here although some general literature on the subject is available [912–916].

Assuming then that the distribution of configurations is indeed given by the distribution at Eq. (105), we can, as in the standard approach with flexible networks, just assume that, having come into being, it then remains permanent. Now the question arises as to how one calculates the elastic constants of such a system. The problem can be solved in principle within the ϕ field formalism. To see this, we must realize that once the network is established, the free energy will be a function of the entire topology of the system. If we now recalculate the free energy in a deformed state, we get the new free energy to be $\tilde{A}$ (topology), which will have an average

$$\langle \tilde{A} \rangle = \int \tilde{A}(\text{topology}) \, P(\text{topology}) \, d(\text{topology}) \tag{106}$$

where P is the probability of finding a topology. For example, if our topology is confined to saying that rod i meets rods k and j, Eq. (106) becomes

$$\langle \tilde{A} \rangle = \iiint \tilde{A}(r_i = r_j = r_k) \, P(r_i = r_j = r_k) d^3 r_i \, d^3 r_j \, d^3 r_k . \tag{107}$$

For simplicity we write the distribution as stemming from an energy (without loss of generality as to other constraints):

$$P = \int_{\text{initial conditions}} e^{(A - H)/kT} d(\text{all}) \tag{108}$$

$$\tilde{A} = - kT \log \int_{\text{deformed conditions}} e^{- H/kT} d(\text{all}). \tag{109}$$

Therefore,

$$\langle \tilde{A} \rangle = - kT \int_{\text{initial}} e^{(A - H)/kT} \log \int_{\text{deformed}} e^{- H/kT}. \tag{110}$$

The rods must meet in an identical way in both integrals. This can be done by writing

$$\log \int e^{- H/kT} = \text{coefficient of a in} \left(\int e^{- H/kT} \right)^a \tag{111}$$

whereupon we can invent a system of $3(a + 1)$ dimensions where the integration is through the initial material for 3 dimensions and the deformed material for the other 3a dimensions. We can now introduce an A for the $a + 1$ replicas and derive

$$e^{- A(a)/kT} = \int_{\text{'0' initial rest}} \int_{\text{deformed}} e^{- \sum_{m=0}^{a} H_m/kT} d(0)(d1. \ldots \ldots da) \tag{112}$$

and the coefficient of a in A is the answer. This can be done provided one invents a new $\emptyset$ where, for instance, $\emptyset(r_0)$ is replaced by $\emptyset(r_0 r_1 \ldots)$. If we now write the same Hamiltonian, Eq. (112), the topology is governed by the mating of $\emptyset$, $\emptyset^*$ and is identical in all the replicas and hence gives us a correct answer. In this formalism, $\emptyset g \emptyset$ (cf. Eqs. (102–105) becomes replaced by $\emptyset(r_0 r_1 \ldots r_a) g(r_0 r_0')$ $g(r_1 r_1'). \ldots g(r_a r_a') \emptyset(r_0' r_1' \ldots r_a')$ and the lattice in the r_0 variables will have the identical topology to that of the $r_1 \ldots r_a$ variables.

For flexible networks this formalism reproduces all the standard results, the calculation being made possible by the mathematical simplicity of the Gaussian function [862, 863]. For rigid rodlike and defect-free networks a solution for arbitrary deformation does not seem possible but such networks can only be deformed by small amounts without fracturing, and an expansion in the strain tensor is always possible [917]. This means that the crude model above can be made quite topologically complete in the behavior of network junction statistics. The intermediate case of defect-free networks created from rigid branchpoints and stiff but not rodlike segments, say, worm-like in character, is more complex. Here, segments may be extended or bent to start with, and under stress various modes of segmental deformation may concomitantly appear and compete with one another on a share from the total population of deformations. The modes of deformation are certain to be intimately dependent on the structure of the network segments, as was shown in Sect. 4.4 above, and change in response to changes in the stress and strain fields. An analytical description of such complex stiff networks and gels is still in the future.

There still remains the issue of the topological constraints of the simple type

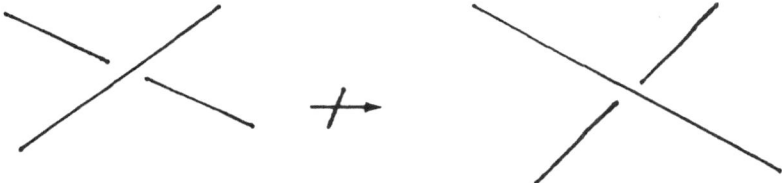

and of course the normal intermolecular forces. The intermolecular forces can be handled in one-step networks by introducing an auxiliary field ψ, which is attached to the monomer $\emptyset(r)g(rr')\emptyset(r')$ by multiplication with $\exp(i\psi[(r + r')/2])$, i.e.,

$$\int\int \emptyset(r)g(rr')\emptyset(r')\exp(i\psi[(r + r')/2]) \tag{113}$$

replacing $\int\int \emptyset g\emptyset$.

If a field weight

$$kT\int\int \psi(r)V^{-1}(r,r')\psi(r')d^3rd^3r' \tag{114}$$

is added to the Hamiltonian, Eq. (105), integration over ψ introduces

$$e^{-V(rr')/kT} \tag{115}$$

between all monomers at r and r' (the monomer is considered to be small so that $(r + r')/2 = r$ or r' without loss of generality).

When replicas are used there has to be a ψ_a, for the interactions are always within particular replicas; it is only crosslinks which are shared between replicas.

Interactions within two-step networks can be expressed by noting that, given the end points of a rod 1, r_1 and r_1', and rod 2, r_2 and r_2', we have enough information to write the mutual energy of the two rods $V(r_1 r_1' r_2 r_2')$. To handle this one needs a $\psi(r_1 r_1')$ and the analysis above goes through. The various cases of stiffness and twist energy can likewise be incorporated in generalized field functions and Hamiltonians.

The various approaches to the entanglement problem, such as the Gaussian winding number, can also be incorporated in the one-step formulae, but at the price of considerable algebraic complexity and a weakness due to finite rod size. One feels that the same such expression must apply to the two-step process, but the finite length of the stiff precursor chains again is a problem. All that can be said at this stage is that the mathematics of topological invariants leaves much to be desired and it is not surprising that rigid networks also suffer.

We have offered an outline of a complete field theoretical formulation, but one whose very precision makes it difficult to operate. We therefore turn to cruder theories which, by intuitive modeling, can give some insight into the overall properties of rigid networks.

5.3 Simpler Theoretical Approaches

5.3.1 Simple Calculations of the Concentrated Rigid Network

The concentrated rigid network, as in Fig. 42, is the simplest to study since it is reasonable to argue that it can be uniform. Network uniformity is defined here as the absence of voids and other flaws whose size is larger than the length l_0 of the average stiff segment in the network. Such a situation implies a concentrated uniformly interpenetrated system before the gelation of the one-step network. It does not mandate that at the gel-point all polymeric material will already be a covalently-bonded integral part of the network, but that the material will be uniformly distributed in space and interpenetrated with the network at the gel-point. The uniformly interpenetrated material continues to polymerize after the gel-point and may or may not be covalently bonded with the network initially created at the gel-point. When the polymerization reaction is completed however, far beyond the gel-point, the resulting concentrated rigid network is uniform for all practical purposes. A typical picture of such a network fragment is shown in Fig. 48.

Following the literature, e.g., the Doi-Edwards [42] discussion, particularly Sects. 6 and 9, we describe the surrounding segments by a tube of radius $a = 1/nl_0^2$ where n is the number concentration of rigid segments and l_0 their average length. Thus Fig. 48a becomes Fig. 48b. Here, the tube has not been drawn sharply to emphasize that it is an average and not a concrete concept. There are two lengths in the problem, l_0 and a. If the junctions in the network are flexible, then there is an entropy in the system of the order of $k_B \log(l_0 - 2a)$. If the tube is deformed by some average λ, the entropy changes to

$$k_B \log(l_0 - 2\lambda a). \tag{116}$$

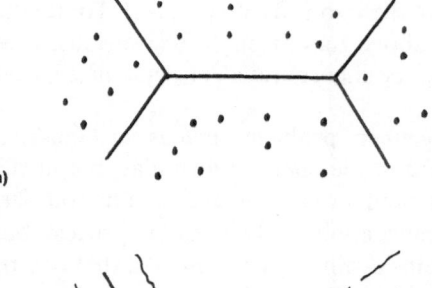

a)

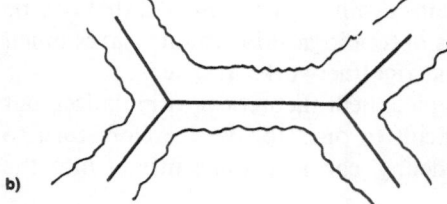

b)

Fig. 48a. Concentrated rigid network. **b** The tube model as applied to concentrated rigid networks

This is catastrophic when λ is about $l_0/2a$, and a further deformation of the tube requires segment bending, i.e., if $\lambda = 1 + e$, the maximum extension of the segment in the tube will be limited to

$$e = (l_0/2a) - 1. \tag{117}$$

If the junctions in the network are rigid, the rodlike segments' bending starts immediately but works up to the value given by Eq. (117). This equation assumes that topological effects bend the rodlike segments into the affine positions if possible, or deform the surrounding material when the rodlike segments are incompressible.

Once one leaves this crude assessment, the behavior is intuitively clear but not easy to handle mathematically. The twisting and bending of rodlike segments, and the coupling between these motions in the segment and through branchpoints, can be incorporated into a refinement of the "affine" calculation above. Likewise, these modes may be inserted into computer programs designed to solve for particular configurations which, by cyclic boundary conditions, can be made characteristic of the homogeneous high concentration case that we consider here. What is not at all simple analytically is to build in the fact that when a threshold of bending energy is reached, the system will jump into a new configuration as in Fig. 49. The same problem arises in one of the properties of liquid crystal polymers possessing statistically flexible spacers between stiff mesogenic groups. Here it pays the molecule to fold over into a "hairpin" where the energy spent on each 360° bend is more than compensated by attractive

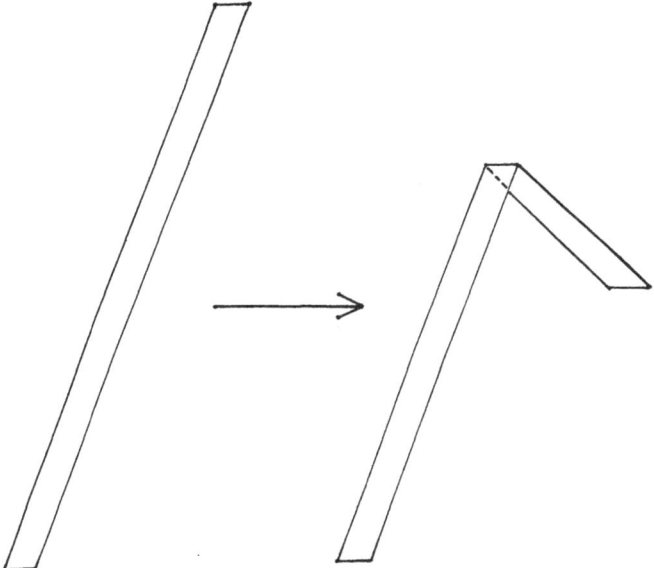

Fig. 49. Sharp bending of rodlike segment upon surpassing a threshold bending energy

energy along the length of the parallel sections of the chain. Although one can easily estimate the numbers involved, it is very difficult to include this in a statistical mechanical treatment. Thus, it seems more appropriate to study the theory of this kind of network deformation by numerical analysis, rather than search for an analytical formula. Clearly a great deal of work would be involved in a full treatment.

5.3.2 Simple Calculations in a Dilute Rigid Network

As we have already indicated, one can expect such networks of all degrees of flexibility to build up as trees. There is always a low probability of intra-tree circuits compared to trees, which increases with the overall flexibility of the network, and one can draw up a hierarchy as follows.

a) Rodlike segments with rigid branchpoints. As is shown in Fig. 50a, the distance A–B is too large and undeformable; there is no way to complete the

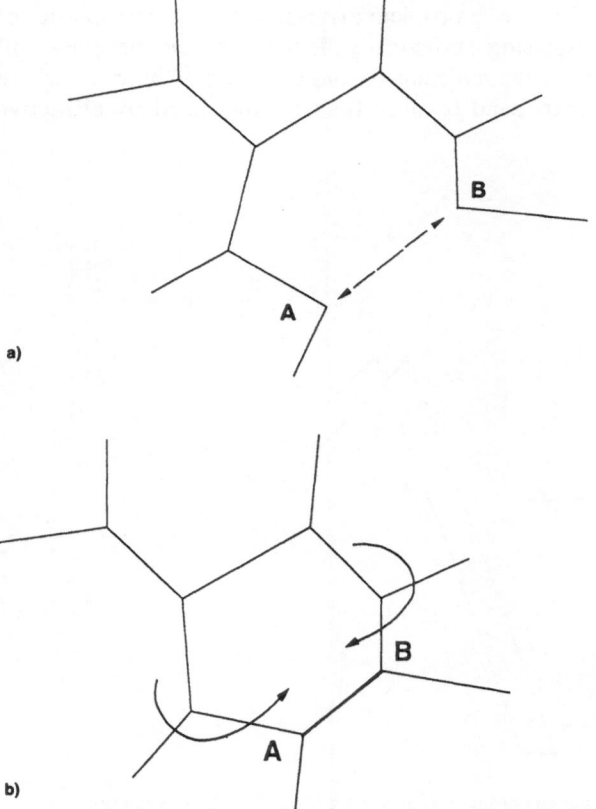

Fig. 50a. Tree fragment with rodlike segments and rigid branchpoints. **b** Tree fragment with rodlike segments but flexible junctions allowing for points A and B to approach one another and be bridged, thus forming an intrafractal macrocycle

circuit for geometric reasons, unless a segment of a perfectly-matched length and appropriate reactive end-groups is added in one or a sequence of steps. In both condensation and addition polymerizations this is statistically highly improbable.

b) Rodlike segments with flexible junctions. As is shown in Fig. 50b, A and B can possibly be joined if angles can bend. Circuit formation is much more likely than in a) above, but limited by segment rigidity and the requirements of the correct angular approach of the reactive groups.

c) Stiff worm-like segments with rigid branchpoints or flexible junctions. These allow more circuits than the respective cases a) and b), but far less than flexible networks. Here, the angular approach restriction is somewhat relaxed relative to rodlike segments, but is still very high relative to flexible segments. Furthermore, the modest average length of both worm-like and rodlike segments places severe restrictions on the volumes swept by their reactive tips and their overall bending capabilities.

d) Flexible segments with flexible or rigid junctions. The combination of, generally, long and flexible segments allows for high translational and rotational motions of all reactive groups with a large increase in the probability of intra-tree circuit formation. This probability is far higher than in any of the cases a), b) or c); it increases with the functionality f of the crosslinking group, with the ratio of such groups to difunctional monomers, and in a very complicated way, with dilution.

In all of these, networks prepared by a one-step method perform about the same as those prepared in two separate steps.

The circuits discussed hitherto in this section are all in the growing or final tree. Because of the fractal nature of these trees and chemical nomenclature, we shall refer to them below as intrafractal rings or intrafractal macrocyclic structures. These should not be confused with interfractal rings or interfractal macrocycles. The latter are formed between fractal polymers as they aggregate to create the network, and each such macrocycle may contain several segments belonging to two or more FPs. Thus, in the final network the number of interfractal macrocyclic structures increases with network perfection and with concentration above C_0^*.

When effects of concentration are considered, one has to differentiate between a situation where fully-grown trees are brought together at concentrations C_0 far above the critical concentration for network formation C_0^*, and a situation where the polymerization starts at $C_0 \gg C_0^*$ and the trees grow interwoven to start with. In the first of these instances, rigid trees made from rodlike segments and rigid branchpoints (case a) above) will not be able to interpenetrate one another, except by deformation. Some interpenetration in the external shells of such trees is expected, however, simply because the material density of the trees rapidly decreases in the exteriors and interpenetrations to depths of about one segment length are not substantially hindered. In case b) above, tree or network deformation is easier and deep interpenetrations of trees is possible by simply pushing rodlike segments out of the way. An undeformed state is possible in principle, especially in cases where the junction functionality is high (f = 6 for instance). As a rule, the interpenetration of fully grown FPs will become easier with increased overall flexibility. The same goes for the ease of disengagement of entangled FPs when pulled apart during, say, dilution.

To demonstrate the above, let us turn to Fig. 51. At the top of the figure, one tree is drawn as a series of dots and the other, with f = 3, as a series of lines. At the bottom of Fig. 51 they are interpenetrated with one another. If one tags on the line tree at the point A and pulls to the right, one can see that with flexible junctions it can slip out back to Fig. 51 (top). The Brownian motion of the flexible junctions means that one must be careful in thinking intuitively of this situation and not confuse it with what would happen if both segments and branchpoints were fully rigid-they would almost certainly jam. In our previous work [311, 312, 350, 413] we did come across situations where FPs of stiff polyamide segments and rigid branchpoints, when synthesized at $C_0 \gg C_0^*$ from

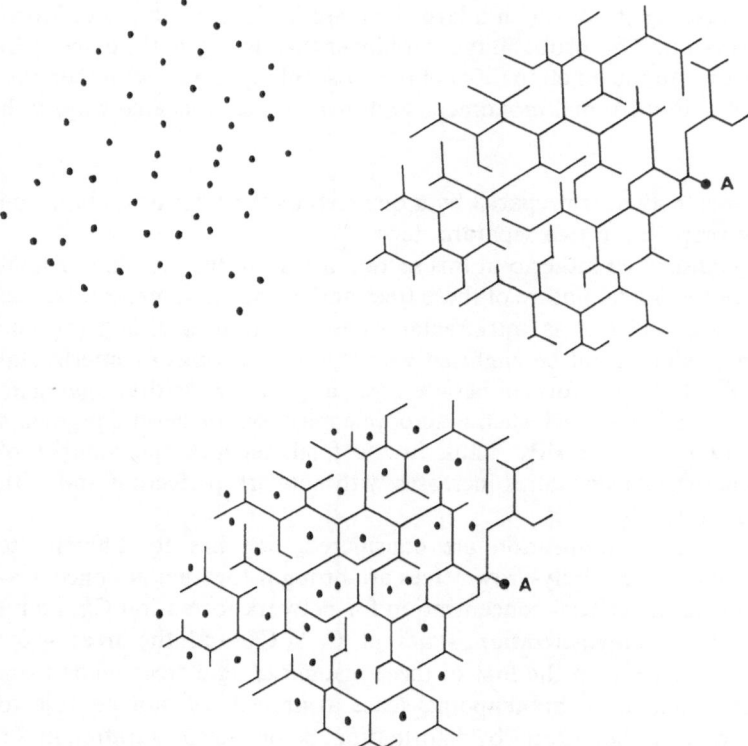

Fig. 51. *Top*: Two fractal polymers with rodlike or stiff segments, unentangled. *Bottom*: The same two fractal polymers entangled. If one pulls to the right at the point A, the FPs with flexible junctions will disentangle back to the top conditions while FPs with rigid branchpoints will remain entangled and their segments jammed together

AB_n-type monomers, got jammed together instead of developing into infinite networks. The resulting systems were amorphous physical gels with 3.0% $\leqslant C_0 \leqslant 10.0\%$, and possessed rather high equilibrium shear modulus. Upon transferring to a good solvent, the gels completely dissolved. Light scattering experiments showed the species in solution to be fractal polymers and not microgels. This clearly indicates that the gelled systems were physical gels in which the applied stresses were supported by the jammed-together segments of the rigid fractal polymers.

The picture that emerges for the dilute solution conditions is that networks can be made in one step from monomers at $C_0 \gtrsim C_0^*$ or in two steps by first allowing the FPs to grow at concentrations lower than C_0^* and then increasing the concentration to C_0^* or slightly higher. In both instances the formed network, at $C_0 \gtrsim C_0^*$, will form Voronoi polyhedra to some extent (see Figs. 31–33). If the FPs are neither too porous not too flexible, the Voronoi picture should be a good one. One can think of the trees as DLA fractals or as a more concentrated internal region with a fractal shell (see Sect. 1.3 and Fig. 13) [311, 312, 918]. The deformation of a Voronoi system is something well explored in the literature both for its intrinsic interest [699–701, 710–712], for the theory of foams [702–707], and for mathematical constructions used as a kind of dynamic finite element method [919–921]. One can expect that this literature will enable the theorist to make a study of the packed FPs. Looking at the various categories listed above, we see that the specifically fractal nature of the trees does not have any major effect on the densely interpenetrated systems, as illustrated in Fig. 42. This stands whether they have rigid or flexible junctions or segments since in this case the free energy of the network is dominated by the local environment. Thus, the simple formulae at Eqs. (87) and (116) will still apply. The interesting new situation arises in the case where the fractal polymers or fractal sub-networks are prepared under dilute conditions and then concentrated until they meet and are packed into Voronoi shapes.

There are two limits here: the case when freely hinged rods interpenetrate a neighboring FP and the case when rigidly hinged rods fail to interpenetrate to any depth and the polyhedra are subject to compression. In the freely hinged case, the freedom of movement of rods pivoted at the junctions is more and more constrained and there is a resulting loss of entropy. It is clearly a theoretical problem of several aspects, but we can make some rough estimate of what happens. Let the number of fractal polymers be N_F, so that the number density of FPs is $N_F/v = C_f$. The average distance between fractal centers, R_a, is then $C_f^{-1/3}$. Suppose the average FP radius is $\langle R^2 \rangle = qL^\beta$ where q has the appropriate dimension and L is the total length of rodlike segments in the fractal, i.e., q has dimension $(\text{length})^{2-\beta}$. Thus, if the length of one segment is l_0, $q = l_0^{2-\beta} \times$ a numerical factor. The actual distribution in space of the segments is itself a problem related to diffusion limited aggregation. We will assume for simplicity that it is

$$P(r) = L(2qL^\beta/3\pi)^{3/2} \exp\left[-(3/2)(r^2/qL^\beta)\right] \qquad (118)$$

i.e., the probability of finding a polymer segment at the point r when the fractal polymer starts at $r = 0$ is $P(r)$ and the total polymer length in the fractal is

$$L = \int P(r)d^3r. \tag{119}$$

Each rod in the isolated fractal can point in any direction if excluded volume is ignored, but as a fractal is interpenetrated by its neighbor the freedom of rotation of each rod is reduced. Thus the angular freedom of a rod is reduced from 4π to a value which will be of the order of the ratio of the density of the intruding fractal polymer to the rod length. A crude extrapolation formula for the entropy could then be

$$\log[4\pi/(1 + \rho(r)/l_0^3)^{1/3}] \tag{120}$$

or even simpler, when r is averaged over the two fractal distributions

$$S = k(\log 4\pi - \gamma e^{-3R_a^2/2qL^\beta}) \tag{121}$$

is a reasonable extrapolation formula where γ is a numerical factor and R_a is the distance between fractal centers. This formula simply says that as fractals interpenetrate the freedom of movement of the rods diminishes. Since R_a is now $C_f^{-1/3}$ the final free energy is

$$F = \gamma kTN_F^2 \exp(-q'C_f^{-2/3}l_0^{-2}) + \text{constants} \tag{122}$$

where q' is now a number. The free energy must be the perfect gas value of N_F fractals in a volume v, so one could imagine van der Waals type equations like

$$P[v - \gamma N_F \exp(-q'C_f^{-2/3}l_0^{-2})] = N_F kT \tag{123}$$

and

$$P[1 - \gamma C_f \exp(-q'C_f^{-2/3}l_0^{-2})] = C_f kT \tag{124}$$

which give the flavor of what can be expected. Full derivations of the equations of state of such systems are clearly possible but at present the state of characterization scarcely justifies the effort involved.

Finally, turning to the case of compressing together FPs which have rigid branchpoints and stiff segments, the formulae derived by Jones and Ball [781] apply directly to this case without further analysis being required. However, it is very possible that the strength of the FPs is such that when brought together many voids will occur (see Sect. 4.2.2) which will then dominate the modulus and failure strength of the resulting material, in the gel as well as in the dry state, as is the case in many other instances in materials science [922]. The propagation of cracks as a consequence of random defects is a problem of much current interest since the resulting fracture surfaces contain interesting power laws [923, 924]. This work may also have applications here since the central position of flaws in the material's strength is well documented [922].

We can further ask if any direct calculation of the strength of the fractal tree is possible. This has been studied in two papers by Jones et al. [732, 781]. Studying the resistance to failure of the minimal path for transport in networks created from fractal trees [925], invasive percolation in systems with fractal void structure [695] and the mechanical and elastic properties of percolation networks in general [854, 926–929] may be additional useful avenues to follow in this quest. Once the strength of fractal trees is calculable, then this in combination with knowledge of the effects of voids, defects and other flaws may afford us the ability to calculate and predict the strength and other mechanical properties of rigid aromatic networks, such as those described in previous Sections of this work.

Finally, the relationships between the shape, size and concentration of monomeric and oligomeric species, and network mesh size and concentration in the reaction bath on one hand and the kinetics of fractal and network formation and the efficiency of the chemical reactions on the other, have not yet been dealt with for rigid networks in any theoretical rigor. Some initial steps in this direction were taken [666, 667] but a lot is yet to be done.

6 References

1. Staudinger H (1920) Ber Chem 53: 1073
2. Carothers WH (1929) J Am Chem Soc 51: 2548
3. Staudinger H (1924) Ber Chem 57: 1203
4. Staudinger H (1929) Angew Chem 42: 67
5. Guth E, Mark H (1934) Monatsh Chem 65: 93
6. Kuhn W (1934) Kolloid Z 68: 2
7. Kuhn W (1936) Kolloid Z 76: 258
8. Flory PJ (1941) J Am Chem Soc 63: 3083
9. Flory PJ (1941) J Am Chem Soc 63: 3091
10. Stockmayer WH (1943) J Chem Phys 11: 45
11. Stockmayer WH (1944) J Chem Phys 12: 125
12. Flory PJ (1946) Chem Revs 39: 137
13. James HM, Guth E (1943) J Chem Phys 11: 455
14. James HM (1947) J Chem Phys 15: 651
15. James HM, Guth E (1947) J Chem Phys 15: 669
16. Wall FT, Flory PJ (1951) J Chem Phys 19: 1435
17. Dusek K, Prins W (1969) Adv Polymer Sci 6: 1
18. Flory PJ (1974) Discuss Faraday Soc 57: 7
19. Graessley WW (1982) Adv Polymer Sci 47: 67
20. Deam RT, Edwards SF (1976) Phil Trans Royal Soc London, A280: 317
21. Edwards SF (1976) Proc Royal Soc London A351: 397
23. Rempp P, Herz E, Borchard W (1978) Adv. Polymer Sci 26: 105
23. Flory PJ (1976) Proc Royal Soc London A351: 351
24. Flory PJ (1977) J Chem Phys 66: 570
25. Freed KF (1971) J Chem Phys 55: 5588
26. Ronca G, Allegra G (1975) J Chem Phys 63: 4990
27. Mark JE (1982) Adv Polymer Sci 44: 2
28. Candau S, Bastide J, Delsanti M (1982) Adv Polymer Sci 44: 27
29. Staverman AJ (1982) Adv Polymer Sci 44: 73
30. Stauffer D, Coniglio A, Adam M (1982) Adv Polymer Sci 44: 103
31. Gordon M (1962) Proc Royal Soc London A268: 240
32. Gordon M, Malcolm GN (1966) Proc Royal Soc London A295: 29
33. Macosko CW, Miller DR (1976) Macromolecules 9: 199
34. Miller DR, Macosko CW (1976) Macromolecules 9: 206
35. Miller DR, Macosko CW (1980) Macromolecules 13: 1063
36. Miller DR, Macoscko CW (1988) J Polymer Sci Polym Phys Ed 26: 1
37. Ilavasky M, Dusek K (1987) Progr. Colloid & Polymer Sci. 75: 11
38. Dušek K, Šomvarsky J (1992) In: Aharoni SM (ed) Synthesis, Characterization, and Theory of Polymer Networks and Gels. Plenum, NY, p 283
39. Flory PJ (1953) Principles of Polymer Chemistry. Cornell University Press, Ithaca, NY
40. Treloar LRG (1958) The Physics of Rubber Elasticity. Clarendon Press, Oxford, UK
41. deGennes P-G (1979) Scaling Concepts in Polymer Physics. Cornell University Press, Ithaca, NY
42. Doi M, Edwards SF (1986) The Theory of Polymer Dynamics. Clarendon Press, Oxford, UK
43. Dickie RA, Bauer RS (eds) (1988) Chemistry, Properties and Structure of Crosslinked Systems. ACS Symposium Series, ACS, Washington, DC
44. Chompff AJ, Newman S (eds) (1971) Polymer Networks, Structure and Mechanical Properties. Plenum NY
45. Mark JE, Lal J (eds) (1982) Elastomers and Rubber Elasticity. ACS Symposium Series, ACS, Washington, DC
46. Mark JE, Erman B (1988) Rubberlike Elasticity: A Molecular Primer. Wiley-Interscience NY
47. Tsvetkov VN, Andreeva LN (1981) Adv Polymer Sci 39: 95
48. Curran SA, LaClair CP, Aharoni SM (1991) Macromolecules 24: 5903
49. Porod G (1949) Monatsh Chem 80: 251
50. Kratky O, Porod G (1949) Rec Trav Chim Pays Bas 68: 1106

51. Flory PJ (1969) Statistical Mechanics of Chain Molecules. Interscience, NY, Appendix G, pp 401
52. Benoit H, Doty PM (1953) J Phys Chem 57: 958
53. Yamakawa H (1971) Modern Theory of Polymer Solutions. Harper & Row, NY, pp 25, 53, 56
54. Aharoni SM (1980) J Appl Polymer Sci 25: 2891
55. Aharoni SM (1981) J Polymer Sci Polym Phys Ed 19: 281
56. Aharoni SM (1987) Macromolecules 20: 2010
57. Strzelecki L, Van-Luyen D (1980) Europ Polymer J 16: 299
58. Jin J-I, Antoun S, Ober C, Lenz RW (1980) Br Polymer J 12: 132
59. Ober CK, Jin J-I, Lenz RW (1984) Adv Polymer Sci 59: 103 and references therein
60. Sirigu A (1991) In: Ciferri A (ed) Liquid Crystallinity in Polymers: Principles and Fundamental Properties. VCH NY pp 261–313 and references therein
61. Aharoni SM (1987) Macromolecules 20: 877
62. Aharoni SM (1988) Macromolecules 21: 185
63. Hoskins BF, Robson R (1989) J Am Chem Soc 111: 5962
64. Duchamp DJ, Marsh RE (1969) Acta Cryst B25: 5
65. Ermer O, Eling A (1988) Angew Chem Int Ed Engl 27: 829
66. Ermer O (1988) J Am Chem Soc 110: 3747
67. Thomas EL, Alward DB, Kinning DJ, Martin DC, Handlin Jr DL, Fetters LJ (1986) Macromolecules 19: 2197
68. Aharoni SM (1989) unpublished observations
69. Strezelecki L, Liebert L (1973) Bull Soc Chim Fr 605
70. Aharoni SM (1988) Macromolecules 21: 1941
71. Aharoni SM (1988) US Patent No. 4,767,839
72. Aharoni SM (1989) Macromolecules 22: 686
73. Aharoni SM (1989) Macromolecules 22: 1125
74. Aharoni SM, Correale ST, Hammond WB, Hatfield GR, Murthy NS (1989) Macromolecules 22: 1137
75. Hatfield GR, Aharoni SM (1989) Macromolecules 22: 3807
76. Murthy NS, Anaroni SM (1992) Macromolecules 25: 1177
77. Fryer JR, Kenney ME (1988) Macromolecules 21: 259
78. Nohr RS, Kuznesof PM, Wynne KJ, Kenney ME, Siebenman PG (1981) J Am Chem Soc 103: 4371
79. Joyner RD, Kenney ME (1960) J Am Chem Soc 82: 5790
80. Metz J, Hanack M (1983) J Am Chem Soc 105: 828
81. Joyner RD, Kenney ME (1962) Inorg Chem 1: 717
82. Orthmann E, Enkelman V, Wegner G (1983) Makromol Chem Rapid Commun 4: 687
83. Takahashi S, Murata M, Kariya K, Sonogashira K, Hagihara N (1979) Macromolecules 12: 1016
84. Fyfe HB, Mlekuz M, Zargarian D, Taylor NJ, Marder TB (1991) J Chem Soc Chem Commun 118
85. Kreuder W, Ringsdorf H, Tschirner P (1985) Makromol Chem Rapid Commun 6: 367
86. Wenz G (1985) Makromol Chem Rapid Commun 6: 577
87. Herrmann-Schoenherr O, Wendorff J, Kreuder W, Ringsdorf H (1986) Makromol Chem Rapid Commun 7: 97
88. Ringsdorf H, Tschirner P, Herrmann-Schoenherr O, Wendorff JH, (1987) Makromol Chem 188: 1431
89. Kreuder W, Ringsdorf H (1983) Makromol Chem Rapid Commun 4: 807
90. Ringsdorf H, Wustefeld R, Zerta E, Ebert M, Wendorff JH (1989) Angew Chem Int Ed Engl 28: 914
91. Allcock HR, Neenan TX (1986) Macromolecules 19: 1495
92. Blumstein A, Hsu EC (1978) In: Blumstein A (ed) Liquid Crystalline Order in Polymers. Academic Press, New York, p 105 and references therein
93. Finkelman H, Rehage G (1984) Adv Polymer Sci 60/61: 99 and references therein
94. Shibaev VP, Plate NA (1984) Adv Polymer Sci 60/61: 173 and references therein
95. Finkelmann H (1991) In: Ciferri A (ed) Liquid Crystallinity in Polymers Principles and Fundamental Properties. VCH Publishers, New York, p. 315 and references therein.
96. Kwolek SL, Morgan PW, Schaefgen JR (1987) In: Kroschwitz JI (ed) Polymers: An Encyclopedic Sourcebook of Engineering Properties. Wiley-Interscience, New York, p. 509
97. Jackson Jr WJ, Kuhfuss HF (1976) J Polymer Sci Polym Chem Ed 14: 2043

98. Ungar G, Feijoo JL, Percec V, Young R (1991) Macromolecules 24: 953
99. Griffin AC, Havens SJ (1980) J Polymer Sci Polym Lett Ed 18: 295
100. Blumstein A, Sivaramakrishnan KN, Clough SB, Blumstein R (1979) Mol Cryst Liq Cryst 49: 255
101. Roviello A, Sirigu A (1982) Makromol Chem 183: 895
102. Antipov EM, Kulichikhin VG, Plate NA (1992) Polym Eng Sci 32: 1188
103. Blumstein A, Maret G, Vilasagar S (1981) Macromolecules 14: 1543
104. Hessel F, Finkelmann H (1986) Polymer Bull 14: 375
105. Zhou QF, Li HM, Feng XD (1987) Macromolecules 20: 233
106. Reck B, Ringsdorf H (1985) Makromol Chem Rapid Commun 6: 291
107. Zentel R, Reckert G (1986) Makromol Chem 187: 1915
108. Zentel R, Reckert G, Reck B (1987) Liquid Cryst 2: 83
109. Bualek S, Zentel R (1988) Makromol Chem 189: 791
110. Zentel R (1989) Angew Chem Adv Mater 101: 1437
111. Peter K, Rätzsch M (1990) Makromol Chem 191: 1021
112. Caruso U, Centore R, Roviello A, Sirigu A (1992) Macromolecules 25: 129
113. Lenz RW (1967) Organic Chemistry of Synthetic High Polymers. Interscience, NY, pp 97–101, 148–151
114. Dimian AF, Jones FN (1987) Proc ACS Div PMSE 56: 640
115. Wang D, Jones FN (1987) Proc ACS Div PMSE 56: 645
116. Braun D, Ritzert HJ (1989) In: Estmond GC, Ledwith A, Russo S, Sigwalt P (eds) Comprehensive Polymer Science, Volume 5. Pergamon, Oxford, pp 649
117. Quirke JME (1984) In: Boulton AJ, McKillop A (eds) Comprehensive Heterocyclic Chemistry, Volume 3. Pergamon, Oxford, pp 457
118. Caruso U, Pragliola S, Roviello A, Sirigu A (1993) Macromolecules 26: 221
119. Finkelmann H, Kock H-J, Rehage G (1981) Makromol Chem Rapid Commun 2: 317
120. Finkelmann H, Kock H-J, Gleim W, Rehage G (1984) Makromol Chem Rapid Commun 5: 287
121. Schätzle J, Finkelmann H (1987) Cryst Liq Cryst 142: 85
122. Gleim W, Finkelmann H (1987) Makromol Chem 188: 1489
123. Schätzle J, Kaufhold W, Finkelmann H (1989) Makromol Chem 190: 3269
124. Hirschmann H, Meier W, Finkelmann H (1992) Makromol Chem Rapid Commun 13: 385
125. Zentel R, Benalia M (1987) Makromol Chem 188: 665
126. Müller H, Müller I, Nuyken O, Strohriegl P (1992) Makromol Chem Rapid Commun. 13: 289
127. Zentel R, Reckert G, Bualek S, Kapitza H (1989) Makromol Chem 190: 2869
128. Zentel R (1988) Liquid Cryst 3: 531
129. Chien LC, Lin C, Fredley DS, McCargar JW (1992) Macromolecules 25: 133
130. Liebert L, Strzelecki L (1973) Compt Rend Acad Sci Paris, C, 276: 647
131. Strzelecki L, Liebert L (1973) Bull Soc Chim Fr 597
132. Liebert L, Strzelecki L (1973) Bull Soc Chim Fr 603
133. Strzelecki L, Liebert L (1981) Eur Polymer J 17: 1271
134. Bouligand Y, Cladis PE, Liebert L, Strzelecki L (1974) Mol Cryst Liq Cryst 25: 233
135. Clough SB, Blumstein A, Hsu EC (1976) Macromolecules 9: 123
136. Toth WJ, Tobolsky AV (1970) Polymer Lett 8: 289
137. Conciatori AB, Choe EW, Farrow G (1984) US Patent No 4,452,993
138. Conciatori AB, Choe EW, Farrow G (1985) US Patent No 4,514,553
139. Stackman RW (1987) US Patent No 4,683,327
140. Broer DJ, Boven J, Mol GN, Challa G (1989) Makromol Chem 190: 2255
141. Broer DJ, Hikmet RAM, Challa G (1989) Makromol Chem 190: 3201
142. Hikmet RAM, Lub J, Maassen vd Brink P (1992) Macromolecules 25: 4194
143. Heynderickx I, Broer DJ, Trevoort-Engelen Y (1992) J Mater Sci 27: 4107
144. Rozenberg BA, Gur'eva LL (1992) In: Aharoni SM (ed) Synthesis, Characterization and Theory of Polymeric Networks and Gels Plenum, New York, p 147
145. Serebryakova II, Gur'eva LL, Tsukruk VV, Shilov VV, Tarasov VP, Erofeev LN, Rozenberg BA (1988) USSR Patent No 1,541,209
146. Tsukruk VV, Gur'eva LL, Tarasov VP, Shilov VV, Erofeev LN, Rozenberg BA (1991) Vysokomol Soed B, 33: 168
147. Barclay GG, Ober CK, Papathomas KI, Wang DW (1992) J Polymer Sci Part A: Polym Chem 30: 1831
148. Barclay GG, McNamee SG, Ober CK, Papathomas KI, Wang DW (1992) J Polymer Sci Part A: Polym Chem 30: 1845

149. Shea KJ, Loy DA, Webster O (1992) J Am Chem Soc 114: 6700
150. Pachter R, Bunning TJ, Socci EP, Farmer BL, Crane RL, Adams WW (1992) Polymer Preprints 33(1): 671
151. Privalko VP (1973) Macromolecules 6: 111
152. Privalko VP, Lipatov YuS (1974) J Macromol Sci Phys B9: 551
153. Privalko VP (1975) Polymer J 7: 202
154. Privalko VP (1980) Macromolecules 13: 370
155. Privalko VP, Lipatov YuS (1974) Makromol Chem 175: 641
156. Boyer RF, Miller RL (1976) Polymer 17: 925
157. Boyer RF, Miller RL (1976) Polymer 17: 1112
158. Boyer RF, Miller RL (1977) Macromolecules 10: 1167
159. Boyer RF, Miller RL (1978) Rubber Chem Technol 51: 718
160. Miller RL, Boyer RF (1978) J Polymer Sci Polym Phys Ed 16: 371
161. Aharoni SM (1974) J Polymer Sci Polym Lett Ed 12: 549
162. Aharoni SM (1976) J Appl Polymer Sci 20: 2863
163. Aharoni SM (1977) J Appl Polymer Sci 21: 1323
164. Graessley WW, Edwards SF (1981) Polymer 22: 1329
165. Aharoni SM (1982) Polymer Preprints 23(1): 275
166. Aharoni SM (1983) Macromolecules 16: 1722
167. He T, Porter RS (1992) Makromol Chem Theory Simul 1: 119
168. Aharoni SM (1986) Macromolecules 19: 426
169. Wool RP (1993) Macromolecules 26: 1564
170. Aharoni SM (1985) Macromolecules 18: 2624
171. Wu S (1989) J Polymer Sci Part B: Polym Phys 27: 732
172. Wu S (1990) Polym Eng Sci 30: 754
173. Wu S (1992) Polym Eng Sci 32: 823
174. Flory PJ (1956) Proc Royal Soc London A234: 60
175. Aharoni SM (1982) J Macromol Sci Phys B21: 105
176. Bedford SE, Yu K, Windle AH (1992) J Chem Soc Faraday Trans 88: 1765
177. Beatty CL, Pochan JM, Froix MF, Hinman DD (1975) Macromolecules 8: 547
178. Froix MF, Beatty CL, Pochan JM, Hinman DD (1975) J Polymer Sci Polym Phys Ed 13: 1269
179. Godovsky YK, Papkov VS (1989) Adv Polymer Sci 88: 129
180. Godovsky YK, Papkov VS (1986) Makromol Chem Macromol Symp 4: 71
181. Godovsky YK (1992) In: Aharoni SM (ed) Synthesis, Characterization and Theory of Polymeric Networks and Gels. Plenum, New York, p 127
182. Neuburger N, Bahar I, Mattice WL (1992) Macromolecules 25: 2447
183. Schneider NS, Desper CR, Beres JJ (1978) In: Blumstein A (ed) Liquid Crystalline Order in Polymers. Academic Press, New YOrk, p 299
184. Aharoni SM (1979) J Polymer Sci Polym Phys Ed 17: 683
185. Aharoni SM (1979) Macromolecules 12: 94
186. Aharoni SM (1979) unpublished observations
187. Shmueli U, Traub W, Rosenheck K (1969) J Polymer Sci PartA-2, 7: 515
188. Clough SB (1975) In: Burke JJ, Weiss V (eds) Characterization of Materials in Research, Ceramics and Polymers. Syracuse University Press, Syracuse, NY, p 417
189. Starkweather HW (1979) J Polymer Sci Polym Phys Ed 17: 73
190. Wunderlich B, Grebowicz J (1984) Adv Polymer Sci 60/61: 1
191. Clark ES, Muus LT (1962) Z Krist 117: 119
192. Aharoni SM, Walsh EK (1979) Macromolecules 12: 271
193. Aharoni SM (1980) Polymer 21: 21
194. Aharoni SM (1981) Macromolecules 14: 222
195. Aharoni SM (1980) J Polymer Sci Polym Phys Ed 18: 1303
196. Aharoni SM (1980) J Polymer Sci Polym Phys Ed 18: 1439
197. Aharoni SM (1980) Polymer Preprints 21(1): 209
198. Berger MN, Tidswell BM (1973) J Polymer Sci Part C: Polym Symp 42: 1063
199. Berger MN (1973) J Macromol Sci Rev Macromol Chem 9: 269
200. Fetters LJ, Yu H (1971) Macromolecules 4: 385
201. Bur AJ, Fetters LJ (1976) Chem Rev 76: 727
202. Burchard W (1963) Makromol Chem 67: 182
203. Tsvetkov VN, Rjumtsev EI, Pogodina NV, Shtennikova IN (1975) Eur Polymer J 11: 37
204. Dobb MG, McIntyre JE (1984) Adv Polymer Sci 60/61: 61 and references therein

205. Ciferri A (1991) In: Ciferri A (ed) Liquid Crystallinity in Polymers: Principles and Fundamental Properties. VCH Publishing, New York, p 209
206. Bechtoldt H, Wendorff JH, Zimmermann HJ (1987) Makromol Chem 188: 651
207. Schwarz G, Kricheldorf (1988) Makromol Chem Rapid Commun 9: 717
208. Gaudiana RA, Minns RA, Sinta R, Weeks N, Rogers HG (1989) Prog Polymer Sci 14: 47
209. Aharoni SM (1983) unpublished results
210. Morgan PW (1977) Macromolecules 10: 1381
211. Preston J (1978) In: Blumstein A (ed) Liquid Crystalline Order in Polymers. Academic Press, New York, p 141
212. Aharoni SM (1982) Macromolecules 15: 1311
213. Zero K, Aharoni SM (1987) Macromolecules 20: 1957
214. Krigbaum WR, Tanaka T, Brelsford G, Ciferri A (1991) Macromolecules 24: 4142
215. Kricheldorf HR, Schmidt B, Burger R (1992) Macromolecules 25: 5465
216. Kricheldorf HR, Schmidt B (1992) Macromolecules 25: 5471
217. Sroog CE (1976) J Polymer Sci Macromol Rev 11: 161
218. Takahashi N, Yoon DY, Parrish W (1984) Macromolecules 17: 2583
219. Vogel H, Marvel CS (1961) J polymer Sci 50: 511
220. Dawans F, Marvel CS (1965) J Polymer Sci Part A, 3: 3549
221. Iwakura Y, Uno K, Imai Y (1964) J Polymer Sci Part A, 2: 2605
222. Iwakura Y, Uno K, Imai Y (1964) Makromol Chem 77: 33
223. Evers RC, Arnold FE, Helminiak TE (1981) Macromolecules 14: 925
224. Wolfe JF, Arnold FE (1981) Macromolecules 14: 909
225. Choe EW, Kim SN (1981) Macromolecules 14: 920
226. Cohen Y, Buchner S, Zachmann HG, Davidov D (1992) Polymer 33: 3811
227. Wolfe JF, Loo BH, Arnold FE (1981) Macromolecules 14: 915
228. Welsh WJ, Bhaumik D, Mark JE (1981) Macromolecules 14: 947
229. Bhaumik D, Welsh WJ, Jaffe HH, Mark JE (1981) Macromolecules 14: 951
230. Zhang B, Mattice WL (1992) Macromolecules 25: 4937
231. Allen SR, Filippov AG, Farris RJ, Thomas EL, Wong CP, Berry GC, Chenevey EC (1981) Macromolecules 14: 1135
232. Martin DC (1992) Macromolecules 25: 5171
233. Wenzel M, Ballauff M, Wegner G (1987) Makromol Chem 188: 2865
234. Helmer-Metzmann F, Ballauff M, Schulz RC, Wegner G (1989) Makromol Chem 190: 985
235. Kovacic P, Kiriakis A (1963) J Am Chem Soc 85: 454
236. Kovacic P, Jones MB (1987) Chem Rev 87: 357
237. Yamamoto T, Hayashi Y, Yamamoto A (1978) Bull Chem Soc Japan 51: 2091
237a. Schlüter A-D, Wegner G (1993) Acta Polymer 44: 59
238. Kern W, Seibel M, Wirth HO (1959) Makromol Chem 29: 165
239. Krigbaum WR, Krause KJ (1978) J Polymer Sci Polym Chem Ed 16: 3151
240. Heitz W (1991) Makromol Chem Macromol Symp 48/49: 15
241. Rehahn M, Schluter A-D, Wegner G, Feast WJ (1989) Polymer 30: 1054
242. Rehahn M, Schluter A-D, Wegner G, Feast WJ (1989) Polymer 30: 1060
243. Noll A, Siegfried N, Heitz W (1990) Makromol Chem Rapid Commun 11: 485
244. Kallitsis JK, Rehahn M, Wegner G (1992) Makromol Chem 193: 1021
245. Kaszynski P, Michl J (1988) J Am Chem Soc 110: 5225
246. Kaszynski P, Friedli AC, Michl J (1988) Mol Cryst Liq Cryst Lett 6: 27
247. Kaszynski P, Friedli AC, McMurdie ND, Michl J (1990) Mol Cryst Liq Cryst 191: 193
248. Kaszynski P, Friedli AC, Michl J (1992) J Am Chem Soc 114: 601
249. Uematsu I, Uematsu Y (1984) Adv Polymer Sci 59: 37 and references therein
250. Samulski ET (1978) In: Blumstein A (ed) Liquid Crystalline Order in Polymers. Academic Press, New York, p 167
251. Werbowyj RS, Gray DG (1976) Mol Cryst Liq Cryst 34: 97
252. Panar M, Wicox OB (1977) German Patent No 2,705,382
253. Aharoni SM (1980) Mol Cryst Liq Cryst Lett 56: 237
254. Aharoni SM (1981) J Polymer Sci Polym Ed 19: 495
255. Werbowyj RS, Gray DG (1980) Macromolecules 13: 69
256. Shimanura K, White JL, Fellers JF (1981) J Appl Polymer Sci 26: 2165
257. Aharoni SM (1982) J Macromol Sci Phys B21: 287
258. Van K, Norisuye T, Teramoto A (1981) Mol Cryst Liq Cryst 78: 123
259. Van K, Teramoto A (1982) Polymer J, 14: 999

260. Itou T, Teramoto A (1984) Macromolecules 17: 1419
261. Southwick JG, Jamieson AM, Blackwell J (1981) Macromolecules 14: 1728
262. Inatomi SI, Jinbo Y, Sato T, Teramoto A (1992) Macromolecules 25: 5013
263. Robinson C (1961) Tetrahedron 13: 219
264. Luzzati V (1963) Prog. Nucleic Acid Res 1: 347
265. Spencer M, Fuller W, Wilkins MHF, Brown GL (1962) Nature 194: 1014
266. Spencer M (1963) Cold Spring Harbor Symp Quant Biol 28: 77
267. Robinson C (1956) Trans Faraday Soc 52: 571
268. Robinson C, Ward JC, Beevers RB (1958) Discuss Faraday Soc 25: 29
269. Brown III FR, Hpfinger AJ, Blout ER (1972) J Mol Biol 63: 101
270. Aharoni SM (1986) Macromolecules 19: 426
271. Tseng SL, Valente A, Gray DG (1981) Macromolecules 14: 715
272. Bhadani SN, Gray DG (1984) Mol Cryst Liq Cryst Lett 102: 255
273. Suto S (1989) J Appl Polymer Sci 37: 2781
274. Song CQ, Litt MH, Manas-Zloczower I (1991) J Appl Polymer Sci 42: 2517
275. Mitchell GR, Guo W, Davis FJ (1992) Polymer 33: 68
276. Song CQ, Litt MH, Manas-Zloczower I (1992) Macromolecules 25: 2166
277. Aharoni SM (1991) Macromolecules 24: 4286
278. Aharoni SM (1981) Polymer 22: 418
279. Kishi R, Sisido M, Tazuke S (1990) Macromolecules 23: 3779
280. Kishi R, Sisido M, Tazuke S (1990) Macromolecules 23: 3868
281. Sisido M, Kishi R (1991) Macromolecules 24: 4110
282. Godovsky YK, Volegova IA, Valetskaya LA, Rebrov AV, Novitskaya LA, Rotenburg SI (1988) Polymer Sci USSR, 30: 329
283. Godovsky YK, Valetskaya LA (1991) Polymer Bull 27: 221
284. Godovsky YK, Volegova IA, Rebov AV, Novitskaya LA, Dolgoplosk SB, Kolokoltseva IG (1990) Polymer Sci USSR 32: 726
285. Papkov VS, Kvachev YuP (1989) Progr Colloid Polym Sci 80: 221
286. des Cloizeaux J, Jannink G (1991) Polymers in Solution: Their Modelling and Structure. Oxford University Press, Oxford, UK
287. Feder J (1988) Fractals. Plenum, NY
288. Avnir D (ed) (1989) The Fractal Approach to Heterogeneous Chemistry. Wiley, Chichester, UK
289. Kaye BH (1989) A Random Walk Through Fractal Dimensions. VCH, Weinheim, Germany
290. DeGennes P-G (1976) J Physique Lett 37: 1
291. Stauffer D (1976) J Chem Soc Faraday Trans II, 72: 1354
292. Daoud M, Martin JE (1989) In: Avnir D (ed) The Fractal Approach to Heterogenous Chemistry. Wiley, Chichester, UK, pp 109–130
293. Daoud M, Lapp A (1990) J Phys: Condens Matter 2: 4021
294. Muthukumar M (1985) J Chem Phys 83: 3161
295. Daoud M (1992) In: Aharoni SM (ed) Synthesis, Characterization and Theory of Polymeric Networks and Gels Plenum, NY, pp 1–12, and references therein.
296. Martin JE, Hurd AJ (1987) J Appl Crystallog 20: 61
297. Schmidt PW, Hohr A, Neumann HB, Kaiser H, Avnir D, Lin JS (1989) J Chem Phys 90: 5016
298. Schmidt PW (1989) In: Avnir D (ed) The Fractal Approach to Heterogeneous Chemistry. Wiley, Chichester, UK, pp 67–79
299. Martin JE (1986) J Appl Crystallog 19: 25
300. Martin JE, Wilcoxon JP (1988) Phys Rev Lett 61: 373
301. Keefer KD (1990) In: Zeigler JM, Fearon FWG (eds) Silicon-Based Polymer Science, a Comprehensive Resource. Advances in Chemistry Series NO. 224. ACS, Washington, DC, pp 227–240
302. Martin JE, Adolf D, Odinek J (1990) Makromol Chem Macromol Symp 40: 1
303. Wijnen PWJG, Beelen TPM, Rummens KPJ, Saeijs HCPL, Van Santen RA (1991) J Appl Crystallog 24: 759
304. Vollet DR, Hiratsuka RS, Pulcinelli SH, Santilli CV, Craievich AF (1992) J Non-Cryst Solids 142: 181
305. Daoud M, Family F, Jannink G (1984) J Physique (Paris) Lett 45: 199
306. Bouchaud E, Delsanti M, Adam M, Daoud M, Durand D (1986) J Physique (Paris) Lett 47: 1273
307. Adam M, Delsanti M, Munch JP, Durand D (1987) J Phys (Paris) 48: 1809

308. Chu B, Wu C, Wu D, Phillips JC (1987) Macromolecules 20: 2642
309. Wu W, Bauer BJ, Su W (1989) Polymer 30: 1384
310. Munch JP, Delsanti M, Durand D (1992) Europhys Lett 18: 577
311. Aharoni SM, Murthy NS, Zero K, Edwards SF (1990) Macromolecules 23: 2533
312. Aharoni SM (1991) Macromolecules 24: 235
313. Cotts PM (1992) In: Aharoni SM (ed) Synthesis, Characterization and Theory of Polymeric
     Networks and Gels Plenum NY pp 41–52
314. McDonnell ME, Zero K, Aharoni SM (1992) In: Aharoni SM (ed) Synthesis, Characterization
     and Theory of Polymeric Networks and Gels. Plenum, NY pp 255–268
315. Kurimura Y (1989) Adv Polymer Sci 90: 105
316. Ross-Murhpy SB (1992) Polymer 33: 2622
317. Bekturov EA, Bimendina LA (1981) Adv Polymer Sci 41: 99 and references therein.
318. Papisov IM, Litmanovich AA (1989) Adv Polymer Sci 90: 139
319. Queslel JP, Mark JE (1984) Adv Polymer Sci 65: 135 and references therein.
320. Petrovic ZS, Ferguson J (1991) Progr Polymer Sci 16: 695 and references therein.
321. Tanaka Y, Mika TF (1973) In: May CA, Tanaka J (Eds) Epoxy Resins Chemistry and
     Technology. Marcel Dekker, NY pp 135–238
322. King AH (1918) Metallurg Chem Eng 18: 243
323. Davankov VA, Tsyurupa MP, Rogozhin SV (1974) J Polymer Sci Polym Symp 47: 95
324. Tsyurupa MP, Lalaev VV, Davankov VA (1984) Acta Polym 35: 451
325. Davanokov VA, Tsyurupa MP (1989) Pure Appl Chem 61: 1881
326. Davankov VA, Tsyurupa MP (1990) Reactive Polymers 13: 27
327. Negre M, Bartholin M, Guyot A (1979) Angew Makromol Chem 80: 19
328. Shea KJ, Loy DA, Webster OW (1989) Chem Mater 1: 572
329. Hagiwara T, Suzuki I, Takeuchi K, Hamana H, Narita T (1991) Macromolecules 24: 6856
330. Japanese Patents JO–4018–408 and JO–4018–443 of Jan. 22, 1992, to Mitsui Toatsu Chem
     Inc.
331. Tomalia DA, Naylor AM, Goddard III WA (1990) Angew Chem Int Ed Engl 29: 138
332. Denkewalter RG, Kolc J, Lukasavage WJ (1981) US Patent No 4,289,872 and (1983) US Patent
     No. 4,410,688
333. Tomalia DA, Hall M, Hedstrand DM (1987) J Am Chem Soc 109: 1601
334. Tomalia DA, Berry V, Hall M, Hedstrand DM (1987) Macromolecules 20: 1167
335. Naylor AM, Goddard III WA (1989) J Am Chem Soc 111: 2339
336. Hawker CJ, Frechet JMJ (1990) J Am Chem Soc 112: 7638
337. Caminati G, Turro NJ, Tomalia DA (1990) J Am Chem Soc 112: 8515
338. Hawker CJ, Frechet JMJ (1990) Macromolecules 23: 4726
339. Uhrich KE, Boegeman S, Frechet JMJ, Turner SR (1991) Polymer Bull 25: 551
340. Maciejewski M (1982) J Macromol Sci Chem A17: 689
341. Miller TM, Neenan TX (1990) Chem Mater 2: 346
342. Newkome GR, Moorefield CN, Baker GR, Johnson AL, Behera RK (1991) Angew Chem Int
     Ed Engl 30: 1176
343. Roovers J, Toporowski PM, Zhou LL (1992) Polymer Preprints 33(1): 182
344. Mourey TH, Turner SR, Rubinstein M, Frechet JMJ, Hawker CJ, Wooley KL (1992)
     Macromolecules 25: 2401
345. Miller TM, Kwock EW, Neenan TX (1992) Macromolecules 25: 3143
346. Morikawa A, Kakimoto M, Imai Y (1992) Macromolecules 25: 3247
347. Uhrich KE, Hawker CJ, Frechet JMJ, Turner SR (1992) Macromolecules 25: 4583
348. Warakomski JM (1992) Chem Mater 4: 1000
349. Bayliff PM, Feast WJ, Parker D (1992) Polymer Bull 29: 265
350. Aharoni SM (1992) US Patent Application No 249,992
350a. Miller TM, Neenan TX, Kwock EW, Stein SM (1993) Polym Prepr 34(1): 58
351. Farin D, Avnir D (1991) Angew Chem Int Ed Engl 30: 1379
352. Kovacic P, Oziomek J (1964) J Org Chem 29: 100
353. Kovacic P, Oziomek J (1966) Macromol Syntheses 2: 23
354. Kovacic P. Feldman MB, Kovacic JP, Lando JB (1968) J Appl Polymer Sci 12: 1735
355. Yamamoto T, Yamamoto A (1977) Chem Lett 353
356. Jones MB, Kovacic P, Lanska D (1981) J Polymer Sci Polym Chem Ed 19: 89
357. Shacklette LW, Eckhardt H, Chance RR, Miller GG, Ivory DM, Baughman RH (1980) J Chem
     Phys 73: 4098
358. Heitz W (1986) Chemiker Zeitung 110: 385

359. Jabloner J, Cessna Jr LC (1976) Polymer Preprints 17: 169
360. Bednarski TM, Delneko JH, Mayer RH, Hagan JA (1975) SPE Ann Techn Papers 21:90
361. Sergeev VA, Shittkov VK, Chernomordik YuA, Korshak VV (1975) Polymer Preprints 16(1): 328
362. Miyaura N, Yanagi T, Suzuki A (1981) Synth Commun 11: 513
363. Kim YH, Webster OW (1990) J Am Chem Soc 112: 4592
364. Kim YH, Webster OW (1992) Macromolecules 25: 5561
365. Kim YH (1989) US Patent No 4,857,630
366. Dill KA, Flory PJ (1981) Proc Natl Acad Sci USA, 78: 676
367. Menger FM, Doll WD (1984) J Am Chem Soc 106: 1109
368. Butcher JA, Lamb GW (1984) J Am Chem Soc 106: 1217
369. Newkome GR, Moorefield CN, Baker GR, Saunders MJ, Grossman SH (1991) Angew Chem Int Ed Engl 30: 1178
370. Aharoni SM, Crosby III CR, Walsh EK (1982) Macromolecules 15: 1093
371. Tomalia DA, Hedstrand DM, Ferritto MS (1991) Macromolecules 24: 1435
372. Tomalia DA, Swanson DR, Hedstrand DM (1992) Polymer Preprints 33(1): 180
373. Tomalia DA, Baker H, Dewald J, Hall M, Martin S, Roeck J, Ryder J, Smith P (1985) Polymer J 17: 117
374. Tomalia DA, Baker H, Dewald J, Hall M, Kallos G, Martin S, Roeck J, Ryder J, Smith P (1986) Macromolecules 19: 2466
375. Newkome GR, Yao Z, Baker GR, Gupta VK (1985) J Org Chem 50: 2004
376. Hawker CJ, Frechet JMJ (1990) J Chem Soc Chem Commun 15: 1010
377. Newkome GR, Lin X, Weis CD (1991) Tetrahedron: Asymmetry 2: 957
378. Uchda H, Kabe Y, Yoshino K, Kawamata A, Tsumuraya T, Masamune J (1990) J Am Chem Soc 112: 7077
379. Morikawa A, Kakimoto M, Imai Y (1991) Macromolecules 24: 3469
380. Webster OW, Gentry FP, Farlee RD, Smart BE (1991) Polymer Preprints 32(1): 412
381. Smart BE, Webster OW (1991) US Patent No 4,987,157
382. Webster OW, Gentry FP, Farlee RD, Smart BE (1991) Polymer Preprints 32(2): 74
383. Webster OW, Kim YH, Gentry FP, Farlee RD, Smart BE (1992) Polymer Preprints 33(1): 186
384. Webster OW, Gentry FP, Farlee RD, Smart BE (1992) Makromol Chem Macromol Symp 54/55: 477
385. Newkome GR, Moorefield CN, Baker GR (1992) Aldrichimica Acta 25: 31
386. Baldwin RA, Cheng MT (1967) J Org Chem 32: 1573
386a. Moore JS, Zhang J (1992) Angew Chem Int Ed Engl 31: 922; Zhang J, Moore JS, Xu Z, Aguirre RA (1992) J Am Chem Soc 114: 2273; Wu Z, Moore JS (1993) Polymer Preprints 34(1): 122 and references therein.
387. Mathias LJ, Reichert VR, Muir AVG (1993) Chem Mater 5: 4
388. Murthy GS, Hassenrück K, Lynch VM, Michl J (1989) J Am Chem Soc 111: 7262
389. Hassenrück K, Murthy GS, Lynch VM, Michl J (1990) J Org Chem 55: 1013
390. Ibrahim MA, Michl J (1991) Polymer Preprints 32(1): 407
391. Friedli AC, Lynch VM, Kaszynski P, Michl J (1990) Acta Crystallog B46: 377
392. Zimmerman HE, Goldman TD, Hirzel TK, Schmidt SP (1980) J Org Chem 45: 3933
393. Aharoni SM, Wertz DH (1983) J Macromol Sci Phys B22: 129
394. Yamazaki N, Matsumoto M, Higashi F (1975) J Polymer Sci Polym Chem Ed 13: 1373
395. Aharoni SM, Hammond WB, Szobota JS, Masilamani D (1984) J Polymer Sci Polym Chem Ed 22: 2579
396. Mitin YuV, Glinskaya OV (1969) Tetrahedron Letters 60: 5267
397. Ogata N, Tanaka H (1971) Polymer J 2: 672
398. Yamazaki N, Higashi F (1974) Tetrahedron 30: 1323
399. Yamazaki N, Higashi F, Kawabata J (1974) J Polymer Sci Polym Chem Ed 12: 2149
400. Aharoni SM, Edwards SF (1989) Macromolecules 22: 3361
401. McMurry J (1976) Organic Reactions, Volume 24. Wiley, NY pp 187–224
402. Aharoni SM, Edwards SF (1992) Polymer Bull 29: 675
403. Schotten C (1882) Ber 15: 1947
404. Schotten C (1884) Ber 17: 2545
405. Schotten C, Baum J (1884) Ber 17: 2548
406. Baumann E (1886) Ber 19: 3218
407. Schotten C (1888) Ber 21: 2238
408. Schotten C (1890) Ber 23: 3430

409. Aharoni SM (1992) Macromolecules 25: 1510
410. Aharoni SM (1992) J Appl Polymer Sci 45: 813
411. Krigbaum WR, Preston J, Jadhav JY (1988) Macromolecules 21: 538
412. Aharoni SM, Hatfield GR, O'Brien KP (1990) Macromolecules 23: 1330
413. Aharoni SM (1993) manuscript in preparation
414. Kwolek SL, Morgan PW, Schaefgen JR, Gulrich LW (1977) Macromolecules 10: 1390
415. Kwolek SL (1971) US Patent No 3,600,350
416. Kim YH (1992) J Am Chem Soc 114: 4947
417. Benicewicz BC, Hoyt AE (1987) US Patent Application No 504,217
418. Melissaris AP, Mikroyannidis JA (1989) J Polymer Sci Part A: Polym Chem 27: 245
419. Hoyt AE, Benicewicz BC (1990) J Polymer Sci Part A: Polym Chem 28: 3417
420. Ballauff M (1986) Makromol Chem Rapid Commun 7: 407
421. Ballauff M, Schmidt GF (1987) Mol Cryst Liq Cryst 147: 163
422. Schrauwen C, Pakula T, Wegner G (1992) Makromol Chem 193: 11
423. Barclay GG, Ober CK, Papathomas KI, Wang DW (1992) Macromolecules 25: 2947
424. Higashi F, Hoshio A, Kiyoshige J (1983) J Polymer Sci Polym Chem Ed 21: 3241
425. Higashi F, Yamade Y, Hoshio A (1984) J Polymer Sci Polym Chem Ed 22: 2181
426. Higashi F, Akiyama N, Takahashi I (1984) J Polymer Sci Polym Chem Ed 22: 1653
427. Higashi F, Takahashi I, Akiyama N (1984) J Polymer Sci Polym Chem Ed 22: 3607
428. Higashi F, Mashimo T, Takahashi I, Akiyama N (1985) J Polymer Sci Polym Chem ed 23: 3095
429. Moore JS, Stupp SI (1990) Macromolecules 23: 65
430. Anon (1992) Polymer News 17: 14
431. Mikolajczyk M, Kielbasinski P (1981) Tetrahedron 37: 233
432. Boden EP, Keck GE (1985) J Org Chem 50: 2394
433. Hawker CJ, Lee R, Frechet JMJ (1991) J Am Chem Soc 113: 4583
434. Frechet JMJ, Hawker CJ, Philippides AE (1991) US Patent No 5,041,516
435. Voit BI, Turner SR (1992) Polymer Preprints 33(1): 184
436. Voit B, Turner SR (1992) Abstract No. 2-P68, 34th IUPAC International Symposium on Macromolecules, Prague, Czechoslovakia, 13–18 July.
437. Carothers WH (1931) Chem Revs 8: 353
438. Jacobson RA (1932) J Am Chem Soc 54: 1513
439. Friedel C, Crafts JM (1885) Bull Soc Chim 43: 53
440. Flory PJ (1952) J Am Chem Soc 74: 2718
441. Rao SP, Jones MB, Baltisberger RJ, Burger VT, Brown CE (1983) J Polymer Sci Polym Lett Ed 21: 551
442. Nystuen NJ, Jones MB (1985) J Polymer Sci Polym Chem Ed 23: 1433
443. Rolando TE, Jones MB (1986) J Polymer Sci Part C: Polym Lett 24: 233
444. Jones MB, Larson JE (1992) J Polymer Sci Part A: Polym Chem 30: 2037
445. Learmonth GS, Osborn P (1968) J Appl Polymer Sci 12: 1815
446. Knop A, Böhmer V, Pilato LA (1989) In: Eastmond GC, Ledwith A, Russo S, Sigwalt P (eds) Comprehensive Polymer Science, Volume 5. Pergamon, Oxford UK, PP 611–647 and references therein.
447. Baekeland LH (1909) US Patent No 942,699
448. Baekeland LH (1910) US Patent No 949,671
449. Lederer L (1894) J Prakt Chem 50: 223
450. Manasse O (1894) Ber. Deutsch Chem Ges 27: 2409
451. Speier A (1897) German Patent No 99,570
452. Luft A (1902) German Patent No 140,552
453. Fyfe CA, Rudin A, Tchir WJ (1980) Macromolecules 13: 1320
454. Bryson RL, Hatfield GR, Early TA, Palmer AR, Maciel GE (1983) Macromolecules 16: 1669
455. Fyfe CA, McKinnon MS, Rudin A, Tchir WJ (1983) Macromolecules 16: 1216
456. Maciel GE, Chuang I-S, Gollob L (1984) Macromolecules 17: 1081
457. Mechin B, Hanton D, Legoff J, Tanneur JP (1984) Europ Polymer J 20: 333
458. Elias HG, Wohwinkel F (1986) New Commerical Polymers 2. Gordon & Breach, NY, pp 171–178, 194–201, 389–392
459. Eg, Japanese Patent No 3,292–322 (1991) to Sumitomo-Durez
     Japanese Patent No. 3,296–521 (1991) to Toshiba Co.
     Japanese Patent No. 3,296–522 (1991) to Toshiba Co.
460. Eg, Japanese Patent No 4,007–313(1992) to Daihachi Co
461. So YH (1992) Macromolecules 25: 516

462. Chu SG, Jabloner H, Nguyen TT (1992) US Patent No 5,095,074
463. Cook M, Katritzky AR, Linda P (1974) Adv Heterocyclic Chem 17: 255
464. Gandini A (1977) Adv Polymer Sci 25: 47
465. Gandini A (1987) in Encyclopedia of Polymer Science and Engineering. Wiley-Interscience, NY, Volume 7, pp 454–473
466. Schmitt CR (1974) Polymer Plast Technol Eng 3(2): 121
467. Fawcett AH, Dadamba W (1982) Makromol Chem 183: 2799
468. Chuang I-S, Maciel GE, Myers GE (1984) Macromolecules 17: 1087
469. Glowinkowski S, Pajak Z (1978) Acta Physiol Polonia A54: 411
470. Eg, Wade RC (1975) US Patent No 3,865,757
    Larsen H-O, Barfoed S, Gent JAG (1976) US Patent No 3,975,318
471. Eg, Wolff PL, Gent JAG (1977) US Patent No 4,016,111
472. Pp 362–363 in reference 39.
473. Hunter WH, Olson AO, Daniels EA (1916) J Am Chem Soc 38: 1761
474. Hunter WH, Woollett GH (1921) J Am Chem Soc 43: 131,135
475. Hunter WH, Seyfried LM (1921) J Am Chem Soc 43: 151
476. Attwood TE, Dawson PC, Freeman JL, Hoy LRJ, Rose JB, Staniland PA (1979) Polymer Preprints 20(1): 191
477. Attwood TE, Dawson PC, Freeman JL, Hoy LRJ, Rose JB, Staniland PA (1981) Polymer 22: 1096
478. Bonner WH (1962) US Patent No 3,065,205
479. Goodman I, McIntyre JE, Russel W (1964) British Patent No 971,227
480. Ueda M, Sato M (1987) Macromolecules 20: 2675
481. Ueda M, Oda M (1989) Polymer J 21: 673
482. Ueda M, Abe T, Oda M (1992) J Polymer Sci. Part A: Polym. Chem 30: 1993
483. Tanabe T, Fukawa I (1989) Japanese Kokai No 64–74223
484. Tanabe T, Fukawa I (1989) Japanese Kokai No 1–153722
485. Fukawa I, Tanabe T (1992) J Polymer Sci Part A: Polym Chem 30: 1977
486. Higuchi J (1963) J Chem Phys 38: 1237
487. Higuchi J (1963) J Chem Phys 39: 1847
488. Ovchinnikov AA (1977) Dokl Acad Sci USSR 236: 928
489. Ovchinnikov AA (1978) Theor Chim Acta 47: 297
490. Mataga N (1968) Theor Chim Acta 10: 372
491. Itoh K (1967) Chem Phys Lett 1: 235
492. Itoh K (1978) Pure Appl Chem 50: 1251
493. Teki Y, Takui T, Itoh K, Iwamura H, Kobayashi K (1986) J Am Chem Soc 108: 2147
494. Korshak YuV, Medvedeva TV, Ovchinnikov AA, Spector VN (1987) Nature 326: 370
495. Ishida T, Inoue K, Koga N, Nakamura N, Iwamura H (1992) Mater Res Soc Symp Proc 247: 407
496. Rajca A (1990) J Am Chem Soc 112: 5889
497. Rajca A (1990) J Am Chem Soc 112: 5890
498. Hellwinkel D, Stahl H, Gaa HG (1987) Angew Chem Int Ed Engl 26: 794
499. Torrance JB, Oostra S, Nazzal A (1987) Synthetic Metals 19: 709.
500. Johannsen I, Torrance JB, Nazzal A (1989) Macromolecules 22: 566
501. Takayanagi M, Kajiyama T, Katayose T (1982) J Appl Polymer Sci 27: 3903
502. Takayanagi M, Katayose T (1981) J Polymer Sci Polym Chem Ed 19: 1133
503. Mercx FPM, Lemstra PJ (1990) Polymer Commun 31: 252
504. Burch RR, Sweeny W, Schmidt H-W, Kim YH (1990) Macromolecules 23: 1065
505. Burch RR, Manring LE (1991) Macromolecules 24: 1731
506. Mailhe-Randolph C, Desilvestro J (1989) J Electroanal Chem 262: 289
507. Michaelson JC, McEvoy AJ, Kuramoto N (1990) J Electroanal Chem 287: 191
508. Genies EM, Penneau JF, Lapkowski M, Boyle A (1989) J Electroanal Chem 269: 63
509. Rodrigue D, Snauwaert P, Demaret X, Riga J, Verbist JJ (1991) Synthetic Metals 41–43: 769
510. Jeon D, Kim J, Gallagher MC, Willis RF (1992) Science 256: 1662
511. Guay J, Dao LH (1989) J Electroanal Chem 274: 135
512. Chandrasekhar P, Gumbs RW (1991) J Electrochem Soc 138: 1337
513. Chandrasekhar P, Thorne JRG, Hochstrasser RM (1991) Appl Phys Lett 59: 1661
514. Kamachi M (1992) Polymer News 17: 198
515. Pekala RW (1989) US Patent No 4,873,218
516. Pekala RW (1989) J Materials Sci 24: 3221

517. Pahl R, Bonse U, Pekala RW, Kinney JH (1991) J Appl Crystall 24: 771
518. Cotts PM, Pekala R (1991) Polymer Preprints 32(3): 451
519. Pekala RW (1992) US Patent No 5,081,163
520. Pekala RW (1992) US Patent No 5,086,085
521. Fricke J (1988) J Non-Cryst Solids 100: 169
522. Pool R (1990) Science 247: 807
523. Grigat E, Putter R (1967) Angew Chem Int Ed Engl 6: 206
524. Wohrle D (1972) Adv Polymer Sci 10: 35
525. Woodburn HM, Fisher JR (1957) J Org Chem 22: 895
526. Anderson R, Holovka JM (1966) J Polymer Sci A-1, 4: 1689
527. Bartlett RK, O'Neill G, Savill NG, Thomas SLS, Wall WF (1970) Brit Polymer J 2: 225
528. Keller TM (1986) J Polymer Sci Part C: Polym Lett 24: 211
529. Keller TM (1988) J Polymer Sci Part A: Polym Chem 26: 3199
530. Keller TM (1987) Polymer Commun 28: 337
531. Keller TM, Moonay DJ (1989) Proc ACS Div PMSE 60(1): 79
532. Keller TM (1992) Polymer Preprints 33(1): 422
533. Gupta AM, Macosko CW (1991) Makromol Chem Macromol Symp 45: 105
534. McCormick FB, Brown-Wensley KA, DeVoe RJ (1992) Proc ACS Div PMSE 66: 460
535. Fang T (1990) Macromolecules 23: 4553
536. Penczek P, Kaminska W (1990) Adv Polymer Sci 97: 41
537. Gupta AM (1991) Macromolecules 24: 3459
538. Hi-Tek Polymers, Product Bulletins for Arocy B-10, B-30, B-50 and M-10 cyanate ester monomers and resins, undated.
539. Rhone-Poulenc, Product Bulletin for Arocy L-10 cyanate ester monomer, undated.
540. Allied-Signal Inc, Preliminary Product Data for PT Resins, undated.
541. Das S, Prevorsek DC, DeBona BT (1990) Modern Plastics, February, P 72
542. Das S, Prevorsek DC (1989) US Patent No 4,831,086
543. Das S, Prevorsek DC (1990) US Patent No 4,970,276
544. Das S, Prevorsek DC (1990) US Patent No 4,978,727
545. Das S (1992) Paper presented at the American Composites International Conference, May 6–8, Orlando, Florida.
546. Zeng S, Ahn K, Seferis JC, Kenny JM, Nicolais L (1992) Polymer Composit 13: 191
547. Prevorsek DC, Chung DC (1979) US Patent No 4,157,360
548. Melissaris AP, Mikroyannidis JA (1988) J Polymer Sci Polym Chem Ed 26: 1885
549. Mikroyannidis JA, Melissaris AP (1988) J Polymer Sci Part A: Polym Chem 26: 1405
550. Melissaris AP, Mikroyannidis JA (1989) Europ Polymer J 25: 455
551. Dolui SK, Maiti S (1991) J Polym Mater 8: 59
552. Hefner Jr RE (1987) US Patent No 4,680,378
553. Hefner Jr RE (1987) US Patent No 4,683,276
554. Hefner Jr RE (1988) US Patent No 4,731,426
555. Hefner Jr RE (1988) US Patent No 4,769,440
556. Hefner Jr RE (1991) US Patent No 5,077,380
557. Gordon M, Ross-Murphy SB (1975) Pure Appl Chem 43: 1
558. Gupta AM, Hendrickson RC, Macosko CW (1991) J Chem Phys 95: 2097
559. Simon SL, Gillham JK, Shimp DA (1990) Proc ACS Div PMSE 62: 96
560. Barton JM, Greenfield DCL, Hamerton I, Jones JR (1991) Polymer Bull 25: 475
561. Osei-Owusu A, Martin GC, Gotro JT (1992) Polym Eng Sci 32: 535
562. Gupta AM, Macosko CW (1992) Proc ACS Div PMSE 66: 447
563. Mirco V, Cao ZQ, Mechin F, Pascault JP (1992) Proc ACS Div PMSE 66: 451
564. Simon SL, Gillham JK (1992) Proc ACS Div PMSE 66: 453
565. Rushton AF, Family F, Herrmann HJ (1985) J Polymer Sci Polym Symp 73: 1
566. Witten Jr TA (1985) J Polymer Sci Polym Symp 73: 7
567. Hoyt AE, Benicewicz BC (1990) J Polymer Sci Part A: Polym Chem 28: 3403
568. Nagai A, Nishimura S, Takahashi A, Maruta M, Fukui A (1992) US Patent No 5,098,971
569. A remarkable number of Unexamined Japanese patents (Kokai) are being issued in this field. For example, during October-November 1991, the following were issued: JP03,223,331, JP03,259,914, JP03,255,067, JP03,252,423, JP03,223,247, JP03,223,248 and JP03,255,128.
570. Yamashita M, Kusumoto K (1991) Japanese Kokai No JP03,292,307
571. Serafini TT, Cheng PG, Ueda KK, Wright WF (1992) US Patent No 5,091,505
572. Hay JM, Boyle JD, Parker SF, Wilson D (1989) Polymer 30: 1032

573. Vijayan TM (1991) J Polym Mater 8: 83
574. Serafini TT, Delvigs P, Lightsey GR (1973) US Panent No 3,745,149
575. Morgan RJ, Jurek RJ, Donnellan T, Yen A (1992) Polymer Preprints 33(1): 426
576. Wilkinson SP, Liptak SC, McGrath JE, Ward TC (1992) Polymer Preprints 33(1): 424
577. Kirchhoff RA (1985) US Patent No 4,540,763
578. Tan LS, Arnold FE (1987) US Patent No 4,711,964
579. Bartmann M (1988) US Patent No 4,719,283
580. Corley LS (1990) US Patent No 4,927,907
581. Arnold FE, Helminiak TE, Tan LS, Hwang WF, Chuah H (1992) US Patent No 5,095,075
582. Kirchhoff RA, Carriere CJ, Bruza KJ, Rondan NG, Sammler RL (1991) J Macromol Sci Chem A28: 1079
583. Kirchhoff RA, Bruza KJ (1993) Prog Polym Sci 18: 85
584. Marks MJ (1992) Proc ACS Div PMSE 66: 362,365
585. Marks MJ (1992) In: Aharoni SM (ed) Synthesis, Characterization, and Theory of Polymeric Networks and Gels Plenum, NY, p 165
586. Upshaw TA, Stille JK, Droske JP (1991) Macromolecules 24: 2143
587. Droske JP, Stille JK (1984) Macromolecules 17: 1
588. Droske JP, Gaik UM, Stille JK (1984) Macromolecules 17: 10
589. Droske JP, Stille JK, Alston WB (1984) Macromolecules 17: 14
590. Sutherlin DM, Stille JK (1986) Macromolecules 19: 251
591. Sutherlin DM, Stille JK, Alston WB (1986) Macromolecules 19: 257
592. Upshaw TA, Stille JK (1988) Macromolecules 21: 2010
593. Meyers RA, Hamersa JW, Green HE (1972) J Polymer Sci B: Polym Lett 10: 685
594. Sivaramakrishnan KP, Samyn C, Westerman IJ, Wong DT-M, Marvel CS (1975) J Polymer Sci Polym Chem Ed 13: 1083
595. Chang DM, Marvel CS (1975) J Polymer Sci Polym Chem Ed 13: 2507
596. Lin S, Marvel CS (1983) J Polymer Sci Polym Chem Ed 21: 1151
597. Hergenrother PM (1980) J Macromol Sci Rev Macromol Chem C19: 1
598. Hergenrother PM (1985) in Encyclopedia of Polymer Science and Engineering, Volume 1. Wiley-Interscience, NY, PP 61–68
599. Kovar RF, Ehlers GFL, Arnold FE (1977) J Polymer Sci Polym Chem Ed 15: 1081
600. Unroe MR, Reinhardt BA (1990) J Polymer Sci Part A: Polym Chem 28: 2207
601. Hergenrother PM (1981) Macromolecules 14: 898
602. Hergenrother PM (1981) Macromolecules 14: 891
603. Samyn C, Marvel CS (1975) J Polymer Sci Polym Chem Ed 13: 1095
604. For example: (a) Japanese Kokai No. 4028–720 (Jan 1992) to Kanegafuchi
     (b) Japanese Kokai No 4027–161 (Jan 1992) to Hitachi
     (c) Pignery AM (1991) US Patent No 4,987,272
     (d) Pignery AM (1992) US Patent No 5,096,987
605. Hay AS, Bolon DA, Leimer KR, Clark RF (1970) J Polymer Sci Part B: Polym Lett 8: 97
606. Landis AL, Bilow N, Boschan RH, Lawerence RE, Aponyi TJ (1974) Polymer Preprints 15(2): 533,537
607. Sefcik MD, Stejskal EO, McKay RA, Shaefer J (1979) Macromolecules 12: 423
608. Glaser C (1870) Ann 137: 154
609. Strauss F (1905) Ann 342: 190
610. Thakur MK, Lando JB (1983) Macromolecules 16: 143
611. Thakur MK, Lando JB (1983) J Appl Phys 54: 5554
612. Baughman RH, Chance RR (1978) Ann New York Acad Sci 313: 705
613. Lomakin SM, Pavlova OV, Oskina OYu, Sivergin YuM, Zaikov GE (1992) Polymer Degradation & Stability 37: 217
614. Erhan S (1989) US Patent No 4,845,278
615. Preston J (1992) personal communication.
616. Meurisse P, Laupretre F, Noel C (1984) Mol Cryst Liq Cryst 110: 41
617. Laupretre F, Noel C (1991) In: Ciferri A (ed) Liquid Crystallinity in Polymers: Principles and Fundamental Properties. VCH, NY pp 3–60
618. Aharoni SM (1993) Polymer Bull 30: 149
619. Aharoni SM (1992) In: Aharoni SM (ed) Synthesis, Characterization, and Theory of Polymeric Networks and Gels Plenum, NY, p 79
619a. Markoski LJ, Walker KA, Deeter GA, Spilman GE, Martin DC, Moore JS (1993) Chem Mater 5: 248

620. David IA (1992) Proc ACS Div PMSE 66: 367
621. Wegner G (1969) Z Naturforsch 24B: 824
622. Baughman RH (1972) J Appl Phys 43: 4362
623. Wegner G (1971) Makromol Chem 145: 85
624. Butera RJ, Simic-Glavaski B, Lando JB (1990) Macromolecules 23: 199,211
625. Neenan TX, Callstrom MR, Scarmoutzos LM, Stewart KR, Whitesides GM, Howes VR (1988) Macromolecules 21: 3525
626. Callstrom MR, Neenan TX, Whitesides GM (1988) Macromolecules 21: 3528
627. Rutherford DR, Stille JK (1988) Macromolecules 21: 3530
628. Rutherford DR, Stille JK, Elliott CM, Reichert VR (1992) Macromolecules 25: 2294
629. Liang R-C, Lai W-YF, Reiser A (1987) Macromolecules 20: 2510
630. Hu X, Stanford JL, Day RJ, Young RJ (1992) Macromolecules 25: 672
631. Wegner G (1970) Makromol Chem 134: 219
632. Day D, Lando JB (1981) J Polymer Sci Polym Ed 19: 227
633. Patil AO, Deshpande DD, Talwar SS, Biswas AB (1981) J Polymer Sci Polym Chem Ed 19: 1155
634. Tiecke B (1985) Adv Polymer Sci 71: 79
635. Neenan TX, Whitesides GM (1988) J Org Chem 53: 2489
636. Rehahn M, Schlüter A-D, Wegner G (1990) Makromol Chem 191: 1991
637. Scherf U, Müllen K (1991) Makromol Chem Rapid Commun 12: 489
638. Scherf U, Müllen K (1992) Synthesis, 23
639. Scherf U, Müllen K (1992) Macromolecules 25: 3546
640. Kroto HW, Allaf AW, Balm SP (1991) Chem Rev 91: 1213
641. Curl RF, Smalley RE (1988) Science 242: 1017
642. Aihara J, Hosoya H (1988) Bull Chem Soc Japan 61: 2657
643. Krätschmer W, Lamb LD, Fostiropoulos K, Huffman DR (1990) Nature 347: 354
644. Kroto HW (1991) Mater Res Soc Sym Proc 206: 611
645. Adams GB, Sankey OF, Page JB, O'Keeffe M, Drabold DA (1992) Science 256: 1792
646. Bausch JW, Surya Prakash GK, Olah GA, Tse DS, Lorents DC, Bae YK, Malhotra R (1991) J Am Chem Soc 113: 3205
647. Wood JM, Kahr B, Hoke SH, Dejarme L, Cooks RG, Ben-Amotz D (1991) J Am Chem Soc 113: 5907
648. Creegan KM, Robbins JL, Robbins WK, Millar JM, Sherwood Rd, Tindall PJ, Cox DM, Smith III AB, McCauley Jr JP, Jones DR, Gallagher RT (1992) J Am Chem Soc 114: 1103
649. Selig H, Lifshitz C, Peres T, Fischer JE, McGhie AR, Romanov WJ, McCauley Jr JP, Smith III AB (1991) J Am Chem Soc 113: 5475
650. Olah GA, Busci I, Lambert C, Aniszfold R, Trivedi NJ, Sensharma DK, Surya Prakash GK (1991) J Am Chem Soc 113: 9385
651. Tebbe FN, Becker JY, Chase DB, Firment LE, Holler ER, Malone BS, Krusic PJ, Wasserman E (1991) J Am Chem Soc 113: 9900
652. Hirsch A, Soi A, Karfunkel HR (1992) Angew Chem Int Ed Engl 31: 766
653. Morton JR, Preston KF, Krusic PJ, Hill SA, Wasserman E (1992) J Am Chem Soc 114: 5454
654. Balch AL, Catalano VJ, Lee JW, Olmstead MM (1992) J Am Chem Soc 114: 5455
655. Fann YC, Singh D, Jansen SA (1992) J Phys Chem 96: 5817
656. Taylor R, Langley GJ, Meidine MF, Parsons JP, Abdul-Sada AK, Dennis TJ, Hare JP, Kroto HW, Walton DRM (1992) J Chem Soc Chem Commun 667
657. Miller GP, Hsu CS, Thomann H, Chiang LY, Bernardo M (1992) Mater Res Soc Symp Proc 247: 293
658. Malhotra R, Narang SC, Nigam A, Ganapathiappan S, Ventura S, Satyam A, Bhardawaj T, Lorents DC (1992) Mater Res Soc Symp Proc 247: 301
659. Loy DA, Assink RA (1992) J Am Chem Soc 114: 3977 and Mater Res Soc Symp Proc 247: 355
660. Huan G, Day VW, Jacobson AJ, Goshorn DP (1991) J Am Chem Soc 113: 3188
660a. Hamada N (1993) Materials Science and Engineering, B19: 181
660b. Baughman RH, Cui C (1993) Synthetic Metals 55–57: 315
660c. Baughman RH, Galvão DS (1993) Chem Phys Lett (in press).
661. Aharoni SM, Hammond WB, Edwards SF (1993) work in progress.
662. Baird DG, Ballman RL (1979) J Rheol 23: 505
663. Hammond WB, Aharoni SM, Curran SA (1992) In: Aharoni SM (ed) Synthesis, Characterization, and Theory of Polymeric Networks and Gels Plenum, NY, p 93

664. Aharoni SM (1992) Intern J Polymeric Mater 17: 35
665. Hummel JP, Flory PJ (1980) Macromolecules 13: 479
666. Spencer CP, Berry GC (1992) Polymer 33: 1909
667. Agarwal US, Khakhar DV (1992) J Chem Phys 96: 7125
668. Warner M (1991) J Chem Soc Faraday Trans 87: 861
669. Aharoni SM (1992) In: Aharoni SM (ed) Synthesis, Characterization, and Theory of Polymeric Networks and Gels Plenum, NY, p 31
670. Stauffer D, Aharony A (1992) Introduction to Precolation Theory. Taylor & Francis, London.
671. Grimmett GR (1989) Percolation. Springer-Verlag, New York.
672. Flory PJ (1941) J Am Chem Soc 63: 3096
673. DeGennes P-G (1979) Scaling Concepts in Polymer Physics. Cornell University Press, Ithaca, NY pp 128–162
674. Klein DJ (1981) J Chem Phys 75: 5186
675. Seitz WA, Klewin DJ (1981) J Chem Phys 75: 5190
676. Chambon F, Winter HH (1985) Polymer Bull 13: 499
677. Muthukumar M, Winter HH (1986) Macromolecules 19: 1284
678. Daoud M, Leibler L (1988) Macromolecules 21: 1497
679. Daoud M (1990) J Phys Condens Matter 2: SA299
680. Lairez D, Adam M, Emery JR, Durand D (1992) Macromolecules 25: 286
681. Martin JE, Adolf D, Wilcoxon JP (1988) Phys Rev Lett 61: 2620
682. Pekala RW (1990) Mater Res Soc Proc 171: 285
683. Ruben GC, Pekala RW, Tillotson TM, Hrubesh LW (1992) J Mater Sci 27: 4341
684. Schaefer DW, Martin JE, Wiltzius P, Cannell DS (1984) Phys Rev Lett 52: 2371
685. Martin JE, Schaefer DW, Hurd AJ (1986) Phys Rev A, 33: 3540
686. Brady RM, Ball RC (1984) Nature 309: 225
687. Gross J, Fricke J, Pekala RW, Hrubesh LW (1992) Phys Rev B, 45: 12774
688. Meakin P (1988) Adv Colloid Interface Sci 28: 249
689. Weitz DA, Oliveria M (1984) Phys Rev Lett 52: 1433
690. Kantor Y, Witten TA (1984) J Phys Lett 45: L-675
691. Katz AJ, Thompson AH (1985) Phys Rev Lett 54: 1325
692. Kaufman JH, Baker CK, Nazzal AI, Flickner M, Melroy OR, Kapitulnik A (1986) Phys Rev Lett 56: 1932
693. Bremer LGB, Van Vliet T, Walstra P (1989) J Chem Soc Faraday Trans I, 85: 3359
694. Allain C, Amiel C (1992) Ann Chim Fr 17: 91
695. Paredes VR, Octavio M (1992) Phys Rev A, 46: 994
695a. Adolf D, Hance B, Martin JE (1993) Macromolecules 26: 2754
696. Voronoi G (1907) J Reine Angew Math 133: 97
697. Voronoi G (1908) J Reine Angew Math 134: 198
698. Dirichlet PGL (1850) J Reine Angew Math 40: 209
699. Aboav DA (1970) Metallography 3: 383
700. Atkinson HV (1988) Acta Metall 36: 469
701. Frost HJ, Thompson CV, Howe CL, Whang J (1988) Script Metall 22: 65
702. Smith CS (1954) Scientific American 190: 58
703. Smith CS (1964) Metall Rev 9: 1
704. Weaire D, Kermode JP (1983) Phil Mag B, 48: 245
705. Kawasaki K, Nagai T, Nakashima K (1989) Phil Mag B, 60: 399
706. Glazier JA, Weaire D (1992) J Phys Condens Matter 4: 1867 and references therein.
707. Magnasco MO (1992) Phil Mag B, 65: 895
708. Lemaitre J, Troadec JP, Gervois A, Bideau D (1991) Europhys Lett 14: 77
709. Noever DA (1992) J Macromol Sci Phys B31: 357
710. Rujan P, Evertsz C, Lyklema JW (1988) Europhys Lett 7: 191
711. Almgren FJ, Taylor JE (1976) Scientific American 235: 82
712. Avron JE, Levine D (1992) Phys Rev Lett 69: 208
713. Arizzi S, Mott PH, Suter UW (1992) J Polymer Sci Part B: Polym Phys 30: 415
714. Ferry JD (1980) Viscoelastic Properties of Polymers. Wiley, New York, Chapters 10, 12, 14 through 19.
715. Volkenstein MV (1963) Configurational Statistics of Polymeric Chains. Interscience, NY, chapters 3, 4, 6 and 8.
716. Candau SJ, Young CY, Tanaka T, Lamarechal P, Bastide J (1979) J Chem Phys 17: 83

717. Boué F, Bastide J, Buzier M, Collette C, Lapp A, Herz J (1987) Progr Colloid & Polymer Sci 75: 152
718. Bastide J, Mendes Jr E, Boué F, Buzier M, Lindner P (1990) Makromol Chem Macromol Symp 40: 81
719. Stein RS (1969) J Polymer Sci Part B, 7: 657
720. Bastide J, Leibler L (1988) Macromolecules 21: 2647
721. Vilgis TA (1992) Macromolecules 25: 399
722. Spurr RA, Erath EH, Myers H, Pease DC (1957) Ind Eng Chem 49: 1839
723. Erath EH, Spurr RA (1959) J Polymer Sci 35: 391
724. Kontou E, Spathis G, Theocaris PS (1985) J Polymer Sci Polym Chem Ed 23: 1493
725. Funke W (1964) Kolloid ZZ Polym 197: 71
726. Gallacher L, Bettelheim FA (1962) J Polymer Sci 58: 697
727. Bergmann K, Demmler K (1974) Kolloid ZZ Polym 252: 204
728. Jacobs PM, Jones FR (1992) Polymer 33: 1418
729. Graessley WW (1975) Macromolecules 8: 186
730. Fischer M, Kreibich UT, Schmid R (1985) In: Batzer H (ed) Polymere Werkstoffe, Vol I Georg Thieme, Stuttgart, Germany, p 233
731. Fischer M (1992) Adv Polymer Sci 100: 313
732. Jones JL, Marques CM (1990) J Phys France 51: 1113
733. Graessley WW (1974) Adv Polymer Sci 16: 1
734. Queslel JP, Mark JE (1985) Adv Polymer Sci 71: 229
735. Labana SS (1986) in Encyclopedia of Polymer Science and Engineering, Volume 4. Wiley-Interscience, NY, pp 350–395
736. Flory PJ (1987) in Encyclopedia of Polymer Science and Engineering, Volume 10. Wiley-Interscience, NY, pp 95–112
737. Queslel JP, Mark JE (1989) in Comprehensive Polymer Science, Volume 2. Pergamon, Oxford, pp 271–309
738. Fox TG, Loshaek S (1955) J Polymer Sci 15: 371
739. Simha R, Boyer RF (1963) J Chem Phys 37: 1003
740. Gibbs JH, Dimarzio EA (1958) J Chem Phys 28: 373
741. Shen MC, Eisenberg A (1970) Rubber Chem Technology 43: 95
742. Tobolsky AV, Katz D, Thach R, Schaffhauser R (1962) J Polymer Sci 62: S176
743. Katz D, Tobolsky AV (1963) Polymer 4: 417
744. Tobolsky AV, Katz D, Takahashi M, Schaffhauser R (1964) J Polymer Sci Part A, 2: 2749
745. Shibayama K, Suzuki Y (1965) J Polymer Sci Part A, 3: 2637
746. Kwei TK (1966) J Polymer Sci Part A-2, 4: 943
747. Smith TL (1974) J Polymer Sci Symp No 46: 97
748. Morawetz H (1975) "Macromolecules in Solution", Wiley-Interscience, New York; pp 96–181
749. Aharoni SM, Walsh EK (1979) J Polymer Sci Polym Lett Ed 17: 321
750. Aharoni SM (1980) Ferroelectrics 30: 227
751. Doi M, Chen D (1989) J Chem Phys 90: 5271
752. Chen D, Doi M (1989) J Chem Phys 91: 2656
753. Potanin AA (1992) J Chem Phys 96: 9191
754. Sonntag RC, Russel WB (1986) J Colloid Interface Sci 113: 399
755. Sonntag RC, Russel WB (1987) J Colloid Interface Sci 115: 378
756. Sonntag RC, Russel WB (1987) J Colloid Interface Sci 116: 485
757. Arnold C, Jr (1979) J Polymer Sci Macromol Rev 14: 265
758. Nguyen TQ, Kausch HH (1992) Adv Polymer Sci 100: 73
759. Popov AA, Zaikov GE (1992) Intern J Polymeric Mater 17: 143
760. Wang X, Gillham JK (1992) J Appl Polymer Sci 45: 2127
761. Wisanrakkit G, Gillham JK (1990) J Appl Polymer Sci 41: 2885
762. Wisanrakkit G, Gillham JK (1991) J Appl Polymer Sci 42: 2453
763. Kiyotsukuri T, Tsutsumi N, Okada H, Nagata M (1992) Polymer 33: 4990
764. Wang X, Gillham JK (1993) J Appl Polymer Sci 47: 425
765. Burfoot JC (1967) Ferroelectrics. Van Nostrand, London
766. Jona F, Shirane G (1962) Ferroelectric Crystals. Pergamon, New York
767. Cady WG (1946) Piezoelectricity. McGraw Hill, New York
768. Daniel VV (1967) Dielectric Relaxation. Academic Press, London
769. Fatuzzo E, Merz WJ (1967) Ferroelectricity. Interscience, New York

770. Gutmann F, Lyons LE (1967) Organic Semiconductors. Wiley, New York
771. Smyth CP (1955) Dielectric Behavior and Structure. McGraw Hill, New York
772. Prasad PN, Williams DJ (1991) Introduction to Nonlinear Optical Effects in Molecules and Polymers. Wiley-Interscience, New York.
773. Vallerien SU, Kremer F, Fischer EW, Kapitza H, Zentel R, Poths H (1990) Makromol Chem Rapid Commun 11: 593
774. Meier W, Finkelmann H (1990) Makromol Chem Rapid Commun 11: 599
775. Meier W, Finkelmann H (1993) Macromolecules 26: 1811
776. Xie S, Natansohn A, Rochon P (1993) Chem Mater 5: 403
777. Kitipichai P, La Peruta R, Korenowski GM, Wnek GE (1993) J Polymer Sci Part A: Polym Chem 31: 1365
778. Mitchel GK, Guo W, Davis FJ (1992) Polymer 33: 68
779. Matsuoka Y, Kishi R, Sisido M (1992) Chemistry Letters 1855
780. Aharoni SM, Edwards SF (1992) unpublished results on LCP networks with stiff polyamide segments and flexible trifurcated junctions.
781. Jones JL, Ball RC (1991) Macromolecules 24: 6369
782. Vilgis TA (1992) Progr Colloid Polym Sci 90: 1
783. Vilgis TA, Heinrich G (1992) Angew Makromol Chem 202/203: 243
784. Vilgis TA (1992) In: Aharoni SM (ed) Synthesis, Characterization, and Theory of Polymer Networks and Gels Plenum, New York.
785. Landau L, Lifshitz IM (1958) Course of Theoretical Physics, Elasticity. Pergamon, New York
786. Sokolnikoff IS (1956) Mathematical Theory of Elasticity. McGraw Hill, New York
787. See Timoshenko SP, Goodier JN (1970) Theory of Elasticity, Sect 99 for an account of the Griffiths-Taylor theory
788. Grosberg AYu, Khokhlov AR (1981) Adv Polymer Sci 41: 53
789. Semenov AN, Khokhlov AR (1988) Sov Phys Usp 31: 988
790. Gupta AM, Edwards SF (1993) J Chem Phys 98: 1588
791. Zhang R, Mattice WL (1992) Macromolecules 25: 4937
792. Wellman MW, Adams WW, Wolff RA, Wiff DR, Fratini AV (1981) Macromolecules 14: 935
793. Suzuki H (1959) Bull Chem Soc Japan 32: 1314
794. Bastiansen O (1949) Acta Chem Scand 3: 348
795. Okuyama K, Sakaitani H, Arikawa H (1992) Macromolecules 25: 7261
796. Martin DC, Thomas EL (1991) Macromolecules 24: 2450
797. Martin DC (1992) Macromolecules 25: 5171
798. Tashiro K, Kobayashi M, Tadokoro H (1977) Macromolecules 10: 413
799. Northolt MG, Van Aartsen JJ (1973) J Polymer Sci Polym Lett Ed 11: 333
800. Northolt MG (1974) Eur Polymer J 10: 799
801. Irwin RS, Vorpagel ER (1993) Macromolecules 26: 3391
802. Chivers RA, Blackwell J, Gutierrez GA (1984) Polymer 25: 435
803. Blackwell J, Gutierrez GA (1982) Polymer 23: 671
804. Blackwell J, Gutierrez GA, Chivers RA (1984) Macromolecules 17: 1219
805. Gutierrez GA, Chivers RA, Blackwell J, Stamatoff JB, Yoon H (1983) Polymer 24: 937
806. Blackwell J, Biswas A, Gutierrez GA, Chivers RA (1985) Faraday Discus Chem Soc 79: 73
807. Coulter P, Windle AH (1989) Macromolecules 22: 1129
808. Tashiro K, Kobayashi M (1991) Polymer 32: 454
808a. Tashiro K (1993) Prog Polym Sci 18: 377
809. Bicerano J, Clark HA (1988) Macromolecules 21: 585
810. Adams JM, Morsi SE (1976) Acta Crystallogr B, 32: 1345
811. Jung B, Schürmann BL (1989) Macromolecules 22: 477
812. Allen RA, Ward IM (1991) Polymer 33: 202
813. Allen RA, Ward IM (1992) Polymer 33: 5191
814. Klunzinger PE, Eby RK (1993) Polymer 34: 2431
815. Welsh WJ, Bhaumik D, Mark JE (1981) J Macromol Sci Phys B20: 59
816. Colquhoun HM, O'Mahoney CA, Williams DJ (1993) Polymer 34: 218
817. Waller JM, Eby RK (1992) Polymer 33: 5334
818. Whittaker ESW (1953) Acta Crystallogr 6: 714
819. Schaefer J, Stejskal EO, Buchdahl R (1977) Macromolecules 10: 384
820. Stejskal EO, Schaefer J, McKay RA (1977) J Magnetic Reson 25: 569
821. Yannas IV, Luise RR (1982) J Macromol Sci Phys B21: 443

822. Luise RR, Yannas IV (1992) Plast Eng Comput Model Polym 25: 191
823. See, for example, Brostow W, Corneliussen RD (1986) Eds, "Failure of Plastics", Hauser, Munich, Germany.
824. Kelly A (1966) Strong Solids. Clarendon Press, Oxford, UK; p 6
825. De Boer J (1936) Trans faraday Soc 32: 10
826. Tobolsky A, Eyring H (1943) J Chem Phys 11: 125
827. Zhurkov SN (1965) J Fract Mech 1: 311
828. Zhurkov SN, Korsukov VE (1974) J Polymer Sci Polym Phys Ed 12: 385
829. Bartenev GM, Zuyev YuS (1968) Strength and Failure of Visco-Elastic Materials. Pergamon, Oxford, UK, pp 1–63
830. Lazurkin YuS, Fogelson RA (1951) Zhur Tekh Fiz 21: 267
831. Lazurkin YuS (1958) J Polymer Sci 30: 595
832. Aharoni SM (1972) J Appl Polymer Sci 16: 3275
833. Aharoni SM (1973) J Polymer Sci Polym Symp 42: 795
834. Aharoni SM (1974) J Macromol Sci Phys B9: 699
835. Aharoni SM (1976) In: Deanin RD, Crugnola AM, (eds) Toughness and Brittleness of Plastics, ACS, Washington DC; pp 123–132
836. Boyer RF (1963) Rubber Chem Technol 36: 1303
837. Sakurada I, Ito T, Nakamae K (1966) J Polymer Sci Part C, 15: 75
838. Sakurada I, Kaji K (1975) Makromol Chem Suppl 1: 599
839. Sakurada I, Kaji K (1970) J Polymer Sci Part C 31: 57
840. Spiess HW (1983) Colloid Polymer Sci 261: 193
841. Inglefield PT, Amici RM, O'Gara JF, Hung CC, Jones AA (1983) Macromolecules 16: 1552
842. Schaefer J, Stejskal EO, McKay RA, Dixon WT (1984) Macromolecules 17: 1479
843. Wehrle M, Hellmann GP, Spiess HW (1987) Colloid Polymer Sci 265: 815
844. Schaefer D, Hansen M, Blumich B, Spiess HW (1991) J Non-Cryst Solids 131–133: 777
845. Nakamae K, Nishino T, Shimizu Y, Matsumoto T (1987) Polymer J 19: 451
846. Treloar LRG (1960) Polymer 1: 95
847. Treloar LRG (1960) Polymer 1: 279
848. Treloar LRG (1960) Polymer 1: 290
849. Tashiro K, Kobayashi M, Tadokora H (1977) Macromolecules 10: 731
850. Phillips JC (1979) J Non-Cryst Solids 34: 153
851. Doi M, Kuzuu NY (1980) J Polymer Sci Polym Phys Ed 18: 409
852. Bastide J, Picot C, Candau S (1981) J Macromol Sci Phys B19: 13
853. Bastide J, Duplessix R, Picot C, Candau S (1984) Macromolecules 17: 83
854. Kantor Y, Webman I (1984) Phys Rev Lett 52: 1891
855. Phillips JC, Thorpe MF (1985) Solid State Commun 53: 699
856. Shaefer DW, Keefer KD (1986) Phys Rev Lett 56: 2199
857. Dumas J, Baza S, Serughetti J (1986) J Mater Sci Lett 5: 478
858. Brown WD (1986) PhD Thesis, Physics, Cambridge University
859. Morris MI (1987) PhD Thesis, Physics, Cambridge University
860. Thorpe MF, Garbocki EJ (1987) Phys Rev B35: 8579
861. Clark AH, Ross-Murphy SB (1987) Adv Polymer Sci 83: 57
862. Edwards SF (1988) J Physique – France 49: 1673
863. Boué F, Edwards SF, Vilgis TA (1988) J Physique – France 49: 1635
864. Ashby MF, Gibson LJ (1988) Cellular Solids, Structure and Properties. Pergamon Press, Oxford, UK
865. Kilb RW (1958) J Phys Chem 62: 969
866. Price FP, Gibbs JH, Zimm BH (1958) J Phys Chem 62: 972
867. Bastide J, Boué F (1986) Physica 140A: 251
868. Bastide J, Leibler L, Prost J (1990) Macromolecules 23: 1821
869. Bastide J, Boué F, Mendes E, Zielinski F, Buzier M, Beinert G, Oeser R, Lartigue C (1992) In: Dusek K, Kuchanov SI, (eds) Polymer Networks '91, VSP, Utrecht, The Netherlands, pp 119–145 and references therein.
870. deGennes PG (1974) J Phys France 35: L–133
871. Marucci G (1981) Macromolecules 14: 434
872. Edwards SF, Vilgis TA (1988) Rep Progr Phys 51: 243
873. Erman B, Monnerie L (1989) Macromolecules 22: 3342
874. Fontaine F, Morland C, Noël C, Monnerie L, Erman B (1989) Macromolecules 22: 3348

875. Fontaine F, Noël C, Monnerie L, Erman B (1989) Macromolecules 22: 3352
876. Erman B, Monnerie L (1992) Macromolecules 25: 4456
877. Cates ME (1985) J Phys France 46: 1059
878. Gaylord RJ (1983) Polymer Bull 9: 181
879. Mooney M (1940) J App Phys 11: 582
880. Rivlin RS (1948) Phil Trans Royal Soc A241: 379
881. Maxwell JC (1870) Trans Royal Soc Edinburgh, 26
882. Stadler R, de Lucca Freitas L (1986) Colloid Polym Sci 264: 773
883. de Lucca Freitas L, Stadler R (1987) Macromolecules 20: 2478
884. Hilger C, Dräger M, Stadler R (1992) Macromolecules 25: 2498
885. Hilger C, Stadler R (1992) Macromolecules 25: 6670
886. Rempp P, Merrill EW (1991) "Polymer Synthesis", Huthig & Wepf, Basel, pp 234–236.
887. Rempp P, Muller R, Gnanou Y (1992) In: Dusek K, Kuchanov SI, (eds) Polymer Networks '91,
     VSP, Utrecht, The Netherlands, pp 25–38 and references therein
888. Annable T, Buscall R, Ettelaie R, Whittlestone D (1993) J Rheology, in press
889. Cox HL (1952) Br J Appl Phys 3: 72
890. Hershey AV (1954) J Appl Mech 21: 236
891. Kroener E (1958) Z Phys 151: 504
892. Dow NF (1963) General Electric Co Report No R635D1
893. Krenchel H (1964) "Fibre Reinforcement", Akademisk Forlag, Copenhagen
894. Lur'e AI (1964) "Three-Dimensional Problems of the Theory of Elasticity", Wiley, New York
895. Hill R (1965) J Mech Phys Solids 13: 213
896. Holister GS, Thomas C (1966) "Fibre Reinforced Materials", Elsevier, London
897. Smith GE, Spencer AJM (1970) J Mech Phys Solids 18: 81
898. Kelly A (1973) "Strong Solids", Clarendon Press, Oxford
899. Korringa J (1973) J Math Phys 14: 509
900. Edwards SF, Miller AG (1976) J Phys D, 9: 535
901. Veldkamp JDB (1979) J Phys D, 12: 1375
902. Gubernatis JE, Krumhansl JA (1975) J Appl Phys 46: 1875
903. Cates ME, Edwards SF (1984) Proc Royal Soc London, A395: 89
904. Batchelor GK (1971) J Fluid Mech 46: 813
905. Edwards SF, Freed KF (1974) J Chem Phys 61: 1189
906. Freed KF, Edwards SF (1974) J Chem Phys 61: 3626
907. Muthukumar M, Edwards SF (1983) Macromolecules 16: 1475
908. Elliott RJ, Krumhansl JA, Leath PL (1974) Rev Mod Phys 46: 465
909. Evans KE, Nkansah MA, Hutchinson IJ, Rogers SC (1991) Nature 353: 124
910. Diederich F, Rubin Y (1992) Angew Chem Int Ed Engl 31: 1101
911. Fixman M (1955) J Chem Phys 23: 1656
912. Friedman B, O'Shaughnessy B (1988) Phys Rev Lett 60: 64
913. Friedman B, O'Shaughnessy B (1991) J Phys France II, 1: 471
914. Friedman B, Levine G, O'Shaughnessy B (1992) Phys Rev A, 46: R7343
915. Friedman B, O'Shaughnessy B (1993) Europhys Lett 21: 779
916. Friedman B, O'Shaughnessy B (1993) Europhys Lett 23: 667
917. Edwards SF, to be published
918. Mansfield ML, Klushin LI (1993) Macromolecules 26: 4262
919. Kumar S, Kurtz SK, Banavar JR, Sharma MG (1992) J Stat Phys 67: 523
920. Wei G, Edwards SF (1992) Computational Polymer Science 2: 44
921. Yang Y, Tobias I, Olson WK (1993) J Chem Phys 98: 1673
922. Aharoni SM (1993) "Interfractal Porosity in Gels of Rigid Aromatic Polyamide Networks and
     its Consequences", and references therein. Manuscript in preparation. Presented in part at the
     35th IUPAC Symposium in Akron, Ohio, on July 11–15, 1994
923. Mandelbrot BB, Passoja DE, Paullay AJ (1984) Nature 308: 721
924. Silberschmidt VV (1993) Europhys Lett 23: 598 and references therein
925. Rossen WR, Mamun CK (1993) Phys Rev B, 47: 11815 and references therein
926. Sahimi M (1992) Modern Phys Lett B, 6: 507
927. Arbabi S, Sahimi M (1993) Phys Rev B, 47: 695
928. Sahimi M, Arbabi S (1993) Phys Rev B, 47: 703
929. Sahimi M, Arbabi S (1993) Phys Rev B, 47: 713

Author Index Volumes 101-118

Subject Index